FIRE SERVICE HYDRAULICS AND WATER SUPPLY

Second Edition

by Michael A. Wieder

Published by
Fire Protection Publications
Oklahoma State University
Stillwater Oklahoma

Project Management: Michael A. Wieder, Managing Editor, FPP
Project Editor: Michael A. Wieder
Graphic Design & Layout: Missy Hannan Sr. Graphic Designer I, FPP
Project Reviewer: Elkie Burnside
Project Reviewer: Gabriel Ramirez

ISBN 978-087939-414-1
Library of Congress 2001012345

Second Edition
First Printing, August 2010
Printed in the United States

10 9 8 7 6 5 4 3 2 1 *Printed in the United States of America*

If you need additional information concerning Fire Protection Publications, contact:
Customer Service, Fire Protection Publications, Oklahoma State University
930 North Willis, Stillwater, OK 74078-8045
800-654-4055 Fax: 405-744-8204

For assistance with training materials, to recommend material for inclusion in an IFSTA or FPP manual, or to ask questions or comment on manual content, contact:
Editorial Department, Fire Protection Publications, Oklahoma State University
930 North Willis, Stillwater, OK 74078-8045
405-744-4111 Fax: 405-744-4112 E-mail: editors@osufpp.org

TABLE OF CONTENTS

ACKNOWLEDGEMENTS

Books do not write themselves. In the same regard, authors who write books do not write them truly through their own efforts. Throughout the course of one's career and personal life, there are many people who are responsible for the author having the ability to write a book. I have been blessed with many positive influences in my career and personal lives that have enabled me to develop the more than thirty books and 100 journal articles that I have published.

Certainly, my parents, Richard and Marie Wieder, as well as my sister Joan laid the foundation for any successes that I have achieved. They, along with all the members of my extended family, never failed to support my educational and professional endeavors, even when they didn't seem to make sense at the time. My only regret is that my parents were unable to live a little longer so that they could see the scope to which their support would grow.

It would be impossible to list all of the people who have contributed to my professional career and successes over the years. I can only highlight some of the major ones:

- The officers and members of the Pennsburg, Pennsylvania Fire Company, who instilled a sense of organizational pride and the importance of training in me from the beginning of my fire service career.
- Jack McElfish, who during his tenure as the Director of the Montgomery County, Pennsylvania Fire Academy in the early 1980's took the time and effort to convince a small-town, young Pennsylvania Dutchman that he needed to pursue a college education in fire protection.
- Gene Carlson and Harold Mace, who as the leadership of IFSTA/Fire Protection Publications at the time hired me as a student employee and offered me permanent employment upon my graduation from Oklahoma State University. Most of what I learned about publishing, fire service politics, and many other work-related topics was greatly influenced by these two gentlemen in the early days of my career. Mr. Mace, though now retired, remains my mentor and sounding board.
- All my fellow employees at Fire Protection Publications and the members of the International Fire Service Training Association, who helped shape me as a writer, editor, supervisor, and person. Most of the little I know, I learned on-the-job with these professionals.
- Professor Pat D. Brock, who in my opinion is the finest educator on the topic of fire protection hydraulics in the United States and beyond. I gained much through the years I spent in his classroom and as his colleague at OSU.

Lastly, my true inspiration for nearly everything, from getting out of bed in the morning to writing this book, comes from my lovely wife Lori, son Luke, daughter Emily, and unimpressed dog Stetson. They make books worth writing and life worth living.

ABOUT THE AUTHOR

Michael A. Wieder currently serves as Assistant Director and Managing Editor of IFSTA/Fire Protection Publications (FPP) at Oklahoma State University, where he has been employed since 1984. In addition to his role with FPP, Mike also serves as the Executive Director of the International Fire Service Training Association (IFSTA). During his tenure at FPP he has written or edited more 30 texts on various aspects of fire fighting and protection. Among these texts were the IFSTA Water Supplies for Fire Protection, Fire Stream Practices, Fire Department Pumping Apparatus, and Pumping Apparatus Driver/Operator Handbook manuals. He has also published over 100 magazine articles and serves as a contributing editor to Firehouse© magazine. He also edited/authored the Fire Streams chapter in the 18th, 19th and 20th editions of the National Fire Protection Association's (NFPA) Fire Protection Handbook. Mike has written numerous government and government-funded research reports on responder rehabilitation, tanker safety, and response and roadway safety for the United States Fire Administration and the International Association of Fire Fighters (IAFF). Mike holds an AAS degree in Fire Technology from Northampton Community College in Bethlehem, PA. He holds an AS in Fire Protection Technology, a BS in Fire Protection and Safety Engineering Technology, and an MS in Occupational and Adult Education from Oklahoma State University. Mike is a Certified Fire Protection Specialist (CFPS) and a member of the Institution of Fire Engineers (MIFireE).

Mike served on the NFPA's Fire Fighter Professional Qualifications committee for more than 20 years. That committee is responsible for the standards for fire fighters, driver/operators, airport fire fighters, and land-based marine fire fighters. Mike chaired the task group that revised the driver/operator standard for 4 editions of that standard. He currently serves as Chair of the NFPA 1026, Incident Management Professional Qualifications committee. He is also a member of the NFPA Fire Service Section Board of Directors and Professional Qualifications Technical Correlating Committee. In 2003 Mike was named the ISFSI George D. Post National Fire Service Instructor of the Year. In 2008 he was awarded the Everett E. Hudiburg award by IFSTA for lifetime contributions to fire service training. Mike has also served as the Secretary of the National Incident Management System Consortium since its inception in the early 1990's.

Mike has served as a firefighter with the Pennsburg, Pennsylvania and Stillwater, Oklahoma Fire Departments. In 1999 he was awarded life membership status with the Pennsburg Fire Company Number 1.

You can contact the author at:

Fire Protection Publications
Oklahoma State University
930 North Willis Street
Stillwater, OK 74078-8045
(800) 654-4055
mwieder@osufpp.org

INTRODUCTION

For most students, in most disciplines, the most unappealing part of their course of study is anything having to do with mathematics. For some unknown reason, beginning with the earliest stages of the secondary education process, students question the need and practicality of courses such as algebra and calculus. "Why do we have to learn this stuff? We'll never do it in real life." These are statements that teachers have come to loathe.

The truth of the matter is that mathematics is all around us in our everyday lives. Whether it is balancing our checkbooks, determining how much grass seed is needed to overseed a portion of our yard, or figuring out how many miles per gallon our vehicle is getting, we use math, in some form, every day.

The fire protection educational process and profession is no different in this respect. I have witnessed this personally, as a student, as a practitioner, and as an instructor. Most students and practitioners love the courses or duties that involve fighting fire, ripping cars apart, and the like. They all disdain applied mathematics courses, such as strengths of materials, statics, fluid mechanics, and hydraulics. However, just as in all other aspects of life, there is no getting away from mathematics in fire protection. Anyone who is going to progress anywhere beyond being an entry-level, "tailboard" firefighter will have to be proficient in mathematics of varying types and degrees.

The first promotional step above firefighter, in most fire departments, is that of becoming a driver/operator. When it is all boiled down, the driver/operator has two primary functions with which they are responsible for: safely driving the apparatus to and from incidents and utilizing the apparatus to it maximum effectiveness while the apparatus is positioned at the emergency scene. For driver/operators who operate fire apparatus equipped with a pump, their primary on-scene duty will be to supply water at a sufficient pressure and volume to achieve the tactical goals of the incident.

Though the fire service has seen enormous advances in technology in recent years, its basic premise has not changed. We still primarily attack fires by discharging an appropriate amount of water onto them until the fire is extinguished. The water may have an additive, such as a wetting agent or foam concentrate, added to it, but we are still pumping water on fires. This process, the movement of water for fire protection, is an applied branch of mathematics known as hydraulics.

Although apparatus and equipment manufacturers would lead us to believe that fire trucks and their pumps basically operate themselves these days, certainly this is not the case. While modern electronic devices, such as pressure regulators and flowmeters, do make the job easier, there is no getting away from the fact that apparatus driver/operators and fire officers still must have a sound working knowledge of fire service hydraulics. These skills will be used in pre-incident planning and on the emergency scene. As long as we are still fighting fires with water, there will be a need to do computations such as pressure loss calculations, required fire flows, and determining pump discharge pressures.

The difference between a "lever-puller" and a true fire service professional is an understanding of the science and theory that support their actions. A lever-puller knows how to get water out of a pump and into a hose line. A professional knows why we are doing it, how the pump works, and how to ensure that the appropriate volume and pressure are being supplied.

It is the purpose of this textbook to allow the student to become a professional in the area of fire service hydraulics. In this text we not only explain how various equipment or equations work, but also *why* they work and why we need to know them. As you will see in the text, as with equipment the theory of hydraulics has benefited from modern technology. Every effort is made by this text to provide the student with the most modern, efficient, and simplest manner of performing the task at hand. In some cases the old methods are explained simply to ensure that the student sees how much better off we are today.

Before getting into the meat of this topic, I would like to address one issue to students and instructors alike. Having previously written books on this subject, having taught hydraulics, and having spent some 25 years answering customer inquiries at IFSTA/Fire Protection Publications, one of the most common issues, or questions I have fielded is from students who work a problem in the book and don't get the same answer as provided by the book. If I had a nickel for every time I heard "I have worked this problem 6 times and every time I get 129.3 psi and the book says it is 129.5 psi. What am I doing wrong?" My standard reply is *nothing*. Calculators all round numbers slightly differently. If you work a problem several times, you get the same answer every time, and that answer is within 1 or 2 gpm or psi of the answer provided in the book, assume that you have performed the calculation correctly and that the deviation can be chalked up to rounding differences.

If you have performed the calculation 6 times and get 3 difference answers or the same answer that is 55 psi or gpm different than what the book says, assume that you need to re-read the section and do it again.

The information contained in this text is designed to meet the objectives put forth in the model course outline for Fire Protection Hydraulics and Water Supply as established by the Fire and Emergency Services Higher Education (FESHE) initiative led by the United States Fire Administration (USFA). This initiative brings together leaders in fire service higher education for the purpose of establishing model core curricula for Associate's, Bachelor's, and Master's level fire service degree programs. It is hoped that this initiative will encourage growth in the fire service higher education field and support commonality of the information that is taught to all of the students.

The fire service has had standards for various levels of vocational competence established for more than a quarter of a century. These standards were initially developed by the National Professional Qualifications Board and today are maintained by the National Fire Protection Association. However, until the USFA started the FESHE initiative, there was little consistency to higher education programs throughout the United States. It is hoped that these institutions of higher education will embrace this initiative in the same manner that training institutions and fire department embrace the professional qualifications standards. The USFA must be commended for taking this important step for the good of the fire service.

Where applicable, this text also addresses some of the job performance requirements contained in NFPA 1002, *Standard for Professional Qualifications for Fire Apparatus Driver/Operators* required of driver/operators who operate fire pumps. For more complete information on meeting that standard, the individual should consult the IFSTA *Pumping Apparatus Driver/Operator Handbook* and *Aerial Apparatus Driver/Operator Handbook.*

For student and instructors seeking more detailed information on fire protection engineering hydraulics and theory, *Fire Protection Hydraulics and Water Supply Analysis* by Pat D. Brock is an excellent reference.

Photo courtesty Ron Jeffers

Principles of Water and Water Flow

A detailed understanding of the physical characteristics of water and the scientific principles of its movement is what sets the true fire protection professional apart from the average firefighter. The first part of this book is intended to give the student a detailed understanding of:

Chapter 1: The basic physical characteristics of water and why it is an effective extinguishing agent.

Chapter 2: The principles of water which is stored or otherwise at rest.

Chapter 3: The important principles that affect the movement of water

Once this foundation has been laid, the student will have a greater appreciation and understanding of the practical information contained in the remainder of the book.

FESHE Course Objectives

The information in this chapter is intended to meet some of the objectives outlined for the Fire Protection Hydraulics and Water Supply course by the United States Fire Administration's National Model Core Curriculum as developed by the Fire and Emergency Services Higher Education (FESHE) initiative.

FESHE Course Objectives

1. Explain the basic extinguishing properties of water that make it useful for fire fighting operations.
2. List the common advantages and disadvantages of water as a fire extinguishing agent.
3. Explain how the Law of Specific Heat and the Law of Latent Heat of Vaporization relate to water as a fire extinguishing agent.
4. Describe how the surface area of water affects its ability to extinguish a fire.

Chapter 1

Water as an Extinguishing Agent

INTRODUCTION

Civilization's progress through time has been charted using mile markers often referred to as ages. These ages are tied to significant discoveries that changed the course of history and human development. The Stone Age, Iron Age, and Bronze Age are examples of these periods. The world we live in today is most likely to be remembered as the Computer Age.

Certainly, one of the most significant breakthroughs in the development of civilization as we know it was the discovery of how to harness fire for useful purposes. This ushered in the Fire Age. No one can say for sure how people first discovered fire. That some curious lad or lass had the foresight to rub two sticks together long enough to produce a combustion reaction is unlikely. More probably, early people witnessed fires that resulted from lightning or volcanic activity. Perhaps these people were a bit chilly at the moment, dressed in the minimalist clothing of the times, and recognized that this newfound discovery provided warmth and comfort. This probably encouraged them to harness the fire for their continued use, and as time went on they developed new ways to create and use it. This whole process could be viewed as one of our earliest technologies.

Of course, discovering fire is one thing; controlling it and using it in a positive manner is entirely something else. We must hope that very quickly after they discovered fire's usefulness, our predecessors also determined ways to control and extinguish it. Otherwise the results must have been disastrous. We can only imagine the conversations that must have occurred before people figured out how to control fire:

Yes, Honey, I realize you are cold. I could make one of the magic warming piles, but we really don't have time to rebuild the entire village right now.

Though history books do not typically show a Water Age in the time chart of civilization, let us hope that it came very soon after the beginning of the Fire Age. Again, we have no record of when early people discovered they could use water to control and extinguish fire. However, this discovery was probably nearly as significant as the discovery of fire itself. We can only assume that this realization resulted when a rain shower dampened one of those early "warming piles."

Our ability to control and extinguish fire remains primarily tied to applying an adequate amount of water to a fire until it is extinguished.

The use of fire and water, however, has evolved differently from most other technologies in that it has remained relatively unchanged through time. Take the ability to count things or compute results, for example. The earliest methods probably involved using small rocks or twigs to signify amounts. Over time we progressed to the abacus, slide rules, electronic calculators, and finally computers. Oddly enough, however, our ability to control and extinguish fire, particularly from a fire service standpoint, remains primarily tied to applying an adequate amount of water to a fire until it is extinguished. Yes, we have more sophisticated equipment today and we may even have additives (such as foaming agents) that enhance water's ability to extinguish fire, but the fact remains that simple water is still the most important element in extinguishing fire.

Every entry-level firefighter learns the basics of how water extinguishes fire and the most effective methods for applying it. Likewise, most entry-level fire apparatus driver/operators learn the basics of water movement. However, the true fire protection professional has a deeper understanding of water's makeup and use. That provides the rationale for this chapter. It examines water's physical characteristics and the basic principles that continue to make it the fire service's most important fire extinguishment tool. This basic information will lay the groundwork for the coming chapters that address the use of water in fire protection.

THE BASIC CHARACTERISTICS OF WATER

Water is a chemical compound that results when two parts of hydrogen (H) combine with one part oxygen (O) to form H_20. Water is most useful for fire fighting purposes when it exists in its liquid state **(Figure 1.1)**. This occurs in a temperature range between about 32°F and 212°F. When water is exposed to temperatures below 32°F it converts to a solid physical state known as ice. At temperatures above 212°F (at sea level) it converts into a gas called water vapor, or steam. The boiling point for water will actually decrease slightly in elevations that rise above sea level, although this is rarely significant in everyday life. When water is initially converted from liquid to vapor, the human eye cannot see it. However, as the vapor rises and cools, it begins to condense (convert back to its liquid state) and becomes visible in the form of condensed steam.

Figure 1.1 At various temperatures water exists as a solid, liquid, or gas.

One of the characteristics that makes water particularly useful for fire fighting is that it is, for all intents and purposes, incompressible in its liquid state. One cubic foot of water contains 7.48 gallons whenever it is in a liquid state. This makes it easy for fire apparatus designers to determine the internal size of an apparatus water tank that is to hold 750 gallons of water (about 100 cubic feet).

While water's incompressibility is a disadvantage from a storage standpoint (you could store more in a given space if it were compressible), it is extremely important from a movement standpoint. Moving water through distribution systems and fire fighting equipment would be terribly difficult if it compressed when handled. That is why we continue to use water in its liquid state for everyday fire fighting. Every firefighter knows that water expanded into steam is most useful in fighting fires within confined spaces (structures, for instance). However, to move and harness steam (which is highly compressible) for fire fighting purposes would be extremely difficult and highly impractical.

Water's incompressibility is one of its most important advantages as a fire fighting agent.

Water's weight varies slightly at different temperatures. Its density (weight per unit of volume) is measured in pounds per cubic foot. Water is densest near its freezing point, weighing approximately 62.4 lb/ft^3. It is least dense near its boiling point, weighing approximately 60 lb/ft^3. Because these differences are fairly insignificant from an everyday use standpoint, for ease of computation fire protection calculations typically assign ordinary fresh water a weight of 62.4 lb/ft^3. The weight of one gallon of water may then be determined as follows:

$$\frac{62.4 \text{ lb/ft}^3}{7.48 \text{ gal/ft}^3} = 8.34 \text{ pounds per gallon}$$

Fire departments who may find themselves using seawater from time to time may note that one cubic foot of sea water weighs on average about 64 pounds per cubic foot or 8.56 pounds per gallon. This will vary slightly in various bodies of water as the salinity is not always the same everywhere.

THE BASIC EXTINGUISHING PROPERTIES OF WATER

In basic, entry-level firefighter training we learn that water can extinguish fire in several ways. The primary way in which water extinguishes fire is by cooling, or removing heat from the fire. Another way water extinguishes fire is by smothering. This works especially well on the surface of flammable and combustible liquids, though adding a foaming agent to the water is generally necessary to enable it to float on the fuel's surface as opposed to mixing with it (polar solvents) or sinking to its bottom (hydrocarbons). Smothering also occurs to some extent when water converts to steam in a confined space. The steam tends to displace the heat, fire gases, and available oxygen from the space, thus extinguishing the fire.

The primary ways in which water extinguishes fire are by cooling and by smothering.

Two natural laws of physics are extremely important from a fire fighting standpoint:

- The Law of Specific Heat
- The Law of Latent Heat of Vaporization

By understanding these laws, you will gain a greater understanding of water's heat-absorbing ability and how that ability is affected by the amount of surface area of the water exposed to the heat.

The Law of Specific Heat

All substances have a given capacity to absorb heat. This capacity is known as the substance's specific heat. Earlier we mentioned the importance of water's incompressibility from a fire fighting standpoint. Of course, that is not water's only useful characteristic. Certainly the fact that it is noncombustible (will not burn) is also very important. While in theory all liquids can smother a fire, achieving extinguishment with a liquid that is flammable or combustible is relatively impractical. That is why gasoline and kerosene make lousy extinguishing agents.

Equally important from a fire protection standpoint is water's ability to absorb large amounts of heat. Heat absorption, or more accurately heat transfer, occurs when heat flows from a warmer object to a cooler object (it never occurs the other way around). Amounts of heat transfer are measured in British thermal units (BTUs) or in joules (J) (1 BTU = 1.055 kJ). A BTU is the amount of heat required to raise the temperature of 1 pound of water 1°F. The joule, also a unit of work, has replaced the calorie in SI (International System of Units) heat measurements (1 calorie = 4.18 joules).

The specific heat of any substance is the ratio between the amount of heat needed to raise the temperature of a specified quantity of that substance and the amount of heat needed to raise the temperature of an identical quantity of water by the same number of degrees. **Table 1.1** shows some common fire extinguishing agents and their specific heat comparison (by weight) with water.

Table 1.1
Specific Heat of Extinguishing Agents

Agent	Specific Heat
Water	1.00
Calcium chloride solution	0.70
Carbon dioxide (solid)	0.12
Carbon dioxide (gas)	0.19
Sodium bicarbonate	0.22

The information in Table 1.1 highlights water's usefulness as an extinguishing agent. For example, divide the specific heat of water (1.00) by the specific heat of carbon dioxide (CO_2) gas (0.19). The result tells us that it takes more than five times as much heat to raise the temperature of 1 pound of water 1°F than it takes for the same amount of carbon dioxide gas. In other words, water absorbs heat 500 percent more effectively than carbon dioxide. Similar comparisons of the other materials listed in Table 1.1 show that water is the best material for absorbing heat.

Water is the best material for absorbing heat.

The Law of Latent Heat of Vaporization

The second scientific principle involving water that is important from a fire protection standpoint involves the amount of heat that water can absorb when it changes from a liquid to a vapor. This is referred to as the latent heat of vaporization. The temperature at which a liquid absorbs enough heat to change to vapor is known as its boiling point.

As mentioned above, at sea level water boils and vaporizes (turns to steam) at 212°F. Vaporization does not occur the instant the water reaches the boiling point, however. Each pound of water requires approximately 970 BTUs of additional heat to complete the conversion to steam **(Figure 1.2 a and b)**. This ability of water to vaporize at a relatively low temperature makes it a very effective extinguishing agent.

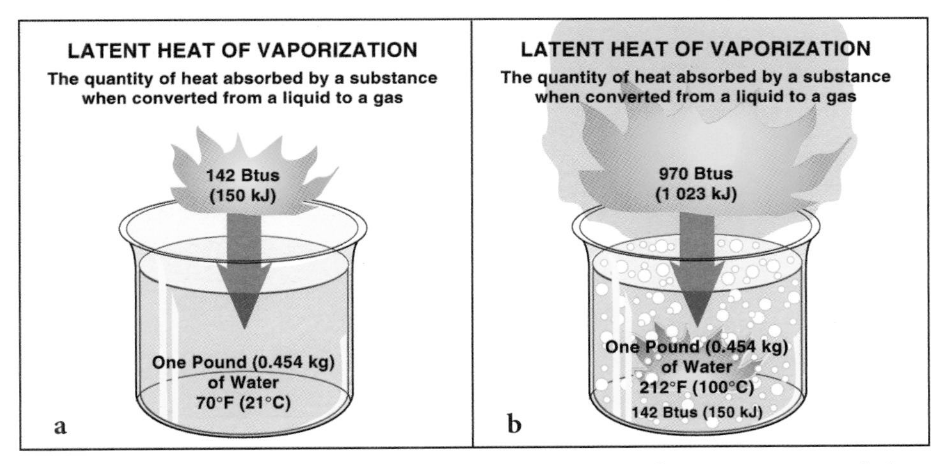

Figure 1.2 (a) To bring water from 70°F to its boiling point at 212°F, 142 BTUs are needed.
Figure 1.2 (b) When water has reached its boiling point, 970 additional BTUs are required to turn it into steam.

Water's specific heat is significant in fighting fire because its temperature does not increase beyond 212° during its absorption of the additional 970 BTUs. A gallon of water at 60°F will therefore absorb 9,357 BTUs of heat if the entire gallon is converted into steam.

Let us apply this information to a fire fighting example. Suppose we are attacking a fire with a 1½-inch hose equipped with a 100-gpm fog nozzle. The temperature of the water is 60°F. Assuming that it is discharged into a highly heated area and completely converts to steam, in theory it will absorb approximately 935,700 BTUs of heat per minute. In reality, complete conversion to steam never occurs, and the amount of heat absorbed will be somewhat less than the theoretical maximum.

The more steam that is produced, the greater will be its ability to displace fire gases and heat and to smother a fire.

The volume of steam produced is also important from a fire fighting standpoint. The more steam that is produced, the greater will be its ability to displace fire gases and heat and to smother a fire. Over the years, the fire service most commonly has used a ratio of 1 to 1,700 for water to steam conversion. Simply put, that means that one part of water will expand to 1,700 parts of steam when it vaporizes. In reality, the actual amount of steam produced will depend on a number of factors, including the amount of heat, how enclosed the space is, the temperature of the water, and the available surface area of the water. In extremely hot atmospheres, water will expand to steam in even greater volumes than the 1 to 1,700 ratio. Using the previous example, for ease of calculation let us say that a 100 gpm nozzle will discharge approximately 10 cubic feet of water per minute (it will actually discharge a little more than that). Using the rule-of-thumb ratio of 1700:1, those 10 cubic feet of water will generate 17,000 cubic feet of steam. This is enough steam to fill a room approximately 10 feet high, 25 feet wide, and 68 feet long **(Figure 1.3)**.

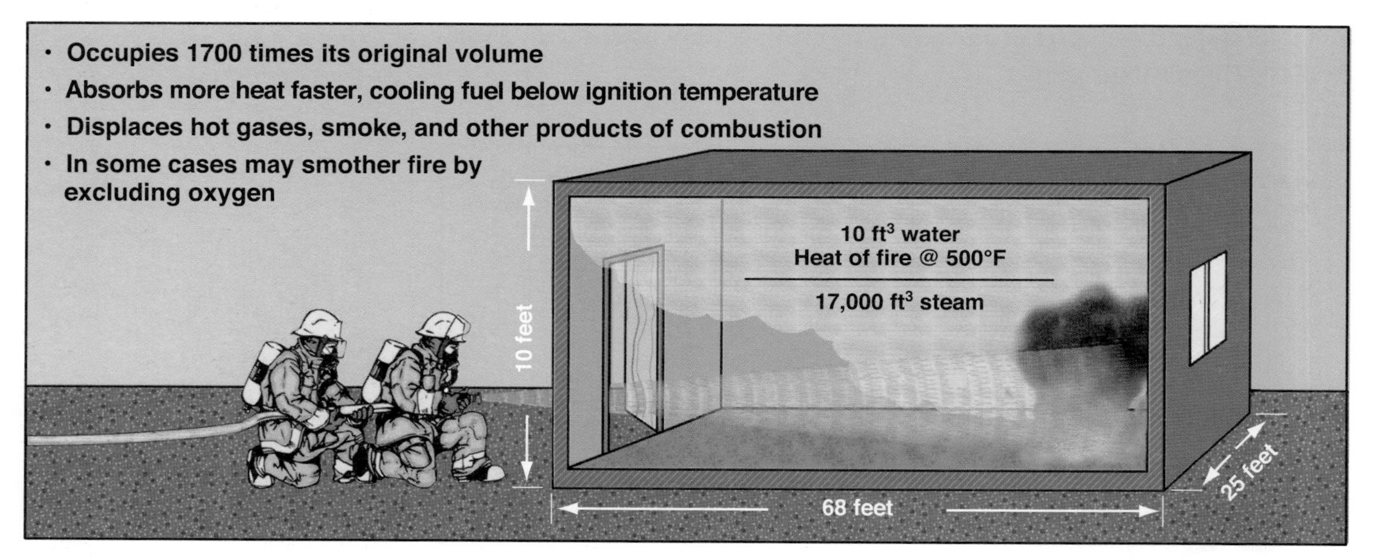

Figure 1.3 A nozzle discharging 100 gallons of water for one minute will generate enough steam to fill a room approximately 10 feet high, 25 feet wide, and 68 feet long.

As we will learn later in this book, determining the necessary amount of water to discharge on a fire to achieve control and extinguishment is an important calculation. In addition to understanding water's ability to absorb heat, the fire protection professional must also understand a fuel's ability to generate heat when it is on fire. The amount of heat a combustible object can produce depends upon the material from which it is composed. The rate at which the object gives off heat depends upon such factors as its physical form, the amount of surface exposed, and the air or oxygen supply. We will examine this topic in greater detail later in this book.

Surface Area of Water

In addition to understanding the Law of Specific Heat and the Law of Latent Heat of Vaporization, the fire protection professional must also realize that the speed with which water absorbs heat increases in proportion to the surface area of the water exposed to the heat. For example, if a 1-inch cube of ice is dropped into a glass of water, the ice cube will take quite a while to absorb its capacity of heat (melt). This is because only 6 square inches of the ice are exposed to the water. If the same cube of ice is divided into 1/8-inch cubes and these cubes are dropped into the water, 48 square inches of the ice are exposed to the water. This principle also applies to water in a liquid state. If water is divided into many drops, its rate of heat absorption increases hundreds of times.

The speed with which water absorbs heat increases in proportion to the surface area of the water exposed to the heat.

ADVANTAGES AND DISADVANTAGES OF WATER

Using the information discussed previously in this chapter, along with other important factors that will be elaborated later in this book, we can develop a simple list of some of the characteristics, or advantages, that make water an excellent extinguishing agent:

- Water has a greater heat-absorbing capacity than other common extinguishing agents.
- Changing water into steam requires a relatively large amount of heat. This will absorb more heat from the fire.
- The greater the surface area of water exposed, the more rapidly it absorbs heat. Using fog streams or deflecting solid streams off objects can expand the exposed surface area.
- As a rule of thumb, water converted into steam occupies 1,700 times its original volume.
- Water is inexpensive and readily available in most jurisdictions **(Figure 1.4)**.
- Water is incompressible and noncombustible.

Figure 1.4 (a) Water supply sources. Water is available from fire hydrants in most urban and suburban jurisdictions. *Courtesy of Dan Evatt*

Figure 1.4 (b) Rural fire departments may have to depend on static water supplies or mobile water supply operations. *Courtesy of Dan Evatt*

On the flip side, water does have some disadvantages as a fire extinguishing agent:

- Water has a considerable amount of surface tension, which limits its ability to soak or penetrate into such combustibles as baled or upholstered materials. Wetting agents are available that when mixed with water will reduce the water's surface tension and increase its penetrating ability.
- Violent reactions can occur when water is applied to certain water-reactive materials, whether they are on fire or not. The majority of these materials are reactive metals (Class D materials) **(Table 1.2, p. 10)**. Do not use water on such substances. In most cases there will be a more suitable extinguishing agent for these materials. Most hazardous materials response personnel will have access to information on appropriate specialized extinguishing agents in these situations.

Table 1.2 Water-Reactive Class D Materials
Aluminum
Magnesium
Titanium
Zirconium
Sodium
Potassium

Figure 1.5 Water will freeze on equipment during fire fighting operations that occur in below freezing weather. *Courtesy of Peter Matthews*

- Water's freezing temperature is a common issue in many jurisdictions **(Figure 1.5)**. This poses a slipping and sticking hazard to firefighters when ice coats equipment, roofs, ladders, and other surfaces. In addition, ice forming in and on equipment may cause it to malfunction. The added weight of ice on ladders and aerial devices may also subject them to possible overloading and failure.
- Water has low viscosity; that is, it flows easily. It does not readily adhere to vertical surfaces. This can be a disadvantage when exposures need protection or when fire involves vertical or rounded surfaces. Again, available additives can convert water to a foaming agent and enable it to stick to vertical surfaces.
- In theory, water from which all impurities have been distilled does not conduct electricity. In reality, the water that a firefighter uses does conduct electricity because it contains chemicals, minerals, and organisms. Firefighters' awareness of this hazard is vitally important during fire fighting operations when the potential for coming into contact with energized electrical equipment is present.

SUMMARY

Despite advances in modern technology, water remains our most common and important fire extinguishing agent. It is inexpensive and readily accessible in most jurisdictions. It is also easy to move, store, and handle. Water's ability to absorb heat and generate large volumes of steam make it particularly useful for fighting fires. In this chapter we have considered important physical characteristics of water. In the next two chapters we will learn additional important facts about water both when it is at a rest and when it is in motion.

CHAPTER ONE REVIEW QUESTIONS

1. What is the most common term used to describe water vapor?

   ____

2. How many gallons of water are in one cubic foot?

   ____

3. How much does a gallon of water weigh?

   ____

4. How much does a cubic foot of water weigh?

   ____

5. Describe the specific heat of a substance.

   ____

6. Define latent heat of vaporization.

   ____

7. Suppose we are attacking a fire with a 2½-inch hose equipped with a 250 gpm fog nozzle. The temperature of the water is 60°F. Assuming that the water is discharged into a highly heated area and that it completely converts to steam, how many BTUs per minute will it absorb?

   ____

8. List at least three characteristics of water that make it an excellent extinguishing agent.
   1. ______________________________
   2. ______________________________
   3. ______________________________

9. List at least three of water's disadvantages as an extinguishing agent.
   1. ______________________________
   2. ______________________________
   3. ______________________________

Answers found on page 423.

FESHE Course Objectives

The information in this chapter is intended to meet some of the objectives outlined for the Fire Protection Hydraulics and Water Supply course by the United States Fire Administration's National Model Core Curriculum as developed by the Fire and Emergency Services Higher Education (FESHE) initiative.

FESHE Course Objectives

1. Explain the basic principles of fluid pressure as they apply to water for fire protection.
2. Explain the relationship between height and density and head pressure.
3. Explain the importance and relevance of potential energy on water in fire protection concerns.

Chapter 2

Water at Rest: Hydrostatics

INTRODUCTION

Most fire protection and suppression personnel's primary concern with water is moving it from one location to another and then applying it to a fire for the purpose of control and extinguishment. The key word in that sentence is moving. Issues involving water at rest are generally not the kind of thing that keeps firefighters awake at night. However, before a true professional can understand the full range of issues involving the movement of water, he or she must have a solid grasp of the physical principles that affect water when it is not moving (static), or in other words, at rest. The study of water at rest and the science behind that study are referred to as hydrostatics. This chapter provides information you will need in order to understand hydrostatics, which will in turn help you to understand the science of water movement.

THE FIVE BASIC PRINCIPLES OF PRESSURE

Before getting too far into the study of water at rest or in motion, it is important to recognize what are commonly referred to as the five basic principles of pressure, sometimes more specifically referred to as the five principles of hydrostatic pressure. You must understand these principles before you can accomplish any practical work in hydraulics.

Some publications, including previous International Fire Service Training Association (IFSTA) and Fire Protection Publications (FPP) documents, recognize six principles of pressure. However, most hydraulics professionals combine two of those six principles into a single principle, thus reducing the number to five. The following section explains this in more detail.

Principle 1: The pressure at a point in a liquid is applied equally in every direction.

This principle is also known as Pascal's Law. The important point to understand here is that pressure created by the weight of the liquid is not only transmitted downward, but rather equally in all directions at any given point. Thus if four pressure gauges were located at the same point in a static body of water, with their intakes each oriented in a different direction, the pressure readings would be exactly the same on all four gauges.

The pressure at a point in a liquid is applied equally in every direction.

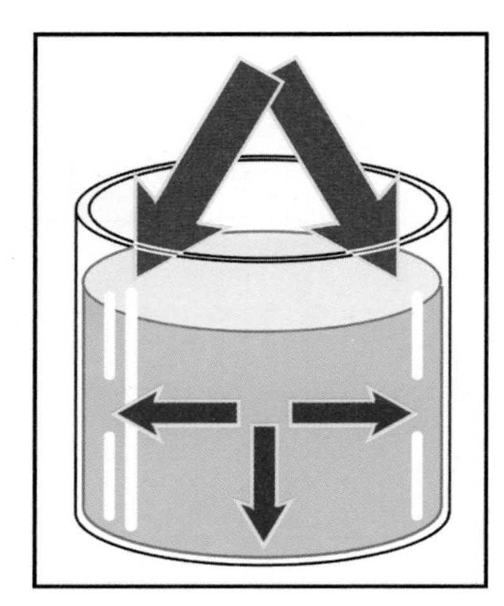

Figure 2.1 The pressure exerted by the weight of the fluid is perpendicular to the walls of the container.

It is within this first principle that some authors derive the previously mentioned "sixth principle." This additional principle typically states that fluid pressure is perpendicular to any surface on which it acts. They illustrate this "principle" using a vessel having flat sides and containing water. The pressure exerted by the weight of the water is perpendicular to the walls of the container **(Figure 2.1)**. If this pressure were exerted in any other direction, as indicated by the slanting arrows, the water would start moving downward along the sides and rising in the center. However, most hydraulics theorists believe that this is simply another way of stating and explaining Pascal's Law.

The pressure applied on a confined liquid from an external source will be transmitted equally in all directions throughout the liquid without a reduction in magnitude.

Principle 2: The pressure applied on a confined liquid from an external source will be transmitted equally in all directions throughout the liquid without a reduction in magnitude.

Before describing Principle 2 in more detail, we need to remember an important fact about water (and most other liquids for that matter) from Chapter 1. That is, one of the properties that make water particularly useful for fire fighting is that it is essentially incompressible. Liquids will generally maintain their original volume when confined and subjected to external pressures. That is not the case for gases, which will contract under pressurization (this process is also known as compression). Because external pressure exerted on a liquid does not result in a volume contraction (compression), the pressure is simply transmitted throughout the liquid in every direction without a reduction in magnitude.

This principle can be illustrated using a hollow sphere with a series of gauges set around its circumference. A water pump is attached to the sphere **(Figure 2.2)**. When the sphere is filled with water and the pump applies pressure, all gauges will register the same pressure, assuming that the gauges are on the same grade line and there exists no change in elevation. If the gauges are located at different elevations, the pressures will not be the same due to the effects of elevation pressure (explained later in this chapter). However, all the gauges will show the same pressure increase as long as the water remains static. This principle, in particular the idea that there is no reduction in magnitude, explains why damaging pressure surges (sometimes called water hammers) can be transmitted over great distances inside a water system.

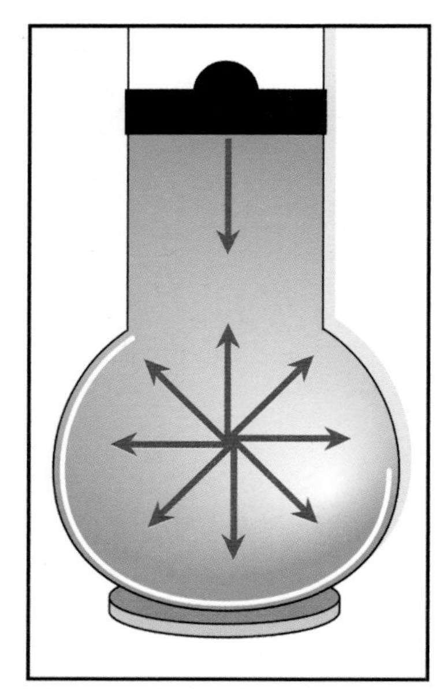

Figure 2.2 Pressure that is transmitted to a confined fluid from without is transmitted equally in all directions.

The pressure created by a liquid in an open container is directly proportional to the depth of the liquid.

Principle 3: The pressure created by a liquid in an open container is directly proportional to the depth of the liquid.

To explain Principle 3, let's first determine the amount of pressure created by a column of water that has a cross-section of one square inch and a height of one foot (a 1-inch by 1-inch by 12-inch column of water). To do this we begin with a container that holds one cubic foot of water. In Chapter One we determined that one cubic foot of water weighs 62.4 pounds. We also know that if the volume of a cube is one cubic foot, the area of each of its sides will be 144 square inches (12 inches x 12 inches) **(Figure 2.3)**.

We must also understand that scientifically pressure is a function of a given force applied over a specific area. The equation that shows this principle is:

Equation 2.1

$$\text{Pressure} = \frac{\text{Force}}{\text{Area}}$$

When a body of liquid is static, the force it generates is equal to its weight. These facts allow us to determine the amount of pressure created by the 1-inch by 1-inch by 12-inch column of water:

$$\text{Pressure} = \frac{62.4 \text{ lbs.}}{144 \text{ in.}^2} = 0.433 \text{ pounds per square inch (psi)}$$

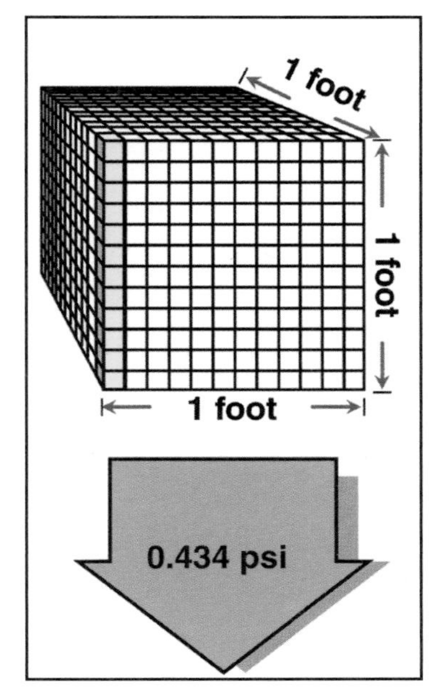

Figure 2.3 The shaded column represents the 0.433 psi created by one foot of water.

If we stacked three cubic feet of water on top of each other, the resulting pressure would be:

$$\text{Pressure} = \frac{62.4 \text{ lbs. x } 3}{144 \text{ in.}^2} = \frac{187.2 \text{ lbs.}}{144 \text{ in.}^2} = 1.3 \text{ psi}$$

Note that 1.3 psi is roughly (through close rounding) 3 times 0.433 psi.

Now visualize three vertical containers, each 1 square inch in cross-sectional area **(Figure 2.4)**. The depth of the water is 1 foot in the first container, 2 feet in the second, and 3 feet in the third. The pressure at the bottom of the second container is twice that of the first, and the pressure at the bottom of the third container is three times that of the first. Thus, the pressure of a liquid in an open container is proportional to its depth.

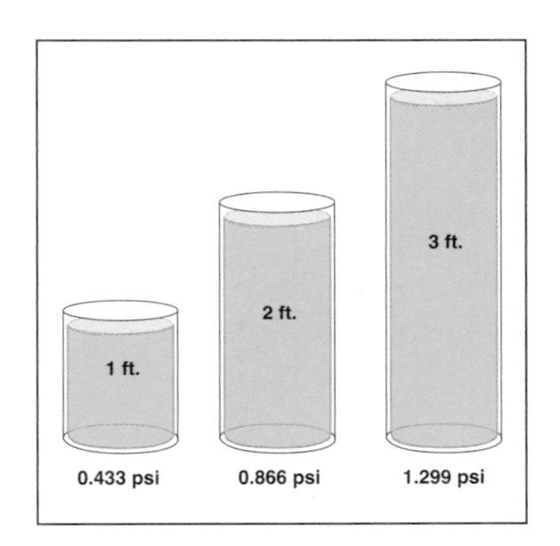

Figure 2.4 The pressure of a fluid in an open vessel is proportional to its depth.

Principle 4: The pressure created by a liquid in an open container is proportional to the density of the liquid.

To understand this principle, we must first understand the relationship between a substance's density and its weight. The denser a substance is, the more it will weigh by volume. For example, we have already established that 1 cubic foot of water weighs 62.4 pounds. Scientifically, water is typically assigned a density, or specific weight, of 1.0. If we filled a similar 1-cubic-foot container with mercury, it would weigh 845.5 pounds. Given those two figures, we can calculate that the density of mercury is about 13.6 times that of water:

$$\frac{\text{Weight of mercury}}{\text{Weight of water}} = \frac{845.5 \text{ lbs./ft}^3}{62.4 \text{ lbs./ft}^3} = 13.6$$

The pressure created by a liquid in an open container is proportional to the density of the liquid.

To illustrate that the pressure created by a liquid is also directly proportional to its density, let's replace the water that was used in the calculation for Principle 3 with mercury:

$$\text{Pressure} = \frac{845.5 \text{ lbs.}}{144 \text{ in.}^2} = 5.87 \text{ psi for a 1 in. x 12 in. column of mercury}$$

This relationship is shown by the following calculation:

$$\frac{\text{Pressure created by mercury}}{\text{Pressure created by water}} = \frac{5.87}{0.433} = 13.6$$

Thus, the ratio of mercury's density to water's is the same as the ratio of the pressures they create.

We can make Principles 3 and 4 more useful for fire protection purposes if we combine them to determine that the pressure created by a static liquid is equal to its weight multiplied by its height:

Equation 2.2 Pressure = weight x height or P = (w) (h)

Given:

P = Pressure in pounds per square inch (psi)
w = Specific weight of the liquid in psi per foot
h = Height of liquid column in feet

Applying Equation 2.2 to water:

Equation 2.3

$$P = \frac{62.4 \text{ lbs./ft}^3}{144 \text{ in.}^2 \text{/ft}^2} (h)$$

$$P = (0.433 \text{ psi/ft.}) (h)$$

The fire service professional will use Equation 2.3 in many contexts and applications, including determining total pressure loss or gain in hose layouts on uneven terrain, determining additional pressure that must be pumped into sprinkler and standpipe systems to overcome elevation pressure loss, and many more.

To demonstrate Equation 2.3 works, let's consider the following example:

Example 2.1

What pressure will be shown on a gauge at the base of an elevated water storage tank in which the water level is 120 feet above the pressure gauge **(Figure 2.5)**?

Solution:

P = (0.433 psi/ft) (h)

P = (0.433 psi/ft) (120 ft) = **52 psi**

For the purpose of entertaining ourselves, let's consider a similar problem substituting oil for water in the tank.

Example 2.2

What pressure will be shown on a gauge at the base of a storage tank containing oil with a specific gravity (S_g) of 0.9 in which the oil level is 70 feet above the pressure gauge?

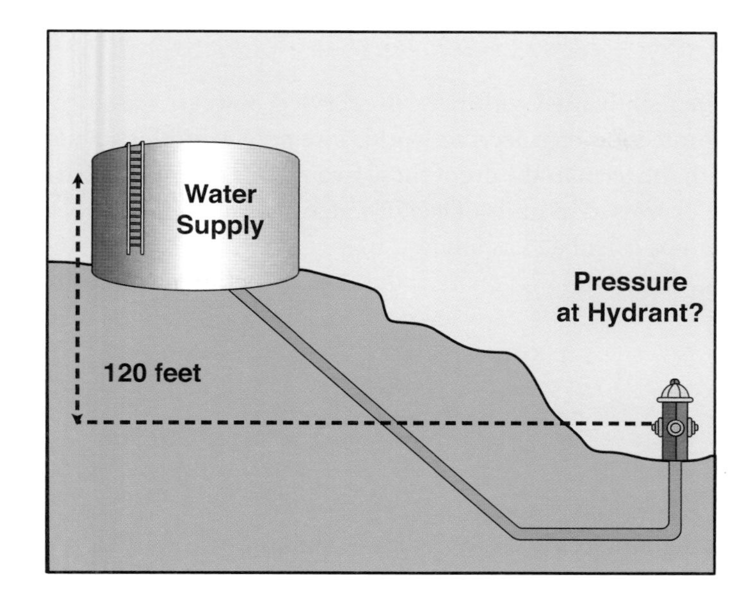

Figure 2.5 What would be the water pressure at this hydrant?

Solution:

$P = [(0.433 \text{ psi/ft})(S_g \text{ of oil})] (h)$

$P = [(0.433 \text{ psi/ft})(0.9)](70 \text{ ft})$

$P = [0.390 \text{ psi/ft}](70 \text{ ft}) = \mathbf{27.3 \text{ psi}}$

Principle 5: The pressure at the bottom of a container is not affected by the shape or volume of the container.

We can illustrate this principle by envisioning water in several different shaped containers, each having the same height **(Figure 2.6)**. The pressure is the same in each container. Though to the eye it may appear otherwise, remember that the pressure at the bottom of the tank depends solely on the depth and specific gravity of the liquid. Keep in mind that this principle applies only to pressure, not to force. The weight of the water will differ if the volume of each tank is not the same; however, all three pressure readings will be the same.

The pressure at the bottom of a container is not affected by the shape or volume of the container.

Figure 2.6 The pressure of a fluid on the bottom of a container is independent of the container's shape.

HEAD

The term head is not frequently used within the fire service; however, it is fairly common in the fire protection engineering world. Fire service professionals should be familiar with the term in the event these two worlds collide. Specifically, head is pressure expressed in units of feet of water instead of pounds per square inch (psi). It can be calculated as follows:

Equation 2.4

$$h = \frac{P}{w}$$

Given:

h = head in feet

P = pressure in psi

w = specific weight in lb/ft^3

Equation 2.4 can be refined for water and fire protection purposes as follows:

$$h = \frac{P}{62.4\ lbs/ft^3}$$

$$h = \frac{(P)\ (144\ in.^2/ft.^2)}{62.4\ lbs/ft^3}$$

$$h = (2.31\ ft./psi)\ (P)$$

Equation 2.5

This determines the inverse relationship of Principle 3 discussed above. To demonstrate this more simply, punch 0.433 into your calculator and hit the inverse (1/X) key. The answer will be 2.31.

Example 2.3

How far above the ground would the level of water in an elevated storage tank need to be in order to create a pressure of 60 psi at the bottom of the tank?

Solution:

h = (2.31 ft/psi)(P)

h = (2.31 ft/psi)(60 psi) = **138.6 feet**

Keep in mind that when we are talking about head, we are talking about height. You are likely to hear the term head pressure used in fire protection discussions. Head and head pressure are two different things. Head pressure is the amount of pressure created by the height of a column of water, more commonly referred to within the fire service as elevation pressure. Head pressure is in reality the inverse of head.

POTENTIAL ENERGY

It would be wrong to assume that just because it is at rest a fluid is has no energy component. Another way of looking at the principles of hydrostatic pressure is to consider hydrostatic pressure as potential energy. Potential energy is stored energy that can perform work once it is released. A static body of water within a water distribution system is subject to two sources of potential energy:

1. Potential energy due to elevation (PE_e)
2. Potential energy due to external pressure sources, such as pumps (PE_p)

To understand potential energy we must first review the definitions of energy and work. Simply defined, energy is the ability to do work. Work is the product of force multiplied by distance, expressed in foot-pounds. Potential energy therefore is the potential or ability to move a given weight (force) a specified distance. Mathematically this can be shown as:

Equation 2.6

$$PE = (W)(h)$$

Given:

W = any unit of weight

h = height in feet

The total potential energy within a water supply system can be expressed as:

$$PE_t = PE_h + PE_p$$

This information may seem theoretical and of little use at this point; however, the concept of potential energy will become more clear and its practicality more evident later in this book when we begin calculating pressure loss (or gain) in a hose or water distribution system.

SUMMARY

Before fire protection professionals can begin to understand the principles of water movement for the purpose of fire extinguishments, they must first master the principles associated with water at rest, also known as hydrostatics. In this chapter we learned that five basic principles are associated with hydrostatics. They are:

Principle 1: The pressure at a point in a liquid is applied equally in every direction.

Principle 2: The pressure applied on a confined liquid from an external source will be transmitted equally in all directions throughout the liquid without a reduction in magnitude.

Principle 3: The pressure created by a liquid in an open container is directly proportional to the depth of the liquid.

Principle 4: The pressure created by a liquid in an open container is proportional to the density of the liquid.

Principle 5: The pressure at the bottom of a container is not affected by the shape or volume of a container.

Furthermore, we explored the concepts of head and potential energy as they relate to water and water movement systems. With these principles covered, we can now move on to studying the basic principles associated with water movement.

CHAPTER TWO REVIEW QUESTIONS

1. Define hydrostatics.

2. Hydrostatic Principle 1, the pressure at a point in a liquid is applied equally in every direction, is also known as what law?

3. Define head.

4. Define potential energy.

5. If an elevated tank containing water is 150 feet high, what pressure will exist at the base of the tank if the tank is full?

6. If a cylindrical water storage tank has a diameter of 120 feet, and the top of the tank is 90 feet above the ground, what pressure will exist at ground level when the tank is full?

7. If a tank of identical height as the tank in question 5 is constructed with twice the volume, how will this affect the pressure available at the base of the tank?

8. If a high-rise building is 580 feet tall, how much pressure must be available at ground level simply to overcome elevation pressure loss in the system?

9. A 750,000-barrel crude oil storage tank is 80 feet high. If the tank is full and the oil has a specific gravity of 0.87, what pressure will exist at the base of the tank.

Answers found on page 424.

FESHE Course Objectives

The information in this chapter is intended to meet some of the objectives outlined for the Fire Protection Hydraulics and Water Supply course by the United States Fire Administration's National Model Core Curriculum as developed by the Fire and Emergency Services Higher Education (FESHE) initiative.

FESHE Course Objectives

1. Explain the importance and relevance of kinetic energy on water in fire protection concerns.
2. Describe the principles of Conservation of Energy and Conservation of Matter.
3. Define the following terms and explain their relevance to fire protection hydraulics:
   - Atmospheric Pressure
   - Head Pressure
   - Static Pressure
   - Normal Operating Pressure
   - Residual Pressure
   - Flow (Velocity) Pressure
4. List and explain the four principles of friction loss.
5. Explain how the Darcy-Weisbach Formula and Hazen-Williams Equation are used to determine friction loss in piping systems.

Chapter 3

Water in Motion: Hydrokinetics

INTRODUCTION

In Chapter 2 we took an extensive look at the basic principles of water at rest. Understanding water at rest provides a knowledge base that the fire protection professional can build upon in order to study more useful applications involving water. Obviously, from a fire protection standpoint it is more useful to understand the principles associated with water when it is moved from one location to another. After all, in its most basic sense, fire suppression simply involves moving water from a place of storage to the location of the fire and then applying it effectively.

The science that studies the characteristics and physical properties of water in motion is hydrokinetics. Fire protection professionals deal with moving water in fixed fire protection systems, fire ground operations, and water supply systems. Later portions of this book will deal directly with water movement in each of those contexts. In this chapter we will examine the theoretical principles of hydrokinetics and lay the foundations for those future, more practical applications.

PRINCIPLES OF KINETIC ENERGY

Two types of energy must be considered when studying hydraulics: potential energy and kinetic energy. Potential energy was discussed in Chapter 2 of this book. We typically associate potential energy with a body of water at rest. Once the water is set into motion, its velocity creates kinetic energy. In reality, it is not as simple as having one or the other type of energy present. Even in a system where water is moving, it is possible that some potential energy will still exist.

Some people find the concept of kinetic energy (expressed scientifically as KE) easier to understand when it is expressed in terms of work being done (expressed scientifically as G). Kinetic energy and work are basically synonymous: the kinetic energy exerted by a moving body of water is the same as the work being done. Mathematically this can be expressed:

Equation 3.1

$$KE = G = \frac{(m)\,(v)^2}{2}$$

Given:

KE = Kinetic Energy

G = Work

m = a mass of water

v = the velocity of the water.

An alternative version of Equation 3.1 can be developed when gravity is causing the movement or acceleration of the water. When acceleration of a particle, in this case water, is due to gravity (g), mass (m) can be expressed as:

$$m = \frac{W}{g}$$

where W is the weight of the particle. This formula may then be combined with Equation 3.1:

Equation 3.2

$$KE = G = \frac{(W)\,(v)^2}{(2)\,(g)}$$

Equation 3.2 is the formula most commonly used to represent the kinetic energy of water. It is important to understand the effect of gravity on water later on in this text when we discuss pumping water up or down hill.

Conservation of Energy

Now that we have established that energy exists in two forms, potential and kinetic, let's see how the two are related. To do this we must accept the fact that energy within a hydraulic system cannot be lost or destroyed; it simply changes forms back and forth between potential and kinetic energy. We may then establish that the total energy (TE) at any point in a system is equal to the sum of the potential energy (PE) and the kinetic energy (KE) at that point. This can be expressed mathematically as:

Equation 3.3

$$TE = PE + KE$$

Principle of Conservation of Energy: The total energy within a system will remain constant.

Related to this concept is the Principle of Conservation of Energy, which states that the total energy within a system will remain constant. What this means is that any change in the potential energy of a system must be matched by a corresponding change in kinetic energy. For example, suppose a prankster is standing on a stepladder holding a bucket of water. Essentially, the water in the bucket is at rest and its energy falls into the potential energy category. Should the prankster turn the bucket over and dump the water onto someone's head as they walked past, all of the water's potential energy will be converted into kinetic energy by the time the water hits the victim and the floor.

When applied to the flow of an incompressible fluid through a hydraulic system, the Principle of Conservation of Energy takes on a derivative form known as Bernoulli's Theorem: in a steady flow without friction, the sum of the velocity head, pressure head, and elevation head is constant for any incompressible fluid particle throughout its course. This is another way of saying that the total pressure is the same at any point within the system.

Bernoulli's Theorem: The total pressure is the same at any point within a hydraulic system.

While Bernoulli's Theorem is important from a scientific standpoint, common sense tells us that it does not reflect real-world conditions. Real-life hydraulic systems, be they sprinkler systems or hose lays, experience pressure losses due to friction and other factors. Bernoulli's Theorem can be applied to real-life conditions through the use of a calculation that is known as Bernoulli's Equation. Though they are mathematically the same, Bernoulli's Equation typically is expressed in one of two forms:

Equation 3.4a

$$h_1 + \frac{P_1}{W} + h_2 + \frac{P_2}{W} + \frac{(v_1)^2}{(2)(g)}$$

Given:

h = head in feet

P = pressure in feet of head

w = the specific weight of the fluid (62.4 pounds/ft^3 for water)

g = the acceleration of gravity

v = velocity in feet of head

Equation 3.4b

$$z_2 + \frac{P_1}{W} + \frac{(v_1)^2}{(2)(g)} + \frac{P_2}{W} + \frac{(v_2)^2}{(2)(g)} + z_2 + h_{1,2}$$

Given:

$h_{1,2}$ = lost head between points 1 and 2 in feet

P = pressure in feet of head

w = the specific weight of the fluid (62.4 pounds/ft^3 for water)

g = the acceleration of gravity (32.2 ft/sec^2)

v = velocity in feet of head

z = head in feet

The difference in these equations is that the components of potential energy attributed to elevation are represented by h in Equation 3.4a and by z in Equation 3.4b.

When applying the equation to water systems, the p/w may be replaced by (2.31)(P) for water as substantiated below:

$$\frac{P}{w} = \frac{P\ (\text{lbs/in}^2)}{62.4\ \text{lbs/ft}^3}$$

$$\frac{P}{w} = \left[\frac{P\ (\text{ft}^3)}{62.4\ (\text{in}^2)}\right]\left[\frac{144\ (\text{in})^2}{(\text{ft}^2)}\right]$$

$$\frac{P}{w} = \frac{144}{62.4} = (P)\ (\text{ft}) = (2.31)(P)$$

In this case P is given in pounds per square inch (psi), and units resulting from multiplying (2.31)(P) are simply feet. Every term in Bernoulli's Equation must have the same units, which are feet of head.

Figure 3.1 illustrates the basic principles of Bernoulli's Equation. Despite the changes in the piping between points 1 and 2, the total energy at each point is the same. Note that this does not mean that the potential energy or the kinetic energy at each point is the same—only the total energy.

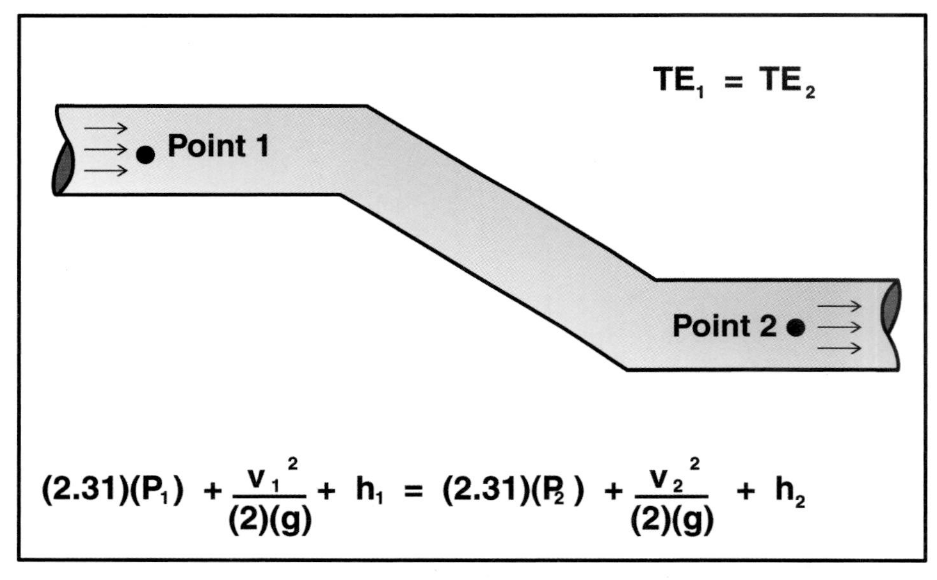

Figure 3.1 Bernoulli's Equation can be used to illustrate the Principle of Conservation of Energy. The total energy at point 1 and point 2 must be the same.

Example 3.1

If the pressure gauge at P_1 in Figure 3.2 reads 70 psi and the gauge at P_2 reads 40 psi, what is the water velocity at P_2 if v_1 is 15 feet per second (fps)?

Solution:

$$(2.31)(P_1) + \frac{(v_1)^2}{(2)(g)} + h_1 = (2.31)(P_2) + \frac{(v_2)^2}{(2)(g)} + h_2$$

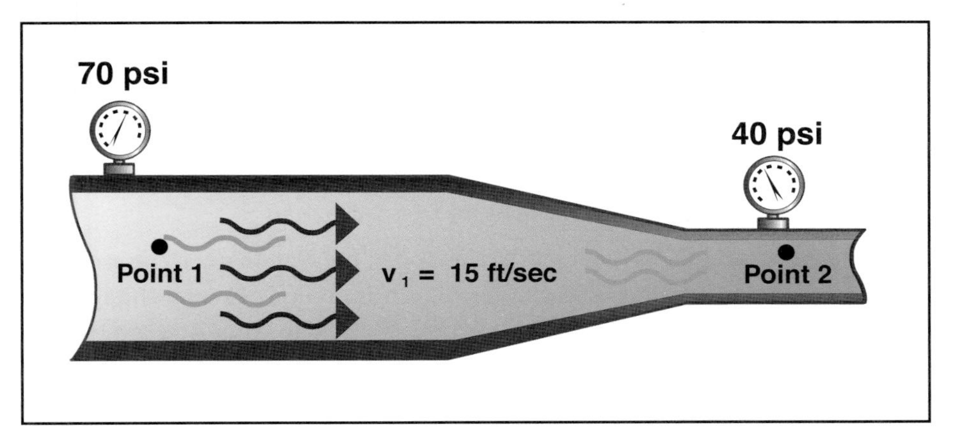

Figure 3.2 Example 3.1.

In this case h_1 and h_2 are the same because there is no change in elevation; thus, through algebra we can synthesize Bernoulli's Equation as follows:

$$(2.31)(P_1\text{-}P_2) + \frac{(v_1)^2}{(2)(g)} = \frac{(v_2)^2}{(2)(g)}$$

$$(2.31)(2)(g)(P_1\text{-}P_2) + (v_1)^2 = (v_2)^2$$

$$v_2 = \sqrt{(2.31)(2)(g)(P_1\text{-}P_2) + (v_1)^2}$$

$$v_2 = \sqrt{(2.31\text{ ft/psi})(64.4\text{ ft/sec}^2)(70\text{-}40\text{ psi})+(15\text{ ft}^2/\text{sec}^2)}$$

$$v_2 = \sqrt{(2.31\text{ ft/psi})(64.4\text{ ft/sec}^2)(30\text{psi})+(15\text{ ft}^2/\text{sec}^2)}$$

$$\mathbf{v_2 = 68.4\ ft/sec}$$

Conservation of Matter

A second important physical law for fire protection professionals concerns the conservation of matter. Very simply, the Principle of Conservation of Matter states that matter can neither be created nor destroyed. In terms of combustion, when a solid fuel such as wood burns, the sum of the resultant ash, smoke, gases, water vapor, and other by-products of combustion will equal the mass of the original fuel. In hydraulics, this principle means that what goes into one end of a piping or hose system must come out the other end. In the immortal words of this author's college hydraulics professor and textbook author Pat D. Brock, "Q_1 equals Q_2 everywhere!" If 750 gpm enters a hose line at one end, it must come out the other end, and the flow at any point within the hose will be the same.

Principle of Conservation of Matter: Matter can neither be created nor destroyed.

The flow of a liquid through a conduit may be determined using the following equation:

Equation 3.5

$$Q = (A)(v)$$

Given:

Q = Some units of volume per unit of time, usually cubic feet per second (cfs) or gallons per minute (gpm).

A = The cross-sectional area of the conduit in square feet (ft2)

v = The velocity of the stream in feet per second.

Having stated that whatever goes in one end of the system must come out the other, we must note that the other variables in Equation 3.5 may change. For example, if 750 gpm are initially pumped into a 5-inch hose line and somewhere down the line the hose is reduced to 4-inch, both the area (A) and the velocity (v) of the stream will change. In this case the area will decrease and the velocity will increase, assuming that 750 gpm is coming out the other end of the hose. Mathematically this is represented as follows:

Equation 3.6

$(A_1)(v_1) = (A_2)(v_2)$

Example 3.2

If the water velocity in the 5-inch hose ($A = 0.136\ ft^2$) is 10 ft/sec, what is the water velocity in the 4-inch ($A = 0.087\ ft^2$) hose, and how many gpm are flowing **(Figure 3.3)**?

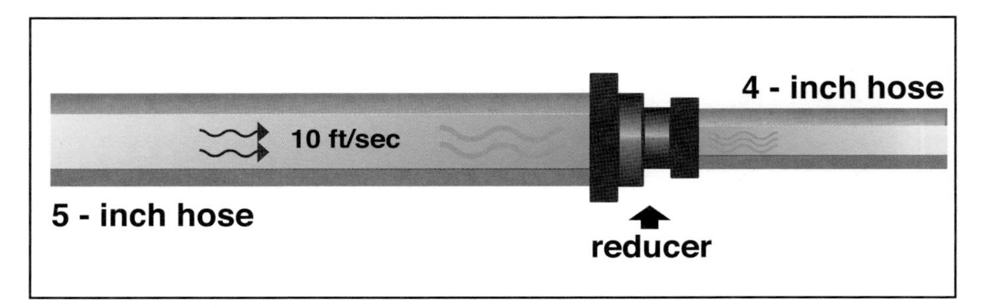

Figure 3.3 Illustration for Example 3.2.

Solution:

$(A_5)(v_5) = (A_4)(v_4)$

$(0.136)(10) = (0.087)(v_4)$

This can be rearranged to find v_2 as follows

$$v_4 = \frac{(0.136)(10)}{(0.087)}$$

$v_4 = 15.6$ feet per second

To determine the flow, you can use either the 4-inch or 5-inch section of hose, because $Q_5 = Q_4$:

$Q = (A_5)(v_5)$

$Q = (0.136\ ft^2)(10\ ft/sec)$

$Q = 1.36\ ft^3$ per second

To convert this to gallons per minute:

GPM = $(1.36\ ft^3/sec)(7.48\ gallons/ft^3)(60\ sec/min)$

GPM = 610 gpm

Once you understand the basics of Equation 3.6, the following principles of water flow in piping or hose systems should become evident:

Principle 1: If the pipe or hose size remains constant, water velocity within a system will be constant.

Principle 2: Within the same system, an increase in pipe or hose diameter will result in a reduction in water velocity.

Principle 3: Within the same system, a reduction in pipe or hose size will result in an increase in water velocity.

Principle 4: If pipe or hose size within a system remains constant, water flowing uphill will travel with the same velocity as water flowing downhill.

These principles will pertain to numerous places throughout the remainder of this book.

THE PRINCIPLES OF PRESSURE

To this point in the chapter, we have focused on hydrokinetic theory. This information is necessary to form a solid platform on which to build the rest of this manual. However, as we progress into this section of the chapter we will begin to explore the more practical aspects of water movement from a fire protection standpoint.

Whenever we talk about pressure, we must remember that this term has a variety of meanings. Ordinarily, we think of pressure as force exerted on one substance by another. In this book, however, pressure is defined as force per unit area. In the U.S. system of measurement, pressure is expressed in units of pounds per square foot (psf) or pounds per square inch (psi), the latter being more common in fire protection.

A common error is to confuse pressure with force. Force is a simple measure of weight and is usually expressed in pounds. This measurement is directly related to the force of gravity, which is the amount of attraction the earth has for all bodies. If several objects of the same size and weight are placed on a flat surface, they each exert an equal force on that surface.

For example, three square containers of equal size (1 x 1 x 1 foot) containing 1 cubic foot of water and weighing 62.4 pounds each are placed next to each other (**Figure 3.4, p. 30**). Each container exerts a force of about 62.4 psf with a total of about 187.2 pounds of force over a 3-square-foot area. If the containers are stacked on top of each other, the total force exerted — 187.2 pounds — remains the same, but the area of contact is reduced to 1 square foot **(Figure 3.5, p. 30)**. The pressure then becomes 187.2 psf.

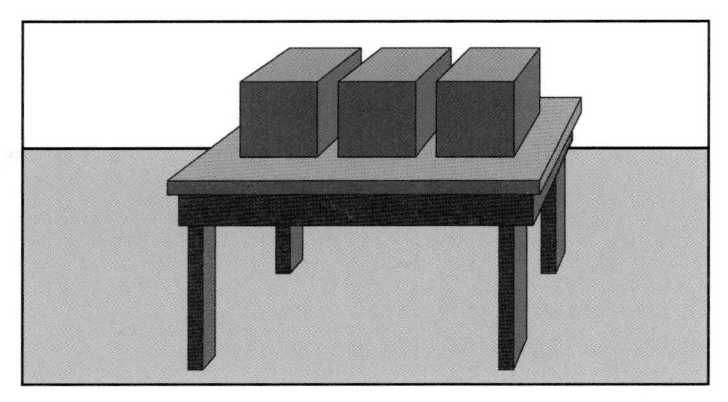

Figure 3.4 When placed side by side, each container exerts a force of 62.5 psf. The total force is 187.5 psf over a 3 ft^2 area.

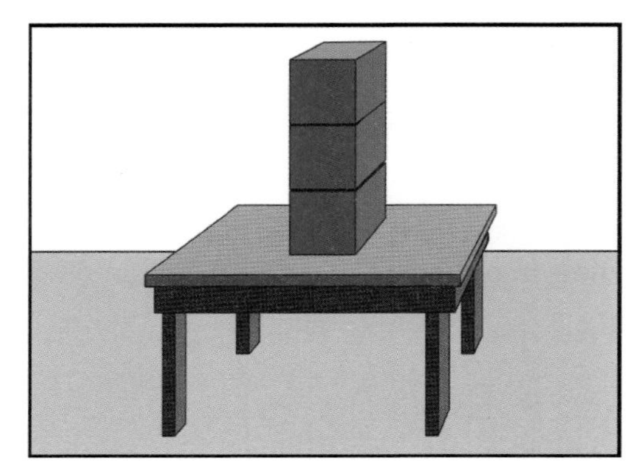

Figure 3.5 When stacked on top of each other the containers exert a force of 187.5 psf.

To understand how force is determined, you must know the weight of water and the height that a column of water occupies. One cubic foot of water weighs approximately 62.4 pounds. Because 1 square foot contains 144 square inches, the weight of water in a 1-square-inch column of water 1 foot high equals 62.4 pounds divided by 144 square inches, or 0.433 pounds. A 1-square-inch column of water 1 foot high therefore exerts a pressure at its base of 0.433 psi. The height required for a 1-square-inch column of water to produce 1 psi at its base equals 1 foot divided by 0.433 psi/ft, or 2.31 feet; therefore, 2.31 feet of water column exerts a pressure of 1 psi at its base. Do these numbers sound familiar? They should, as this is another way of stating the same information that we calculated in Principle 3 of Chapter 2.

Throughout this manual we will refer continually to different types of pressure in water supply systems and fire service applications. The fire protection professional must be acquainted with each of these terms in order to use them in their proper context.

Atmospheric Pressure

The earth's atmosphere has both depth and density, and it exerts pressure upon everything on earth. Atmospheric pressure is greatest at low altitudes, particularly those below sea level as at Death Valley, California. Atmospheric pressure is lowest at very high altitudes, such as on top of Pike's Peak in Colorado. The measurement most commonly associated with atmospheric pressure is 14.7 psi. This is the atmospheric pressure at sea level, and it is considered the standard atmospheric pressure.

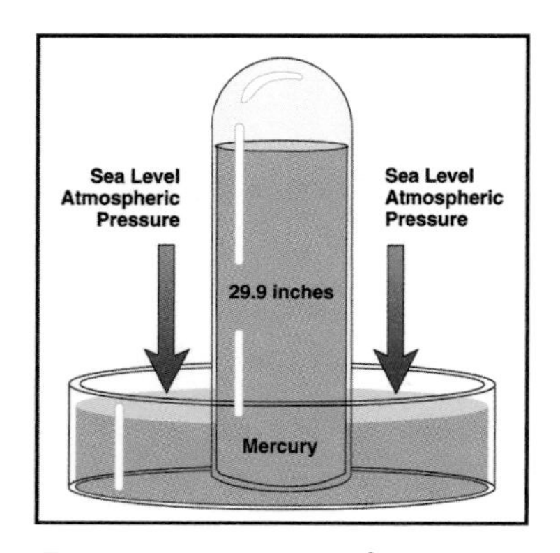

Figure 3.6 A pressure of 14.7 psi causes the mercury column in this barometer to rise 29.9 inches.

Standard atmospheric pressure is 14.7 psi.

A common method of measuring atmospheric pressure is by comparing the weight of the atmosphere with the weight of a column of mercury: the greater the atmospheric pressure, the taller the column of mercury. A pressure of 1 psi makes the column of mercury about 2.04 inches tall. At sea level, then, the column of mercury is 2.04 x 14.7, or 29.9 inches tall **(Figure 3.6)**.

Above sea level, atmospheric pressure decreases approximately 0.5 psi for every 1,000 feet. In fire service applications this pressure drop is of little consequence to about 2,000 feet. Above 2,000 feet, however, the lower atmospheric pressure can be of concern, particularly when operating from a draft. Because the less dense atmosphere at high altitudes reduces the pump's effective lift when drafting, fire pumps must work harder to produce the pressures required for effective fire streams.

Always keep in mind that the readouts from most pressure gauges show the pressure in addition to the existing atmospheric pressure. For example, if a discharge pressure gauge on a fire department pumper shows 120 psi at sea level, it is actually indicating, 134.7 psi (14.7 + 120). To distinguish between a gauge reading and actual atmospheric pressure, engineers use the annotations psig (pounds per square inch gauge) and psia (pounds per square inch absolute). The psi above a perfect vacuum is absolute zero. Any pressure less than atmospheric pressure is called vacuum. When a gauge reads –5 psig, it actually indicates 5 psi less than the existing atmospheric pressure. Absolute zero pressure is called a perfect vacuum. Though scientists have come very close, no one has ever successfully created a perfect vacuum in a practical setting. Anytime this manual uses the term psi, it refers to psig.

Head Pressure

We discussed the concept of head and its application to water supply and pressure in Chapter 2. In fire protection terms, head refers to the height of a water supply above the discharge orifice. If a water supply is 200 feet above the hydrant discharge opening that it supplies, it is said to have 200 feet of head **(Figure 3.7)**. Head in feet may be converted to head pressure by dividing the number of feet by 2.31. The result is the number of feet that 1 psi raises a column of water. Thus, if the water supply were located 200 feet above the discharge opening, then the head pressure would be 200 divided by 2.31, or about 87 psi. **Table 3.1, p. 32** lists head pressure for water supplies at various heights.

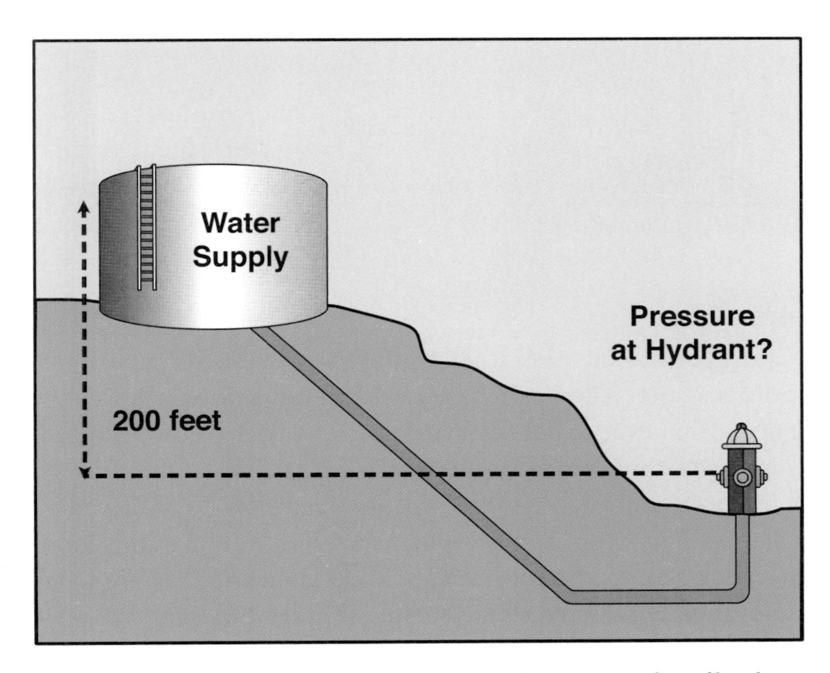

Figure 3.7 In this example the water supply is said to exert 200 feet of head.

Table 3.1
Head in Feet and Head Pressure

Feet of Head	Pounds per Square Inch	Pounds per Square Inch	Feet of Head
5	2.17	5	11.50
10	4.33	10	23.00
15	6.50	15	34.60
20	8.66	20	46.20
25	10.83	25	57.70
30	12.99	30	69.30
35	15.16	35	80.80
40	17.32	40	92.30
50	21.65	50	115.40
60	26.09	60	138.50
70	30.30	70	161.60
80	34.60	80	184.70
90	39.00	90	207.80
100	43.30	100	230.90
120	52.00	120	277.00
140	60.60	140	323.20
160	69.20	160	369.40
200	86.60	180	415.60
300	129.90	200	461.70
400	173.20	250	577.20
500	216.50	275	643.00
600	259.80	300	692.70
800	346.40	350	808.10
1,000	433.00	500	1,154.50

Static Pressure

Static means at rest, or without motion. In real-life water system applications, true static conditions rarely if ever exist because there is always some flow in the pipes due to normal domestic or industrial use. Thus, the water-flow definition of static pressure is the stored potential energy available to force water through pipe, fittings, fire hose, and adapters. Pressure on water may be produced by the elevation of the water supply, by atmospheric pressure, or by a pump. From an actual application standpoint, the pressure in a water system before water flows from a hydrant is considered static pressure **(Figure 3.8)**.

Normal Operating Pressure

Normal operating pressure is that pressure found in a water distribution system during normal consumption demands. As soon as water starts to flow through a distribution system, static pressure no longer exists. The demands for water consumption fluctuate continuously, causing water flow to increase or decrease in the system. The difference between a system's static pressure and its normal operating pressure results from the friction caused by water flowing through the various pipes, valves, and fittings in the system.

Figure 3.8 The static pressure is measured with no water flowing from the discharge.

Residual Pressure

In its basic sense, the term residual means a remainder, or that which is left. With that understood, then residual pressure is that part of the total available pressure not used to overcome friction loss or gravity while forcing water through pipe, fittings, fire hose, and adapters. The concept of residual pressure is important, as in essence, if you do not have any residual pressure you do not have a water supply. As we will see later in this manual when we discuss water flow tests, we begin by taking a static pressure reading off of the test hydrant. Once that static pressure is recorded, one or more hydrants in the area of the test hydrant are opened and allowed to flow. This will reduce the pressure reading on the test hydrant. The lowered pressure is the residual pressure for that hydrant. In a water distribution system, residual pressure varies according to the amount of water flowing from one or more hydrants, water consumption demands, and the size of the pipe. Remember that residual pressure must be identified at the test hydrant (where the static pressure reading was taken), not at the hydrants that are being flowed.

Flow Pressure

The flow pressure (velocity pressure) is that forward velocity pressure created at a discharge opening while water is flowing (**Figure 3.9, p. 34**). Using a device called a pitot tube and gauge, the velocity of flow pressure can be measured and converted to psi. If the size of the discharge opening is known, the measurement of flow pressure can be used to calculate the volume of water flowing in gpm.

THE PRINCIPLES OF FRICTION LOSS

In our earlier discussion about the Principle of Conservation of Energy and Bernoulli's Theorem, we stated that in a steady flow without friction, the total energy in a hydraulic system is constant. Unfortunately this is not the case in water supply systems or fire service hose layouts. Friction created by hose linings, appliances, and other parts of the system will reduce pressure at the discharge end of the system. In scientific terms, the forces of friction are converting some of the kinetic energy back to potential energy. From a fire protection stand-

Figure 3.9 The flow pressure is typically measured by inserting a pitot tube and gauge in the stream from an open hydrant discharge.

point, friction loss can be defined as that part of the total pressure lost while forcing water through pipe, fittings, fire hose, and adapters. Friction loss is caused by movement of water molecules against each other, the linings in fire hose or the inside of a pipe, couplings, sharp bends, changes in hose, pipe, or orifice size by adapters, and improperly sized gaskets. Anything that affects water movement may cause additional friction loss. The smoother the inside of the hose or pipe, the less friction loss will result. Good quality fire hose has a smoother inner surface and causes less friction loss than lower quality hose. The friction loss in old hose may be as much as 50 percent greater than that in new hose.

Friction loss can be demonstrated by inserting in-line gauges in a hose line or pipe run. The difference in the residual pressures between gauges when water is flowing through the hose is the friction loss over that section. The more common fire service example of friction loss is the drop in pressure in a hose line from the time it leaves the fire pumper until it reaches the nozzle.

We will spend considerable time and effort calculating friction loss and explaining its consequences throughout this manual. But first it will be helpful to understand the four basic principles that govern friction loss in fire hose and pipes. For the fire protection professional, these are as important as any of the hydrokinetic principles discussed to this point.

First Principle of Friction Loss

The amount of friction loss is directly proportional to the length of the hose or pipe.

Given that all other conditions are the same, the amount of friction loss is directly proportional to the length of the hose or pipe. In other words, given the same flow, the longer the hose or pipe is, the more friction loss there will be in the system. This principle can be illustrated using one 2½-inch hose line that is 100 feet long and another 2½-inch hose line that is 200 feet long. If a constant flow of 200 gpm is maintained in each hose, the 100-foot hose line will have a friction loss of 8 psi and the 200-foot hose line will have a friction loss of 16 psi.

Second Principle of Friction Loss

When hoses or pipes are the same size, friction loss varies approximately with the square of the increase in the velocity of the flow.

When hoses or pipes are the same size, friction loss varies approximately with the square of the increase in the velocity of the flow. In other words, friction loss not only increases as the flow velocity increases, but it also increases at a much higher rate. Using the 200 foot 2½-inch hose line from Principle 1, we saw that at a flow of 200 gpm friction loss was 16 psi. If that flow were doubled to 400 gpm, the friction loss would be 64 psi. Thus, doubling the flow created four times the friction loss. If the original flow were tripled from 200 to 600 gpm, friction loss would be 144 psi, or nine times the loss at 200 gpm.

Third Principle of Friction Loss

Given the same discharge volume, friction loss varies inversely as the fifth power of the diameter of the hose. In basic terms, given the same flow, the larger the hose is, the less friction loss will occur. This principle will become very important later in the manual when we discuss water supply operations, particularly relay pumping. To illustrate Principle 3 mathematically, consider 4-inch and 5-inch supply hose as follows:

$$\frac{(4)^5}{(5)^5} = \frac{1024}{3125} = 0.328$$

Thus, given the same flow, 5-inch hose will have approximately 33 percent as much friction loss as 4-inch hose.

Given the some discharge volume, friction loss varies inversely as the fifth power of the diameter of the hose.

Fourth Principle of Friction Loss

For a given flow velocity, friction loss is approximately the same, regardless of the pressure on the water. This principle explains why friction loss is the same when hose lines or pipes at different pressures flow the same amount of water. For example, if 100 gpm passes through a 3-inch hose line within a certain time, the water must travel at a specified velocity (feet per second). For the same rate of flow to pass through a 1 1/2-inch hose line, the velocity must be greatly increased. Four 1 1/2-inch hose lines are needed to flow 100 gpm at the same velocity as the single 3-inch hose line. Because water is practically incompressible, the same volume of water pumped into a hose or pipe under pressure at one end will be discharged at the other end. The size of the hose or pipe determines the velocity at which that volume of water will be discharged. The smaller the hose, the greater the velocity needed to deliver the same volume. Some fire hose tends to expand to a larger inside diameter under higher pressures. This will decrease the velocity and, therefore, the friction loss.

For a give flow velocity, friction loss is approximately the same regardless of the pressure on the water.

As mentioned in Friction Loss Principle 1, the friction loss will increase as the length of piping or hose increases. Flow pressure will always be greatest near the supply source and lowest at the farthest point in the system. The arrangement in Figure 3.10 illustrates this. Note that the readings are lower at each successive pressure gauge in the hose lay as the water travels from the source to the discharge.

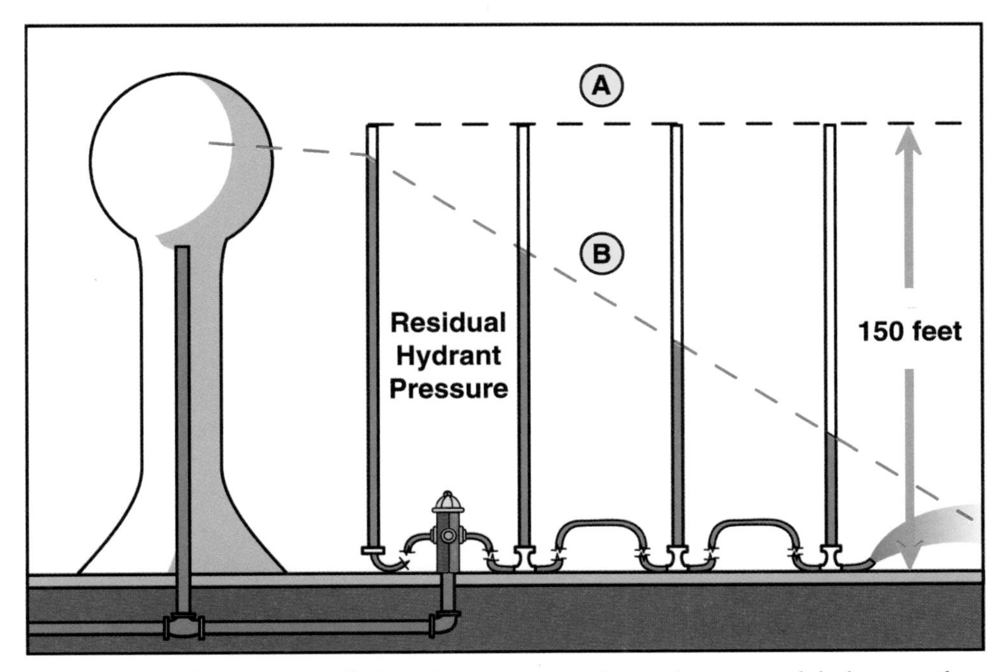

Figure 3.10 The flow pressure will always be greatest near the supply source and the lowest at the most distant point in the system.

DETERMINING FRICTION LOSS IN PIPING SYSTEMS

Most of the hydraulic calculations covered in this book will involve determining pressure loss and necessary pump discharge pressures in various layouts of fire hose. We will cover the formulas and other information necessary to perform those hydrokinetic calculations later in this text.

The true fire service professional should recognize that other applications within the broader fire protection field require similar calculations. For example, engineers who design underground water supply systems or fixed, water-based fire suppression systems must also determine pressure loss within those systems. Granted, they perform much of this work using computer-aided design programs. However, those programs tend to be based on one or both of two classic engineering friction loss formulas: the Darcy-Weisbach formula and the Hazen-Williams formula.

Most fire service professionals will never have occasion to use either of these formulas. However, you should be aware of them as during the course of your career you may encounter other fire protection professionals who use them in their work.

The Darcy-Weisbach Formula

Of the two formulas that we will discuss in this section, the Darcy-Weisbach formula is the more accurate, as well as the more widely used by the engineering profession in general. However, it is rarely used in fire protection applications, although it is recommended for testing foam proportioning systems according to NFPA® 11, Standard for Low-Expansion Foam. The basic Darcy-Weisbach formula is:

Equation 3.7

$$h_f = \frac{(f)(v)^2(L)}{(2)(g)(D)}$$

Given:

h_f = head loss due to friction in feet.

f = a dimensionless friction factor.

L = pipe length in feet.

v = fluid velocity in fps.

D = pipe diameter in feet.

g = acceleration due to gravity (32.2 ft/sec^2).

Note that the friction factor (f) depends on whether the flow of water through the pipes is laminar (in parallel layers) or turbulent (with no definite pattern). Which of these flows is present has historically been determined using a mathematical formula that calculates a factor known as the Reynolds number:

Table 3.2
Kinematic Viscosity Of Certain Liquids
(Kinematic Viscosity = Tabular Value x 10^{-5})

Temp.	Water	Medium Lubricating Oil	Medium Fuel Oil	Gasoline
°F	Kin. Visc. ft²/sec	Kin. Visc. ft²/sec	Kin. Visc. ft²/sec	Kin. Visc. ft²/sec
40	1.664	477	6.55	0.810
50	1.410	280	5.55	0.765
60	1.217	188	4.75	0.730
70	1.059	125	4.12	0.690
80	0.93	94	3.65	0.660
90	0.826	69	3.19	0.630
100	0.739	49.2	2.78	0.600

Equation 3.8

$$R_E = \frac{(v)(D)}{\mu}$$

Given:

R_E = Reynolds number

v = Fluid velocity in feet per second

D = Pipe diameter in feet

μ = Kinematic viscosity of the fluid in ft²/second **(Table 3.2)**

A variety of opinions exist regarding the point at which a laminar flow becomes turbulent. Most fire protection circles recognize a Reynolds number of 2,100 as the transition point from a laminar flow ($R_E \leq 2{,}100$) to a turbulent flow ($R_E > 2{,}100$). Seldom, if ever, will you encounter a true laminar flow in a fire protection setting.

Numerous derivatives of the original Darcy-Weisbach formula have been developed for specific applications. Because fire protection professionals rarely use them, they are not detailed in this text. For more information on the Darcy-Weisbach formula and sample calculations, see *Fire Protection Hydraulics and Water Supply Analysis* by Pat D. Brock.

The Hazen-Williams Formula

The fire protection industry uses the Hazen-Williams formula more commonly than the Darcy-Weisbach formula. In fact, NFPA® 13, Standard for the Installation of Sprinkler Systems, requires the use of the Hazen-Williams formula in hydraulic friction loss calculations, and NFPA® 24, Standard for the Installation of Private Fire Service Mains and Their Appurtenances, recommends its use. While its result may not be quite as accurate as the Darcy-Weisbach formula, it is generally easier to understand and use. This formula dates to the beginning of the twentieth century. In those days it was difficult to use because of the multiple exponential factors that had to be calculated. Now electronic calculators and computers have made its use much more routine.

The original Hazen-Williams formula was:

Equation 3.9

$$v = (1.318)(C)(R)^{0.63}(S)^{0.54}$$

Given:

v = water velocity in ft/sec.

R = the hydraulic radius in inches.

S = the slope of the hydraulic gradient in feet of head loss per foot of pipe.

C = Hazen-Williams coefficient of roughness.

This formula allows us to calculate the water velocity given a specified head loss. However, that information is of little practical use from a fire protection standpoint. Fire protection professionals are more interested in determining the friction loss in a pipe when the pipe's size and roughness and the flow in gallons per minute are known. To accomplish this, the original formula can be revised, changing water velocity in feet per second to gallons per minute, and feet of head to psi. This yields the more commonly used fire protection version:

Equation 3.10

$$P_f = \frac{(4.52)(Q)^{1.85}}{(C)^{1.85}(D)^{4.87}}$$

Given:

P_f = the pressure lost to friction in psi/ft of pipe.

Q = the flow rate in gpm.

C = the Hazen-Williams coefficient of roughness **(Table 3.3)**.

D = the internal pipe diameter in inches.

Before we begin to calculate some sample problems, a simple examination of the Hazen-William formula will provide some important facts concerning friction loss:

Pipe diameter is the single variable with the greatest impact on friction loss.

- The flow (Q) is raised to the 1.85 power in the equation. Thus, when the flow rate is doubled, if all other things remain constant, the friction loss will be about four times greater. If the flow triples, friction loss will be almost nine times greater. This is similar to the concepts of the Second Principle of Friction Loss.

Table 3.3 Values Of C	
Kind of Pipe	**C-Factor**
Unlined Cast or Ductile Iron	100
Asbestos Cement, Cement Cast or Ductile Iron, Steel Cement-Lined Steel and Concrete	140
Copper, Fiberglass Filament Wound Epoxy, Polyethylene and Polyvinyl Cloritde (PVC)	150

- The pipe diameter is raised to the 4.87 power in the denominator of the equation. Thus, any increase in pipe size will reduce friction loss if all other factors remain the same. If the diameter is doubled, the friction loss will be reduced by a factor close to 1/32. If the diameter is tripled, the friction loss will be about 1/243 of its original value. Thus, pipe diameter is the single variable with the greatest impact on friction loss.

The actual internal diameter of any given pipe will differ slightly from the nominal size of the pipe. For example, Schedule 40, steel, 6-inch aboveground pipe will have an actual internal diameter of 6.065 inches. Using the actual internal diameter of the pipe is important to achieving accurate results with the Hazen-Williams formula; **Tables 3.4 and 3.5, p. 40-41** list nominal and actual internal diameters for various common pipe sizes and materials.

The coefficient of roughness (C) is a reflection of the interior surface condition of the pipe. Large values of C indicate smooth pipe and thus less friction loss. Smaller values of C reflect rough or deteriorating pipe, which in turn will have greater friction loss. See Table 3.3 for sample values of C in commonly encountered types of pipe.

Example 3.3

How much pressure is lost to friction as 1,250 gpm travel through 800 feet of 6-inch Schedule 40 steel pipe having a C of 100?

Solution:

$$P_f = \frac{(4.52)(Q)^{1.85}}{(C)^{1.85}(D)^{4.87}}$$

$$P_f = \frac{(4.52)(1250)^{1.85}}{(100)^{1.85}(6.065)^{4.87}}$$

$P_f = 0.074$ psi per foot

Table 3.4
Nominal and Actual Internal Diameters Aboveground Piping

	Nominal											
	0.75	1	1.25	1.5	2	2.5	3	3.5	4	5	6	8
Steel Pipe												
Sch 10		1.097	1.442	1.682	2.067	2.469	3.068	3.548	4.026	5.047	6.065	7.981
Sch 30												8.071
Sch 40		1.049	1.38	1.61	2.067	2.469	3.068	3.548	4.026	5.047	6.065	7.981
XL		1.104	1.452	1.687	2.154							
Copper Pipe												
Type K	0.745	0.995	1.245	1.481	1.959	2.435	2.907	3.385	3.857	4.805	5.741	7.583
Type L	0.785	1.025	1.265	1.505	1.985	2.465	2.945	3.425	3.905	4.875	5.845	7.725
Type M	0.811	1.055	1.291	1.527	2.009	2.495	2.981	3.459	3.935	4.907	5.881	7.785
CPVC Plastic Pipe												
	0.884	1.109	1.400	1.602	2.003	2.423	2.951					

Then the total loss over the 800 feet length is:

$P_T = (0.074 \text{ psi/ft})(800 \text{ ft})$

$\mathbf{P_T = 59.2 \text{ psi}}$

Many derivatives of the Hazen-Williams formula may be used for many different fire protection engineering applications. Again, to learn more about these more hard-core engineering applications, consult Fire Protection Hydraulics and Water Supply Analysis by Pat D. Brock.

SUMMARY

Most fire protection professionals, whether they work on the engineering side of the industry or on the fire service side, are most concerned with water when it is in motion. The study of fluids in motion is called hydrokinetics. As in most scientific fields, the various principles associated with hydrokinetics have been researched and studied extensively. Most fire service professionals will never pursue this subject to that depth, or for that matter, will never even use the formulas explained in this chapter. However, to be considered a true member of the fire protection profession, every fire service professional should at least be familiar with these concepts and formulas for one of two reasons:

Table 3.5
Nominal and Actual Internal Diameters Underground Piping

	Nominal									
	4	6	8	10	12	14	16	18	20	24
Ductile Iron Pipe										
CL 50		6.4	8.51	10.52	12.58	14.64	16.72	18.8	20.88	24.04
Cl 51	4.28	6.34	8.45	10.46	12.52	14.58	16.66	18.74	20.82	24.78
CL 52	4.155	6.275	8.385	10.4	12.46	14.52	16.4	18.68	20.76	24.72
CL 54		6.16	8.27	10.28	12.34					
Cast-Iron Pipe Class 150										
Unlined	4.1	6.14	8.23	10.22	12.24	14.28	16.32	18.34	20.36	24.34
Enamel	3.98	6.02	8.11	10.1	12.12	14.09	16.13			
Cement	3.85	5.99	7.98	9.97	11.99	13.9	15.94			
Plastic Underground Pipe										
CL 150	4.24	6.09	7.98	9.79	11.65					
PVC CL 200	4.08	5.96	7.68	9.42	11.2					
Permastran	4.28	6.32	8.23	10.23	12.12					
Blue Brute										
CL 150	4.27	6.13	8.04	9.87	11.73					
CL 200	4.11	5.91	7.76							
Transite										
CL 100	4.0	6.0	8.0	10.0	12.0	13.59	15.5			
CL 150	4.0	5.85	7.85	10.0	12.0	14.0	16.0			
CL 200	4.0	5.7	7.6	9.63	11.56	13.59	15.5			

So that they can speak relatively intelligently on the subject when they encounter other fire protection professionals.

So that they will appreciate how much simpler the fire service hydraulic calculations that we will learn later in this manual are than the pure engineering calculations!

CHAPTER THREE REVIEW QUESTIONS

1. Define hydrokinetics.

 ____

2. What are the two types of energy?

 ____

3. Briefly explain the concept behind the Principle of Conservation of Energy.

 ____

4. Explain Bernoulli's Theorem.

 ____

5. Match the type of pressures listed on the right with the descriptions listed on the left. Write the appropriate letters on the blanks.

 ____

Description	Type of Pressure
______ a. Pressure created by the height of a water supply above the discharge orifice.	1. Atmospheric Pressure
______ b. Pressure created at a discharge opening while water is flowing.	2. Head Pressure
______ c. Pressure found in a water distribution system during normal consumption demands.	3. Normal Operating Pressure
______ d. Pressure created by earth's atmosphere.	4. Residual Pressure
______ e. The stored potential energy available to force water through pipe, fittings, fire hose, and adapters.	5. Static Pressure
______ f. That part of the total available pressure not used to overcome friction loss or gravity while forcing water through pipe, fittings, fire hose, and adapters.	6. Flow (Velocity) Pressure

6. If the water velocity in a 3-inch hose ($A = 0.049\ ft^2$) is 20 ft/sec, what is the water velocity in the 2½ -inch ($A = 0.034\ ft^2$) hose that is connected to the end of the 3-inch hose, and how many gpm are flowing?

 Velocity ______________________________

 GPM ______________________________

7. If the water velocity in a 10-inch hose is 5 ft/sec, what is the water velocity in the 8-inch hose that is connected to the end of the 10-inch hose, and how many gpm are flowing?

 Velocity ______________________________

 GPM ______________________________

8. How much pressure is lost to friction as 2,500 gpm travel through 1,800 feet of 8-inch, Schedule 40 pipe having a C of 55?

   Velocity ______________________________

   GPM ______________________________

9. How much pressure is lost to friction as 300 gpm travel through 200 feet of 4-inch, Schedule 10 pipe having a C of 90?

   ____

Answers found on page 424.

Part 2

Water Supply Systems and Water Flow Analsis

If you ask the average municipal firefighter where the water for fire fighting comes from, he or she is likely to respond "From a hydrant." While in essence correct (unless you are talking to a rural firefighter who counts on static water supply sources for fire protection), this answer is also very shortsighted. The fire hydrant is the proverbial "tip of the iceberg" from a water supply system standpoint. Much of the average municipal water supply system goes unseen by the public and firefighters alike. However, fire service professionals should have a working knowledge of their community's water supply system so that they will be able to use it to its fullest capabilities.

If you then ask the average municipal firefighter the capacity of the community's water supply system, again you are likely to get a well-intentioned but poorly informed answer. Most commonly, firefighters will respond in terms of how much pressure they usually get out of their hydrants once a pumper is connected to them. The true fire service professional understands that many variables are involved with water supply analysis and that simply knowing the flow pressure out of a hydrant is not a reliable indicator of the system's capability.

Let's give our average firefighter one last attempt at giving a correct answer and ask how much water will be required to successfully extinguish a fire in the building he or she is currently sitting in. Some may give a baseless ballpark figure and others are likely to respond that "this is what the chiefs get paid big bucks to figure out."

The true fire service professional should have a fundamentally sound working knowledge of the following important topics:

Chapter 4: The basic components and operation of the municipal water supply system used in their jurisdiction.

Chapter 5: The capabilities of the water supply system in their jurisdiction and the methods used to evaluate that system.

Chapter 6: The various methods that can be used to determine roughly how much water will be required to successfully extinguish a fire in a specified building within their jurisdiction.

The three chapters in Part 2 of this manual provide the information needed to reliably answer the above questions on each of these important topics.

FESHE Course Objectives

The information in this chapter is intended to meet some of the objectives outlined for the Fire Protection Hydraulics and Water Supply course by the United States Fire Administration's National Model Core Curriculum as developed by the Fire and Emergency Services Higher Education (FESHE) initiative. The information in this chapter also allows the student to meet the requirements of NFPA® 1002, *Standard for Fire Apparatus Driver/Operator Professional Qualifications* (2009 edition) listed below.

FESHE Course Objectives

1. List the sources of water used to supply water supply systems.
2. Describe the function of water treatment facilities in a water supply system.
3. List and describe the three basic mechanisms for moving water through a water supply system.
4. Describe the piping system used to distribute water throughout a water supply system.

NFPA® 1002 Requirements

15.2.1 (A) and 5.2.2 (A) Requisite Knowledge. Problems related to small-diameter and dead-end mains, low-pressure and private water supply systems, and hydrant coding systems.

Chapter 4

Water Distribution Systems

INTRODUCTION

We like to think that firefighters, fire trucks, and other tools of our trade are the most important elements in a community's overall fire protection "toolbox." Being a firefighter myself, I would never wish to understate the importance of firefighters, fire trucks, or the other highly visible elements of the overall fire protection effort. Realistically, however, none of those elements would be of much use if we did not have a reliable source of water for fire fighting when the need arose. With this fact established, we must realize that a community's water supply system often has as big an impact on fire protection as do the firefighters who use it. That is why a community's water supply system always accounted for such a significant percentage of the fire protection ratings that were assigned by the Insurance Services Office (ISO) and other rating bureaus.

While some unique features are likely in every community's water supply systems, the fundamental components of these systems are generally the same. In this chapter we will examine those components so that you will have a better-than-average understanding of how these systems get the water to the fire hydrant. In general any municipal, or for that matter private, water supply system has four basic components:

1. A water supply source
2. Water treatment facilities
3. A mechanism for forcing water through the system
4. A system of piping to transport the water through the community

Each of these components and its major subcomponents are detailed in this chapter.

WATER SYSTEM FUNDAMENTALS

Prior to discussing the four major components of water supply systems, we must understand a few important water system fundamentals. First and foremost, any community usually has only one water supply system, and it is used for domestic consumption, industrial applications, and fire protection. In simple terms, the hydrants that you use for fire fighting are hooked to the same pipes as the faucet in your kitchen. This knowledge is important to the fire service

Any given community usually has only one water suppy system for domestic consumption, industrial applications, and fire protection.

professional's understanding of the impact of factors like widespread lawn sprinkling and high industrial demands on the amount of water available from hydrants when a fire occurs.

A few major cities (Philadelphia and Chicago come to mind) have had secondary water supply systems intended solely for fire department use. These systems typically utilize high-volume, low-pressure supplies of untreated (nonpotable) water that can be used during large-scale fire fighting operations. They may not even be continually pressurized. If that is the case, the authority operating the system must be notified to pressurize the system when it is needed. These systems generally are used only during large-scale fire fighting operations. Routine fires are handled from the standard water supply system.

Some industrial complexes may have special, separate water supply systems for fire fighting emergencies. In particular these are common in refineries, paper manufacturing facilities, and other high-fire-hazard operations. Industrial complexes also tend to use untreated water for supplying these systems.

We also should establish who has responsibility for the operation of the water supply system. Even though the fire department relies heavily on these systems for its success, rarely does it have a major hand in their construction, maintenance, repair, or operation. In most municipalities a specifically designated water department manages the water supply system. This department may be an actual division of the municipal government, a separate trust or authority, or in a few cases a privately owned company that contracts with the community.

The fire department must have an excellent working relationship with the water system operator.

In any case, it is important that the fire department have an excellent working relationship with the water system operator. This will help to ensure that the fire department's concerns will be addressed when the system is constructed or modified. For example, if a new factory is being constructed within the jurisdiction, the fire department needs to consult with the water department on how much water must be available in the vicinity of the new occupancy in the event of a fire. The fire department also needs the water department to be responsive during critical periods when the system is down or when additional water supply is needed for a major fire fighting operation.

WATER SYSTEM CAPACITY

One of the most important things for a fire protection professional to understand about a water supply system is its capacity. Water supply system designs typically take into consideration three basic rates of consumption. These form an established base to which the fire flow requirements can be added during the design process. They also allow the engineers and fire protection people alike to determine the adequacy of the water distribution system. Fire protection professionals should be familiar with these basic rates of consumption, which they will encounter often during water supply testing.

The average daily consumption (ADC) is the average of the total amount of water used daily in a water distribution system over a period of one year. The maximum daily consumption (MDC) is the maximum total amount of water that was used during any 24-hour interval within a three-year period. Unusual situations, such as refilling a reservoir after cleaning, should not be considered in determining the maximum daily consumption. The peak hourly consumption (PHC) is the maximum amount of water used in any one-hour interval over the course of a day.

The maximum daily consumption is normally about one and one-half times the average daily consumption. The peak hourly rate normally varies from two to four times the normal hourly rate. The effect of these varying consumption rates on the system's ability to deliver required fire flows varies with system design. Both maximum daily consumption and peak hourly consumption should be considered to ensure that the water supplies and pressure do not reach dangerously low levels during these periods. Adequate water must be available at all times in the event that a fire occurs.

SOURCES OF WATER SUPPLY

The water supply system begins with a reliable source for water that can be drawn into the system. Depending on the size of the community and its demand for water, the system may have a single source or multiple sources. Experts generally recommend that at least one back-up source be available in the event that the main source is compromised. Water department officials must monitor and plan constantly for the community's water needs. This should incorporate an engineering estimate of the total amount of water needed both for domestic and industrial use and for fire fighting use. In large communities, the domestic/industrial requirements usually far exceed fire protection needs. In small towns, the requirements for fire protection may exceed other demands.

The two basic categories of water supply sources are surface water and groundwater. Most communities utilize only one type of source, although a few rely on a combination of surface and groundwater.

Surface Water Supplies

Of the two types of water supply sources, surface water is by far the more commonly used. Surface water supply sources include rivers, streams, lakes, reservoirs, coastal waters, and ponds **(Figure 4.1, p.50)**.

Rivers and streams are very common sources of domestic water supply due to their close proximity to many communities. However, rivers and streams do present a number of challenges from a reliability standpoint. Of all the types of surface water supplies, their available volume is most affected by variations in rainfall. During periods of drought or even minor decreases in rainfall, these bodies of water can drop to relatively low levels, which can also result in reduced water quality. During periods of heavy rain and flooding, the spreading water can swamp treatment facilities and make the water dirtier. Water officials also have to be constantly on the lookout for any pollution that has entered the source upstream of the treatment facility and deal with it accordingly. Because of all of these factors, treatment and purification processes for water from rivers and streams must be constantly reviewed and adjusted as necessary.

Figure 4.1 Lakes commonly serve as a surface water supply source. *Courtesy of Dan Evatt*

Lakes (naturally occurring bodies of water), reservoirs (man-made impoundments), and ponds (natural or man-made) are all generally similar from a water supply standpoint. They tend to be fairly stable in water quality and not as quickly affected by periods of decreased rainfall as rivers and streams. The water quality may decrease somewhat near the shore and in shallower areas of the body of water. The rate at which a surface water source will decrease in volume depends on the size of the body of water, the capabilities of the springs or streams feeding the body, and recent rainfall.

Many water departments utilize man-made impoundments that are constructed close to the water treatment facility. Often these impoundments are supplied by raw, untreated water pumped from a remote surface water supply source. As the water sits in these impoundments, many impurities filter out naturally, making the treatment process somewhat easier. Most jurisdictions try to keep these impoundments as full as possible at all times so that maximum water is available in the event of a major fire or emergency.

Oceans and coastal waters generally are not used as domestic water supplies because the process of desalinating and purifying them in useful quantities is difficult and expensive. Some municipalities and industries use these sources for dedicated fire protection water supply systems and for other purposes such as cooling water for power plants. In these roles they provide an almost limitless supply of water.

In general, surface water supply sources tend to:

- Have softer water (less minerals) than ground water
- Have more suspended solids and color than ground water
- Have more bacterial contaminants than ground water
- Deplete more rapidly during periods of low rainfall

Ground Water Supplies

Two primary types of ground water sources may be used as water supplies for domestic systems: wells and springs. The quality of the water coming from both of these sources is generally very high. This is because the process of traveling down through the earth to the underground source filters the water and substantially purifies it. Well or spring water generally requires little if any treatment for purification. We should note that wells might be drilled into the same stratum that supplies a spring. This may eventually deplete the flow volume of a spring.

In some cases, a stratum of water may actually be under sufficient pressure from the earth's surface or subterranean formation to force the water to the ground surface or above. Wells of this type are called flowing or artesian wells. These wells in particular may deplete the available water in the subsurface strata if they are pumped excessively.

In general, ground water supply sources tend to:

- Have harder water (more minerals) than surface water
- Be clearer (have less color) than surface water
- Have less bacteria than surface water and require less treatment
- Be less quickly affected by periods of decreased rainfall than surface water

WATER TREATMENT FACILITIES

The treatment of water for the water supply system is important from a health and safety standpoint **(Figure 4.2)**. It improves the water's clarity and color and removes odors, contaminants, bacteria, and other disease-carrying organisms. The impetus for water departments to properly treat water dates back to federal regulations imposed by the Safe Drinking Water Act of 1974. This law specified requirements for purification and limits for contamination in drinking water.

Figure 4.2 Water treatment and pumping plants are the heart of the water supply system.

Water treatment generally involves two basic processes: adding elements that may improve the quality of the water and removing contaminants. Some of the additives that may be used include fluoride (which improves dental health) and oxygen (which improves water quality). Contaminants can be removed by four basic processes:

1. Sedimentation
2. Filtration
3. Coagulation
4. Chemical treatment

Sedimentation uses gravity to cause particles to fall out of the water. This is accomplished by impounding water in settling basins or filter beds and keeping it relatively still. Over time the solid contaminants fall to the bottom of the bed and the clean water is pumped off the top. Periodically the sludge that forms at the bottom of the basin must be removed so that additional water can be treated.

Filtration is accomplished by passing the water through a series of filters that catch and remove suspended matter. The media used to filter the water include sand, diatomaceous earth, metal or composite screens, and similar materials. The filters must be periodically cleaned or replaced for the system to work effectively.

Coagulation typically is not used as a sole means of purification but in combination with one of the other methods. This method introduces a chemical to the water that causes solid particles to bond together forming larger pieces. This allows the particles to be more easily caught by filters in a filtration system. It also increases the fallout speed in a sedimentation system.

Chemical treatment can be used to remove or add elements to the water. Living organisms, such as bacteria, can be removed from the water by chlorination (the addition of chlorine). Other chemicals can be added to remove some of the elements such as calcium and magnesium that make the water hard. These chemicals are referred to as water softeners. Again, fluoride and oxygen are also commonly added to the water to make it healthier to drink. In particular, fluoride greatly improves the strength and overall health of teeth.

In general, fire departments and fire protection professionals have little to do with water treatment. Their main concern regarding treatment facilities is that a maintenance error, natural disaster, loss of power supply, or fire that could disable the pumping station(s) or severely hamper the purification process. Any of these situations would drastically reduce the volume and pressure of water available for fire fighting operations. Another problem would be the treatment system's inability to process water fast enough to meet the demand. Fire officials must have a plan to deal with either of these potential shortfalls. A related concern for fire protection personnel regarding water treatment facilities is the bulk storage and usage of hazardous chemicals at these locations (**Figure 4.3**). Large quantities of chlorine, fluorine, various acids, and other hazardous materials are commonly on site. Preincident plans for these facilities must account for these chemicals.

Fire departments must have plans to deal with potential water supply shortfalls.

Figure 4.3 The treatment of water necessitates the storage of some hazardous chemicals, such as chlorine, at water treatment facilities.

MEANS OF MOVING WATER

Locating and treating water is only half the battle of operating a water supply system. The other half involves moving the water and distributing it through the community. For a water supply system to work, water must first be moved from its supply source to the treatment facility. Once treated, the water must be introduced into the water distribution system with sufficient pressure to make it useable. Three basic types of water movement systems are in common use:

1. Gravity
2. Direct pumping
3. Combination

Gravity Systems

Gravity systems use a primary water source located at a higher elevation than the distribution system **(Figure 4.4, p. 54)**. The head pressure created by the higher elevation provides the energy to move the water through the entire system. You will recall from our earlier discussions about head or elevation pressure that water gains 0.433 psi for every foot of elevation drop. Therefore, in order to obtain sufficient pressure to supply a reliable municipal water distribution system, the water source generally must be located at least several hundred feet higher than the highest point in the water distribution system. The most common examples include a mountain reservoir that supplies water to a city below or a system of elevated tanks in a city itself. Total gravity systems are the least common of the three types of systems.

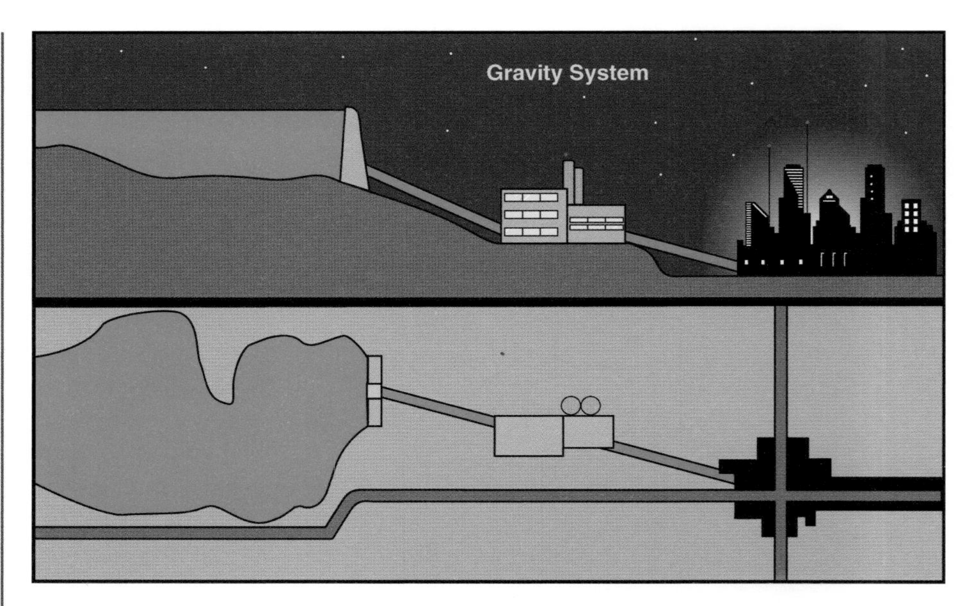

Figure 4.4 Gravity systems are only practical in jurisdictions where the water supply source is located substantially above the area being served.

Direct Pumping Systems

Direct pumping systems are most commonly found in jurisdictions where the system's water supply source is located at the same elevation as or lower than the system it supplies. Direct pumping systems use one or more pumps that take water from the primary source and discharge it through the treatment processes **(Figure 4.5)**. From there, another series of pumps forces the water into the distribution system. Most direct pumping systems make little or no use of elevated storage.

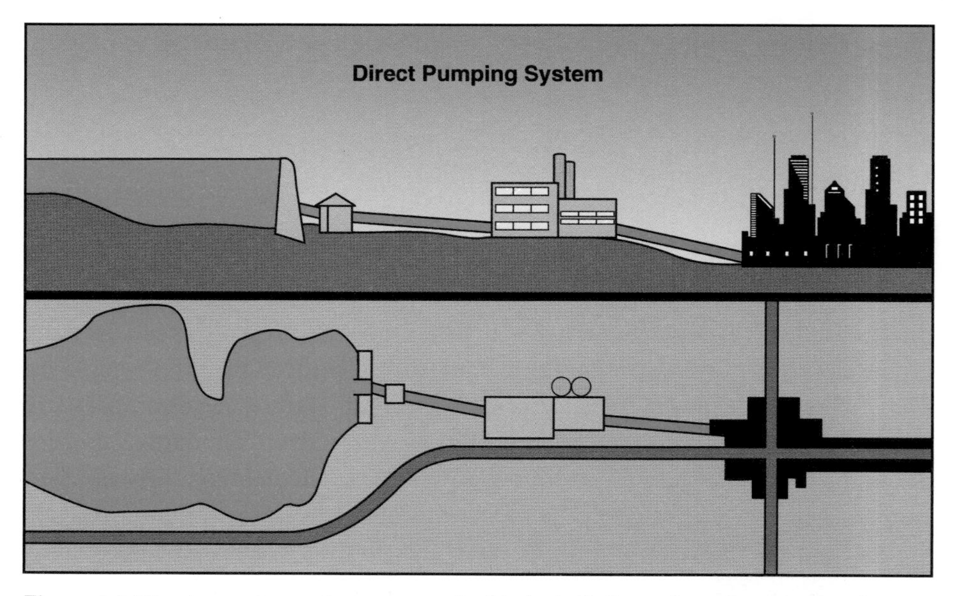

Figure 4.5 Direct pumping systems are required in jurisdictions where the water supply source is at the same level or below the area being served.

The types of pumps and drivers to power the pumps will vary depending on the local system. The most commonly used drivers are electric motors, followed by diesel engines. The most reliable systems have available both types of pump drivers. Thus, if a power failure knocks out the electric motors, the diesel pumps may be used to supply the system. Most electric pumps are also equipped with back-up electrical generators that may be used for power if the normal supply source is interrupted.

Most direct pumping systems are computer controlled. The computers regulate the pumps, valves, and other components of the system. They also monitor the demand on the system and make adjustments to ensure that an adequate flow of water is maintained. As with the pump drivers, it is important that this computer equipment be redundant and have sufficient backups in the event that the primary equipment fails. Booster pumps may be located at strategic points in large water distribution systems to compensate for pressure losses due to friction or elevation changes.

Combination Systems

The most common type of water supply system in North America is the combination system. Combination systems utilize elements of both the direct pumping and gravity systems **(Figure 4.6)**. In most cases, elevated storage tanks supply the gravity flow. These tanks serve as emergency storage and provide adequate pressure through the use of gravity. When the system pressure is high during periods of low consumption, automatic valves open and allow the elevated storage tanks to fill. When the pressure drops during periods of heavy consumption, the elevated tanks provide extra water by feeding it back into the distribution system. A good combination system requires reliable, duplicated equipment and properly sized, strategically located elevated tanks.

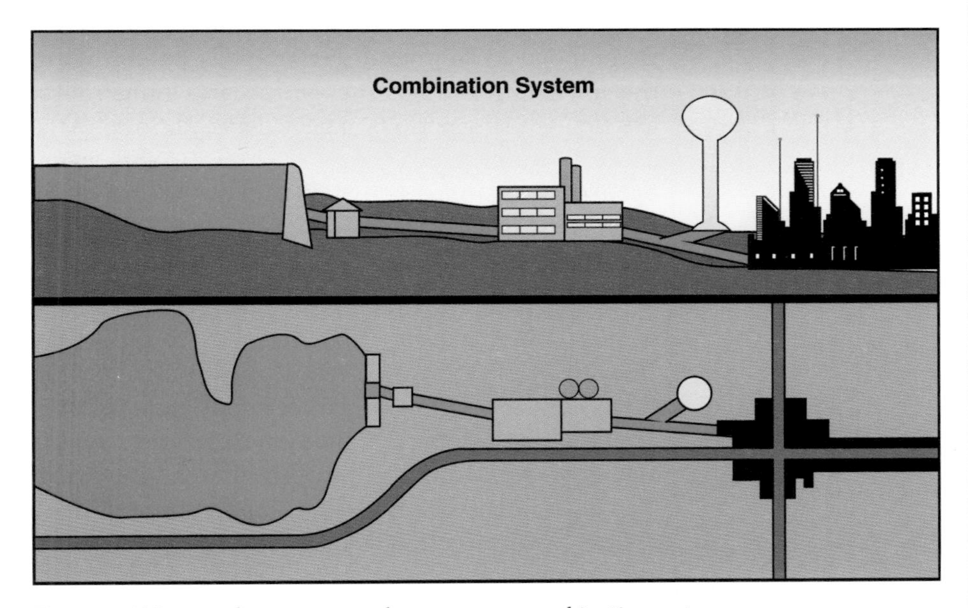

Figure 4.6 Most modern water supply systems are combination systems.

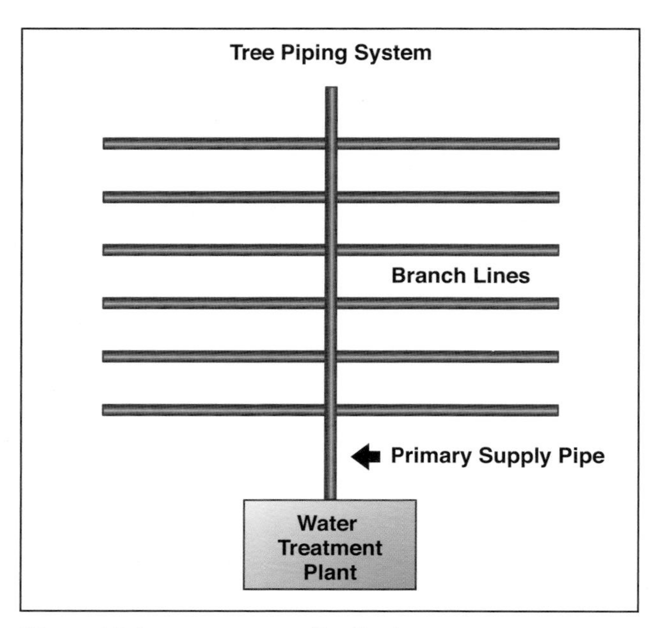

Figure 4.7 A tree-type water distribution system.

Figure 4.8 Circle water distribution systems are common in industrial facilities.

One advantage of a combination system is that the water stored in elevated reservoirs can ensure continued supply when the pumping portion of the system becomes inoperative. Elevated storage should be sufficient to provide domestic and industrial demands plus the demands expected in fire fighting operations. Such elevated storage also should be sufficient to permit making most repairs, alterations, or additions to the system. The locations of the elevated storage tanks and the capacity of the mains leading from them are also important. The elevated storage tanks should be placed where the system is most likely to require additional pressure and flow such as at the ends of long feeder lines, at the outermost boundaries of the system, and near areas that are subject to frequent high demands.

Water Distribution System

The distribution component of the water supply system receives the water from the pumping station and delivers it throughout the area served. In addition to the pumps and elevated storage tanks, the distribution system includes pipes, valves, meters, and fire hydrants.

The ability of a water system to deliver an adequate quantity of water relies upon the carrying capacity of the system's network of pipes. There are three common designs for water supply system piping arrangements:

1. **Tree system.** These systems contain a central primary supply pipe that feeds branch distribution lines throughout the jurisdiction (**Figure 4.7**). Their primary disadvantage is that most places in the system receive water supply in only one direction. Thus, points toward the ends of the branch lines can receive dangerously low flows during periods of peak demand closer to the center of the system. As well, if a break in the system occurs, all customers downstream of the break may completely lose their water supply.
2. **Circle, or belt, system.** These systems usually are found in small jurisdictions or in private systems that supply an industry or small complex (**Figure 4.8**). They form a loop that serves all of the customers on the system. They generally have a consistent, normal direction of flow through the system. However, in the event of a break on one side of the system, the direction of flow may be reversed to supply the rest of the system, assuming that there are no automatic check valves or backflow preventers. These systems are not commonly used for municipal domestic consumption water supply systems. However, they are frequently used for the secondary fire protection water supply system in large cities.

3. **Grid system.** Most modern municipal water distribution systems are grid systems **(Figure 4.9)**. These systems utilize an interlooped system of pipes connected at standard intervals to provide a reliable, multidirectional flow of water through the system. They are the most reliable type of water distribution. A break in almost any part of the system can be isolated using valves to maintain water supply throughout the rest of the system.

A grid system generally consists of the following piping components:

- **Primary feeders** —Primary feeders are the largest pipes (or mains) in the system. They convey large quantities of water to various points of the system for local distribution to the smaller mains.
- **Secondary feeders** —Secondary feeders comprise the network of intermediate-sized pipes that reinforce the grid within the various loops off the primary feeder system. They aid in the concentration of the required fire flow at any point.
- **Distributors** —Distributors are the smaller mains serving individual fire hydrants and blocks of consumers throughout the system.

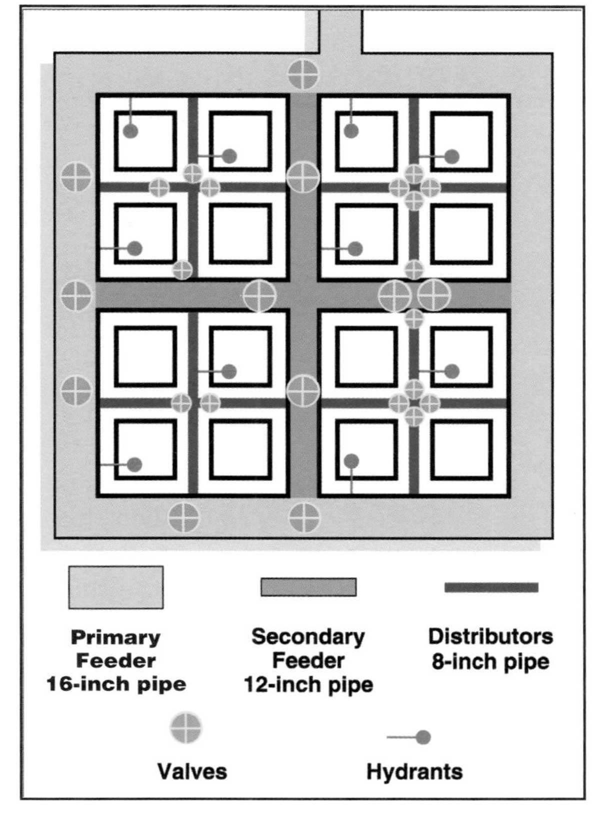

Figure 4.9 Most modern municipal water supply systems are grid systems.

From a fire protection standpoint, one of the grid system's primary advantages is that most of the fire hydrants can receive water from two directions. This factor greatly reduces the amount of friction loss in the system and increases the volume of available water. Fire hydrants that receive water from one direction are referred to as dead-end hydrants, and they generally should be avoided if at all possible.

In well-designed grid systems, two or more primary feeders run by separate routes from the supply source to the community's high-risk and industrial districts. This helps to ensure an adequate water supply at all times. Similarly, secondary feeders should be arranged in loops as far as possible to give two directions of supply to any point. This increases the capacity of the supply at any given point and ensures that a break in a feeder main will not completely cut off the supply to the system.

The earliest water supply systems utilized piping made from hollowed logs. Most piping in modern distribution systems is cast iron, ductile iron, asbestos cement, steel, plastic, or concrete, and it is almost always installed underground. (Asbestos cement pipe is no longer being installed, and most previously installed asbestos pipe has been replaced or relined to encase the asbestos.) Pipe should always be the proper type for the soil conditions, pressures, and temperatures to which it will be subjected. When water mains are installed in unstable or corrosive soils or in difficult access areas, steel or reinforced concrete pipe may be used to give the strength needed. Some locations that may require extra protection include areas beneath railroad tracks and highways, areas close to heavy industrial machinery, areas prone to earthquakes, or areas of rugged terrain.

The valves in a water distribution system have a tremendous impact on the system's operation.

The internal surface of the pipe, regardless of the material from which it is made, offers resistance to water flows. Some materials, however, offer considerably less resistance to water flow than others. Personnel from the engineering division of the water department should determine the type of pipe best suited to the conditions at hand.

The amount of water able to flow through a pipe and the amount of friction loss created can also be affected by other factors. Frequently, friction loss is increased by encrustation of minerals on the interior surfaces of the pipe. Another problem is sedimentation that settles out of the water. Both of these conditions result in a restriction of the pipe size, increased friction loss, and a proportionate reduction in the amount of water that can be drawn from the system.

The sizes of pipes used in systems will vary depending on the size of the community they serve. Primary feeders may range from 8 inches to 72 inches in diameter or more. Secondary feeders generally fall in the range of 6 to 36 inches. Distributors may be as small as 2 inches, but are most commonly between 4 and 20 inches. Any of these dimensions may be smaller or larger than those stated, depending on local need. The American Waterworks Association currently recommends a minimum diameter of 8 inches for any new pipe installation.

WATER MAIN VALVES

The valves in a water distribution system have a tremendous impact on the system's operation. For this reason, the fire protection professional should have a basic understanding of the different types of valves and their operation.

The function of a valve in a water distribution system is to provide a means for controlling the flow of water through the distribution piping. Valves should be located at frequent intervals in the grid system so that only a minimum length of pipe is out of service at one time and only small districts are cut off if it becomes necessary to stop the flow at any specific point. They should be operated at least once a year to keep them in good condition. The actual need for valve operation in a water system rarely occurs, sometimes not for many years.

One of the most important factors in a water supply system is the water department's ability to operate the valves promptly during an emergency or equipment breakdown. A well-run water utility has records of the locations of all valves. Valves should be inspected and operated on a regular basis. If each fire company is informed of the locations of valves in the distribution system, their condition and accessibility can be noted during fire hydrant inspections. The fire department will then inform the water department if any valves need attention.

Figure 4.10 Indicating valves are more common on private water supply systems.

Valves for water distribution systems are broadly divided into indicating and nonindicating types. An indicating valve visually shows whether the gate or valve seat is open, closed, or partially closed **(Figure 4.10)**. Indicating valves are not common in municipal water distribution systems but are more frequently found in private fire protection systems. Two common types of indicator valves are the post indicator valve (PIV) and the outside screw and yoke valve (OS&Y valve). Post indicator valves have a hollow metal post attached to the valve housing. The words OPEN and SHUT printed on the valve stem inside this post indicate the position of the valve. PIVs are commonly used on private water supply systems. The OS&Y valve has a yoke on the outside with a threaded stem that controls the gate's opening or closing. The threaded portion of the stem is out of the yoke when the valve is open and inside the yoke when the valve is closed. These are most commonly used on sprinkler systems but may be found in some water distribution system applications.

Figure 4.11 Most municipal water supply systems use nonindicating valves.

Most valves in municipal water distribution are of the nonindicating type **(Figure 4.11)**. As the name suggests, it is not possible to tell the position of a nonindicating valve by simply looking at it. Nonindicating valves in a water distribution system typically are buried or installed in manholes. If a buried valve is properly installed, it can be operated aboveground through a valve box using a special socket wrench on the end of a reach rod.

Control valves in water distribution systems may use either gate mechanisms or butterfly mechanisms. Both mechanisms can be used in either indicating or nonindicating valves. Gate valves may be of the rising stem or the nonrising stem type. The rising stem type is similar to the OS&Y valve. On the nonrising stem type, a valve nut is turned by a valve key (wrench) to raise or lower the gate and control the water flow. Non-rising-stem gate valves should be marked with a number indicating how many turns are necessary to completely close the valve. If a valve resists turning after fewer than the indicated number of turns, it usually means debris or other obstructions are in the valve. Butterfly valves are tight closing, and they usually have a rubber or a rubber-composition seat that is bonded to the valve body. The valve disk rotates 90 degrees from the fully open to the tight-shut position. The nonindicating butterfly type also requires a valve key. Its principle of operation provides satisfactory water control after long periods of inactivity.

The advantages of proper valve installation in a distribution system are readily apparent. If valves are installed according to established standards, only one or perhaps two fire hydrants will have to be closed off from service while a single break is being repaired. The advantage of proper valve installation is reduced, however, if all valves are not properly maintained and kept fully open. Valves that are only partially open cause turbulence resulting in high friction loss. Closed or partially closed valves may not noticeably affect ordinary domestic flows of water. As a result, the impairment will go undetected until a fire occurs or until detailed inspections and fire flow tests are made. A fire department will experience difficulty obtaining water in areas of the distribution system where valves are closed or partially closed.

Control valves must be properly maintained and kept fully open to ensure adequate water flows for fire fighting.

Figure 4.12 A standard fire hydrant.

Figure 4.13 This nonstandard hydrant located in a metropolitan area has two 2½-inch connections.

Report any problems with fire hydrants to the water department as soon as possible.

FIRE HYDRANTS

Access to the underground water distribution network for fire protection purposes is made through a fire hydrant. The American Water Works Association (AWWA) has adopted specifications for a national standard hydrant for ordinary waterworks service in the United States. These specifications are designed to produce a hydrant that is free from difficulties such as trouble in opening and closing, interior mechanical parts that can work loose, leakage, excessive friction loss, failure to drain properly, or loose nipples.

The two main types of modern fire hydrants are dry-barrel hydrants and wet-barrel hydrants. Regardless of the design or type, a standard hydrant will have at least one large outlet (4 or 4½ inches) for pumper supply and two outlets for 2½-inch couplings **(Figure 4.12)**. However, individual communities may specify different connections on their fire hydrants based on their standard operating procedures. Common deviations include hydrants that have only one or two small-diameter connections or hydrants that have two large-diameter connections and no small-diameter connections **(Figure 4.13)**.

Standard hydrants usually are equipped with a 5-inch valve opening and a 6-inch connection to the water main. The threads on all hydrant outlets must conform to those used by the local fire department. The principal specifications covered by the standard are the number of threads per inch and the outside diameter of the male thread. For exact details, refer to NFPA® 1963, Standard on Screw Threads and Gaskets for Fire Hose Connections. Hydrant location is usually determined by the type, size, and location of the protected occupancy(ies), and local code requirements.

Dry-barrel hydrants are used in areas that have freezing temperatures **(Figure 4.14)**. The dry-barrel hydrant has a base valve below the frost line; the stem nut to open and close the base valve is on the top of the hydrant. Any water remaining in a closed dry-barrel hydrant drains through a small valve that opens at the bottom of the hydrant when the main valve approaches a closed position. During freezing conditions, a firefighter should visually check a dry barrel hydrant after use to ensure that it drains completely. If the hydrant does not drain completely, the firefighter should use a dewatering pump to remove the remaining water and report the hydrant to the water department for repair.

The wet-barrel hydrant usually has a compression valve at each outlet but may have another valve in the bonnet that controls the water flow to all outlets **(Figure 4.15)**. This type of hydrant features the valve at the hose outlet and is used in mild climates where prevailing weather conditions are above freezing.

In addition to draining dry barrel hydrants, firefighters should be alert for leaks in the following areas anytime a fire hydrant is in use:

- At the main valve when the valve is closed
- At the drain valve when the main valve is open but the outlets are capped
- In the water mains near the hydrant

Any problems should be reported to the water department as soon as possible. The actual flow of water from a hydrant may vary due to such conditions as feeder main location, encrustation, deposits, and totally or partially closed supply valves. Firefighters can make better tactical decisions if they know at least the relative available water flow of different hydrants in the vicinity. To address this problem, either the fire or water department, depending on local policy, should conduct regular flow testing of hydrants. Following these tests, the hydrants should be color-coded according to a local system or to NFPA® 291, *Recommended Practice for Fire Flow Testing and Marking of Hydrants* (**Table 4.1**).

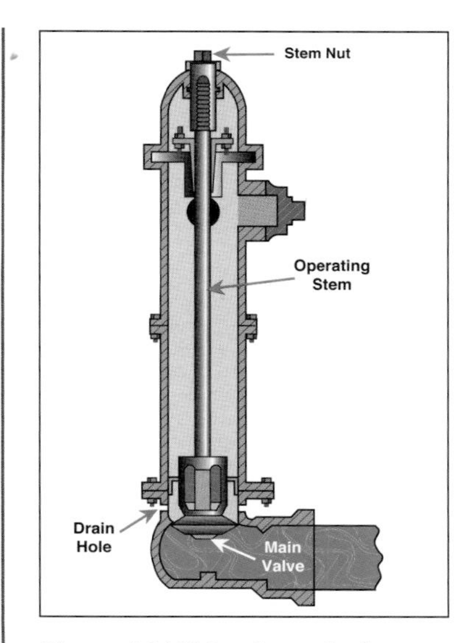

Figure 4.14 This schematic shows the important parts of a dry barrel hydrant.

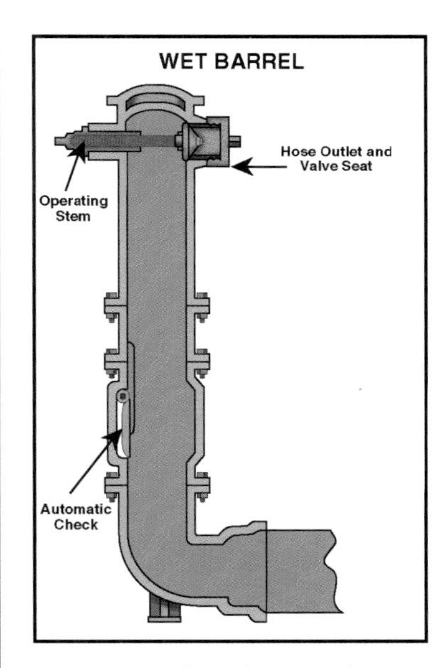

Figure 4.15 This schematic shows the important parts of a wet barrel hydrant.

Table 4.1
Hydrant Color Codes

Hydrant Class	Color	Flow
Class AA	Light Blue	1,500 gpm or greater
Class A	Green	1,000 -1,499 gpm
Class B	Orange	500-999 gpm
Class C	Red	Less than 500 gpm

PRIVATE WATER SUPPLY SYSTEMS

In addition to the public water supply systems that service most communities, fire protection professionals must also be familiar with the basic principles of any private water supply systems within their response jurisdiction. Private water supply systems are most commonly found on large commercial, industrial, or institutional properties. They may service one large building or a series of buildings on the complex. In general, the private water supply system exists for one of three purposes:

- To provide water solely for fire protection
- To provide water for sanitary purposes, consumption, and fire protection
- To provide water for fire protection and manufacturing processes

The design of private water supply systems is typically similar to that of the municipal systems described earlier in this chapter. Most commonly, private water supply systems receive their water in bulk from a municipal water supply system. Some private systems may have their own water supply source and treatment facility independent of the municipal water distribution system.

Some properties are served by two sources of water supply for fire protection: one from the municipal system and the other from a private source. In many of these cases, the private source of water for fire protection provides nonpotable (not for drinking) water. When this is the case, adequate measures must be taken

Fire protection professionals must be familiar with the design and reliability of private water supply systems in their jurisdiction.

to prevent contamination caused by the backflow of nonpotable water into the municipal water supply system. A variety of backflow prevention measures can prevent this problem. Some jurisdictions do not allow the interconnection of potable and nonpotable water supply systems. This means that the protected property is required to maintain two completely separate systems. Keep in mind that private water supply systems that rely on the municipal water distribution system as their sole water supply source are subject to service interruptions in the event that the municipal system fails.

Even if they have their own water supply source, most private water supply systems maintain separate piping for fire protection and domestic/industrial services. This is in direct contrast to most municipal water supply systems, in which fire hydrants are connected to the same mains that supply water for domestic/industrial use. Separate systems are cost prohibitive for most municipal applications but are economically practical in many private applications. Having separate piping arrangements in a private water supply system offers a number of advantages, including the following:

- The property owner has control over the water supply source.
- Either system (fire protection or domestic/industrial) is unaffected by service interruptions to the other system.

Fire protection professionals must be familiar with the design and reliability of private water supply systems in their jurisdiction. Large, well-maintained systems may provide a reliable source of water for fire protection purposes. Small-capacity, poorly maintained, or otherwise unreliable private water supply systems should not be relied upon to provide all the water necessary for adequate fire fighting operations. Historically, many significant fire losses can be traced, at least in part, to the failure of a private water supply system that was being used by municipal fire departments working the incident. Problems such as the loss of electrical service to a property whose fire protection system is supplied by electrically driven fire pumps have resulted in disastrous losses.

If the fire department has any question about the reliability of a private water supply system or of its capacity to provide an adequate amount of water for a large-scale fire fighting operation, it should arrange to augment the private water supply. This may be accomplished by relaying water from the municipal water supply system or by drafting from a reliable static water supply source close to the scene.

SUMMARY

A community's water supply system is one of the most important elements in its overall fire protection system. The importance of providing a source of clean, reliable water for normal daily consumption goes without saying. However, unless this source can be extended to meet the needs of fire fighting operations, firefighters will have virtually no ability to effectively fight fires.

Modern water supply systems incorporate four basic elements:

1. A water supply source
2. Water treatment facilities
3. A mechanism for forcing water through the system
4. A system of piping to transport the water through the community

Fire protection professionals need to understand the basic design and operation of the municipal water supply system in their jurisdiction, as well as any private water supply systems within their response area, so that they can use these systems to their maximum effectiveness and be able to spot problems that could affect fire protection concerns when they occur.

CHAPTER FOUR REVIEW QUESTIONS

1. Name the four primary elements of a municipal water supply system.

____

____

____

____

2. Describe the major features of the independent fire protection water supply systems used in some large cities.

____

____

____

3. Match the type of water system consumption with the appropriate definition. Write the appropriate numbers on the blanks.

Description	Type of Consumption
______ a. The maximum total amount of water used during any 24-hour interval within a 3-year period.	1. Average daily consumption (ADC)
______ b. The average of the total amount of water used daily in a water distribution system over a period of one year.	2. Maximum daily consumption (MDC)
______ c. The maximum amount of water used in any 1-hour interval over the course of a day.	3. Peak hourly consumption (PHC)

4. Name at least three surface water supply sources.

____

____

____

5. List at least two general characteristics of surface water.

____

____

6. List at least two general characteristics of ground water.

____

____

7. Name the four methods by which contaminants are removed from water.

____

____

8. Describe the main concerns regarding water treatment facilities.

____

____

9. Name the three basic types of water movement systems and describe the basic principles of each.

____

____

____

____

____

____

10. Match the type of water distribution piping system with the appropriate definition.

Description	Type of Piping System
______ a. An interlooped system of pipes that are connected at standard intervals to provide a reliable, multidirectional flow of water through the system.	1. A circle or belt system
______ b. Contains a central primary supply pipe that feeds branch distribution lines throughout the jurisdiction.	2. Grid system
______ c. These systems form a loop that serves all of the customers on the system.	3. A tree system

11. Match the descriptions of mains in a grid piping system with the appropriate type.

Description	Type of Piping System
______ a. Smaller main serving individual fire hydrants and blocks of consumers throughout the system.	1. Primary feeder
______ b. Largest pipe (or main) in the system; conveys large quantities of water to various points of the system for local distribution to smaller mains.	2. Secondary feeder
______ c. Network of intermediate-sized pipes that reinforce the grid within the various loops off the primary feeder system.	3. Distributor

12. What two basic categories of valves are used in a water supply system?

____

____

13. What two basic operating designs for fire hydrants are in use today?

____

____

____

14. What are the advantages to having separate piping arrangements in a private water supply system?

____

Answers found on page 424.

FESHE Course Objectives

The information in this chapter is intended to meet some of the objectives outlined for the Fire Protection Hydraulics and Water Supply course by the United States Fire Administration's National Model Core Curriculum as developed by the Fire and Emergency Services Higher Education (FESHE) initiative.

FESHE Course Objectives

1. Explain the importance of conducting water supply testing on the water supply system.
2. List and demonstrate the operation of equipment used to test a water supply system.
3. Demonstrate the procedures for determining the flow pressure and volume from a fire hydrant.
4. Explain the effect of the discharge opening on the flow testing process.
5. Perform a flow test on a water supply system.
6. Demonstrate the ability to compute flow tests results obtained during water supply testing.

Chapter 5

Water Flow Analysis

INTRODUCTION

In Chapter 4 we surveyed the basic elements of water supply distribution systems and their importance in the overall fire protection system of any given community. While all fire protection professionals can benefit from knowing how the water supply distribution system is designed and operated, the information they need most is a realistic picture of the system's capacity and performance capabilities. While engineering estimates or calculations may provide some data on the flow characteristics of a water supply distribution system, the only truly reliable method for determining the amount of available flow at any chosen point in the system is to perform water flow analysis testing at that point.

Water flow analysis testing is the only truly reliable method for determining the amount of available flow at any chosen point in a distribution system.

In this chapter we will examine the need to conduct water flow analysis, the equipment necessary to perform this testing, the proper procedures for conducting a water flow test, and methods for using the collected data to develop an overall picture of the water flow capability.

WHY WATER FLOW ANALYSIS IS NECESSARY

Water flow analysis is necessary for a number of reasons, including:

- *To determine whether a water supply system is operating as designed.* Many cases can be cited where a closed valve or significant blockage in the system came to light only as a result of a flow test. It is much more desirable to determine these types of problems during testing than during a fire emergency.
- *To determine whether a properly functioning water supply system can provide an adequate flow of water for the target hazards in the area of the test.* Again, learning of an insufficient flow only at the time of a working fire can have disastrous results. If the fire department can determine ahead of time that a potential water supply deficiency exists, it may develop special operating procedures to compensate for this problem. Fire officers familiar with fire flow test results are better prepared to position pumpers at strong hydrants and avoid weak locations. The water department also may choose to use this information to plan on upgrading the system in that area. Tests repeated at the same locations year after year may reveal a loss in the water mains' carrying capacity, increases in consumption, and a need for strengthening certain arterial mains. Flow tests should be run at least every five years, as well as after any extensive water main improvements or extensions have been made.

Learning of an insufficient flow only at the time of a working fire can have disastrous results.

Not only can water supply differ from location to location, but it also may change over time at the same location.

Although it is possible to mathematically calculate the water flow in the distribution system, even with modern computer programs these calculations are difficult, time-consuming, and in the end yield only approximations of the actual flow. The calculations may not be accurate for a number of reasons. First, sediment or mineral deposits within the system's pipes may reduce their actual internal diameter and thus reduce their flow capability. Secondly, piping corrosion and deterioration can result in greater pressure losses to friction and reduced carrying capacity of the pipe. Lastly, the mathematical calculations cannot forecast unforeseen valve closures or debris in the system. Each of these conditions can have a significant impact on the actual flow through the system.

In addition to problems with the piping system, the magnitude of water consumption near the test area also can affect the available water supply and fire protection. For example, if an industrial occupancy is using large volumes of water, the pressures and flows remaining for fire protection purposes may be reduced. Not only can water supply differ from location to location, but it also may change over time at the same location. This can result from population increases or industrial/commercial growth that places additional demands on the water supply system. For these reasons, the best way to establish the quantities and pressures available at a particular location is through actual flow testing.

As mentioned in Chapter 4, in most jurisdictions the repair, maintenance, and testing of public water systems is the water department's responsibility. The water department is better equipped to do this work than any other agency, including the fire department. Although the water department technically may be responsible for these functions, fire departments depend on the system to be in top operating condition during a fire. This requires the fire department to perform, or at least monitor, periodic testing of the hydrants and water supply system.

Testing of private water supply systems is the property owner's responsibility. Insurance providers often specify the frequency and extent of the tests to be conducted. Again, however, fire department officials may still wish to witness water supply testing at these sites to ensure that they are adequate for fire fighting operations. When performing or requiring testing of a private water supply system, fire department officials need to remember that water flowed from private hydrants may be metered by the municipal water department. Extensive testing can increase the property owner's water bills.

WATER SUPPLY ANALYSIS EQUIPMENT

To conduct a complete water supply analysis, the fire protection professional will need a variety of equipment. As we will discuss later in this chapter, each water flow test should begin with a thorough examination of the hydrant(s) being used for the test. Some of the equipment necessary to do this includes:

- **Thread gauging device**: to check the condition and type of thread used on the hydrant discharges.
- **Can of lubrication oil**: to lubricate the valve stem or the hydrant discharge caps.

- **Small, flat brush**: to clean debris from the hydrant discharge threads or from the discharge caps.
- **Gate valve key**: to check the operation of underground valves near the hydrant or to shut the valve in the event of a problem during testing.
- **Large pail**: to hold any debris removed from the hydrant for proper disposal at a later time.
- **Hydrant wrench**: to open the hydrant valve and to loosen tight discharge caps **(Figure 5.1)**.

Figure 5.1 A hydrant wrench.

Conducting the flow portion of the test will require several different items. The first is a static pressure gauge **(Figure 5.2)**. The static pressure gauge is essentially a discharge cap outfitted with two appendages. One is a small discharge orifice that is typically equipped with a petcock valve. This is used to bleed air bubbles from the hydrant before taking a static pressure. Another item is a pressure gauge that records the pressure bearing against the cap. Static pressure gauges used for water system testing should be capable of measuring at least 200 psi.

Figure 5.2 Static pressure cap gauge.

A third item is a pitot tube and gauge, often simply referred to as a pitot gauge. This device consists of a blade with a small opening through it (the pitot) and a pressure gauge **(Figure 5.3)**. The blade is inserted into a flowing stream of water to measure the stream's velocity pressure. Pitot gauges used for water system testing should be capable of measuring at least 200 psi.

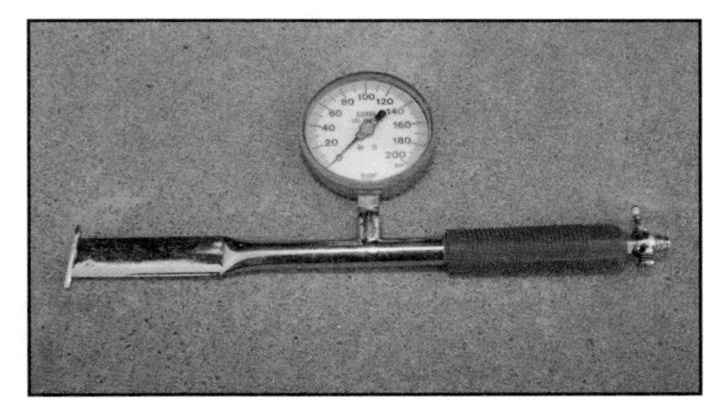

Figure 5.3 A pitot tube and gauge.

Using a pitot gauge to take a flow reading is not difficult, but it must be done properly to obtain accurate readings. **Figure 5.4** illustrates a good method of holding a pitot gauge at a hydrant outlet or nozzle. Note that the pitot gauge is grasped just behind the blade with the first two fingers and thumb of the left hand while the right hand holds the air chamber. The little finger of the left hand rests upon the hydrant outlet or nozzle tip to steady the instrument. Unless some effort is made to steady the pitot tube, the movement of the water will make getting an accurate reading difficult.

Figure 5.4 For this method, the pinkie finger is used to position the pitot tube and gauge.

Figure 5.5, p. 72 illustrates another method of holding the pitot gauge. The left hand's fingers are split around the gauge outlet, and the heel of the fist is placed on the edge of the hydrant orifice or outlet. The blade can then be sliced counterclockwise into the stream (**Figure 5.6 p. 72**). The right hand once again steadies the air chamber.

The procedure for using a pitot gauge is as follows:

Step 1: Open the petcock on the pitot tube and make certain the air chamber is drained. Then close the petcock.

Step 2: Edge the blade into the stream, with the small opening or point centered in the stream and held away from the butt or nozzle a distance of approximately one-half the diameter of

Figure 5.5 Place the heel of the hand so that the blade of the pitot tube is parallel to the stream of water.

Figure 5.6 Rotate the hand to insert the pitot blade into the stream.

Figure 5.7 The pitot blade should held a distance away from the discharge equal to one-half the diameter of the discharge.

the opening (**Figure 5.7**). For a 2½-inch hydrant butt, this distance is 1¼ inches. The pitot tube blade should now be parallel to the outlet opening, with the air chamber kept above the horizontal plane passing through the center of the stream. This increases the efficiency of the air chamber and helps avoid needle fluctuations.

Step 3: Take and record the velocity pressure reading from the gauge. If the needle fluctuates, read and record the value halfway between the high and low extremes.

Step 4: After the test is completed, open the petcock and be certain all water is drained from the assembly before storing.

The stream of water from a hydrant can have tremendous force. It has the potential to damage lawns, shrubs, vehicles, and other objects in the area. To avoid such damages, a water stream diffuser may be placed on the hydrant discharge. The diffuser reduces the stream's pressure to a point that will not cause any damage. It has an opening through which the pitot gauge may be inserted to get a reading.

In recent years some personnel have used flowmeters to measure the flow from the hydrant. Flowmeters are electronic devices that automatically convert the velocity pressure to a flow quantity readout (in gallons or liters per minute). This eliminates the need for some of the volume calculations discussed later in this chapter. However, flowmeters must be calibrated carefully to ensure reliable results.

The tester will also need some means of recording all of the data determined during the course of the test. The age-old clipboard or notebook works fine for this task. Portable computer equipment also may be used. Programs that contain electronic worksheets for recording information and calculating results may be useful. If a computer is not being used to calculate the results automatically, a scientific calculator also will be required.

DETERMINING AVAILABLE WATER SUPPLY

Water flow tests are the only reliable way for fire service personnel to determine the quantity of water available for fire protection in the water supply system. These tests are also commonly referred to as fire flow tests, but that term should not be confused with the required fire flow for a given occupancy. Required fire flow is the calculated amount of water needed to extinguish a fire in a given occupancy. Required fire flows are covered in detail in Chapter 6.

Fire flow tests are performed to determine the water flow available for fire fighting at various locations within a distribution system. First the static (normal operating) and residual pressures are measured, and then these measurements are used to calculate available water. A hydrant's flow and pressure readings can be applied to calculations or graphical analysis to determine the flow available at any pressure or the pressure available at any flow.

Before the fire department conducts a flow test, it should notify the responsible water department official because opening hydrants may upset the water supply system's normal operating conditions. Notification is also important because water service personnel may be performing maintenance work in the immediate vicinity; therefore, the results of the flow test would not be typical for normal conditions. Proper notification also will promote a better working relationship between the water department and fire service personnel.

Before the fire department conducts a flow test, it should notify the responsible water department official or, in the case of private water supply systems, the owner.

If a private water supply system is being tested, not only should the owner of the system be notified, but the owner or a representative of the owner should accompany the fire department personnel and assist in conducting the test. At the very least, the owner's representative should open and close all valves and perform any required maintenance functions. This releases the fire department from any responsibility for damages or omissions that may occur during the process.

Flow Test Procedures

Many fire service personnel operate under the mistaken impression that testing the water flow is as simple as opening a hydrant and measuring the flow. In reality that does not give an accurate picture of the system's true range of performance capabilities. For example, a hydrant may appear to have a strong flow when simply opened and flowed. However, if a pumper were to hook up to the hydrant and start pumping, the residual pressure could quickly drop and minimize the amount of water actually available for fire fighting. On the other hand, a hydrant may appear to have a low velocity pressure, but that pressure may be maintained regardless of how much water a connected pumper takes from the hydrant. Only a properly conducted flow test can accurately predict how the system will perform once the pumper is connected to the hydrant.

Only a properly conducted flow test can accurately predict how the water supply system will perform once the pumper is connected to the hydrant.

Before beginning the flow test, the personnel should inspect the condition of the hydrants they are using for the test. If any hydrant appears to have serious mechanical problems, the test should be postponed until it is repaired. Minor deficiencies may be corrected on the spot if possible. Personnel should note the following conditions during this inspection:

- Check for any obstructions near the hydrant, such as sign posts, utility poles, shrubbery, or fences **(Figure 5.8)**.
- Check the hydrant outlet(s) to ensure that they face the proper direction and allow adequate clearance between the outlet and the surrounding ground or other obstacles **(Figures 5.9 a and b)**. The clearance between the bottom of the butt and the grade should be at least 15 inches.
- Check for mechanical damage to the hydrant, such as dented outlets, damaged discharge threads, or rounded (stripped) stem nuts **(Figure 5.10)**.
- Check the paint for rust or corrosion.
- Check for foreign objects that may have been inserted into the hydrant discharges.
- Check water flow by fully opening the hydrant, then closing it, and finally checking its ability to drain once closed. You may check drainage visually or by placing the palm of your hand over the discharge opening and feeling for a slight sucking action **(Figure 5.11)**.

Figure 5.8 The hydrant should be free of obstructions.

Figures 5.9 a and b The hydrant discharges should be an appropriate distance above the ground and free of obstacles.

Figure 5.10 This damaged discharge would prevent a hose from being connected to the hydrant.

Certain precautions must be taken before and during flow tests to avoid injuries to personnel or passersby. Efforts must also be made to minimize damage to property from the flowing stream. These may include the use of stream diffusers. Both pedestrian and automobile traffic must be controlled during all phases of the testing. This may require assistance from the local law enforcement agency. Conducting flow tests in busy areas at off hours, such as very early in the morning, may also be advisable.

Other safety measures include tightening caps on hydrant outlets not being used, not standing in front of closed caps, and not leaning over the top of the hydrant when operating it. Property damage control measures include opening and closing hydrants slowly to avoid water hammer, not flowing hydrants where drainage is inadequate, and always remembering to check downstream to see where the water will flow. Because flowing water across a busy street could cause an accident, take proper measures beforehand to slow or stop traffic. Performing these tests during freezing weather generally is not a good idea. A good rule to follow is: When in doubt, do not flow! If conducting a flow test presents difficulties, seek solutions that will allow you to complete the test without disruptions or property destruction.

Figure 5.11 Suction on the hand indicates that the hydrant is draining properly.

Take precautions to avoid injuries to personnel or passerby and to minimize damage to property before and during flow tests.

When in doubt, do not flow!

A proper water supply test involves the simultaneous use of at least two fire hydrants. These hydrants are referred to by specific names:

- **Test hydrant**. This is the hydrant for which the test is being conducted. The calculations that we will learn later in this chapter are relevant only to this particular hydrant. Ironically, no water is discharged from this hydrant during the test.
- **Flow hydrant(s)**. This hydrant is opened and flowed to create a pressure drop at the test hydrant. Depending on the characteristics of the water distribution, opening more than one hydrant may be necessary to achieve the required amount of pressure drop in the system to compute accurate results.

The test hydrant must be located between the flow hydrant and the water supply source. In other words, the flow hydrant should be downstream from the test hydrant (**Figure 5.12**). The direction of flow can be determined by reviewing water maps supplied by the water department. When the test involves flowing multiple hydrants, the test hydrant should be centrally located relative to the flow hydrants.

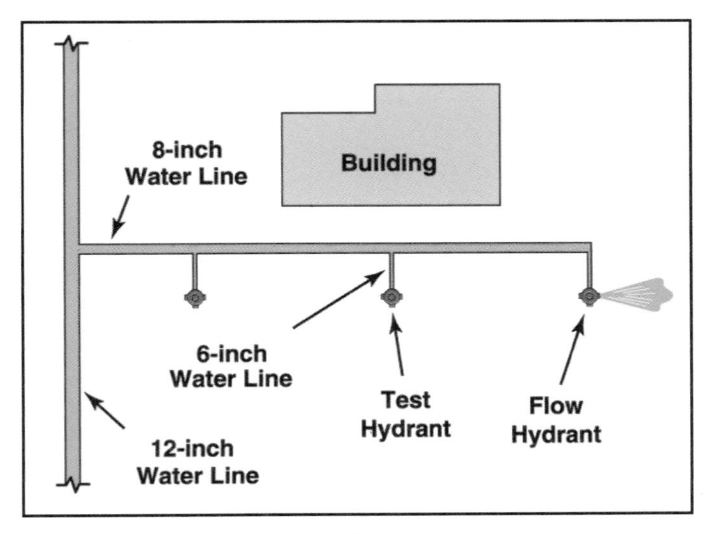

Figure 5.12 The test hydrant should be between the water supply source and the flow hydrant.

The following procedure should be used for conducting an available water test:

Step 1: Locate personnel at the test hydrant and at all flow hydrants to be used.

Step 2: Remove a hydrant cap from the test hydrant, check for debris inside the hydrant, and then snugly attach the static pressure gauge with the petcock valve in the open position. After checking the other caps for tightness, slowly open the hydrant several turns. Once all of the air has escaped and a steady stream of water is flowing, close the petcock and fully open the hydrant (**Figure 5.13, p. 76**).

Figure 5.13 Close the petcock when a steady stream of water is flowing.

Step 3: Read and record the static pressure as seen on the static pressure gauge.

Step 4: The individual(s) at the flow hydrant(s) removes the cap(s) from the outlet(s) to be flowed. Generally, 2½-inch outlets should be used to conduct hydrant flow tests. When using a hydrant outlet, check and record the hydrant coefficient and the actual inside diameter of the orifice. If a nozzle is placed on the outlet (the reasons for which are discussed later in this section), check and record its discharge coefficient and diameter. Use a ruler with a scale that measures to at least sixteenths of an inch to determine the actual diameter of the outlet or nozzle opening.

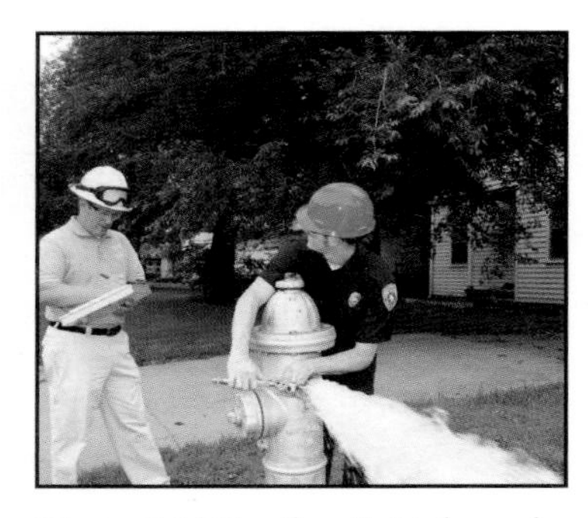

Figure 5.14 Use the pitot tube and gauge to record the flow pressure.

Step 5: Fully open the flow hydrant(s) as necessary, and take and record the pitot reading of the velocity pressures (**Figure 5.14**). The individual at the test hydrant simultaneously reads and records the residual pressure shown on the static pressure gauge. The residual pressure should not drop below 20 psi during the test. If it does, the number of flow hydrants must be reduced.

Step 6: Slowly close the flow hydrant(s) to prevent water hammer in the mains. After checking for proper drainage, replace and secure all hydrant caps. Report any hydrant defects.

Step 7: Check the test hydrant for a return to normal operating pressure; then close the hydrant. Open the petcock valve to prevent a vacuum on the static pressure gauge. Remove the static pressure gauge. After checking for proper drainage, replace and secure the hydrant cap. Report any hydrant defects.

Generally, 2½-inch outlets should be used to conduct hydrant flow tests. This is because the stream from a large hydrant outlet (4 to 4½- inches) contains voids; that is, the entire stream of water is not solid. Thus, the formulas for calculating flow will not give accurate results. If you must use large outlets, you can apply a correction factor to achieve more accurate results. Multiply the flow (as determined by $gpm = 29.83 \text{ x } C_d \text{ x } d^2 \text{ x } \sqrt{P}$) by the appropriate factors in Table 5.2 corresponding to the velocity pressure measured by the pitot tube and gauge.

Table 5.1 indicates a flow of 6 psi through a 4-inch outlet as 1,050 gpm. However, tests have shown that only 84 percent of this quantity is actually flowing due to voids in the water stream. Accordingly, actual flow is 1,050 x 0.84 = 882 gpm.

In some cases, opening more than one hydrant may be necessary to achieve accurate results. The number of hydrants to be opened depends on the estimated flow available in the area. For example, if you suspect that the system is very strong in the test area, you may have to open several hydrants for a more accurate test. You should open enough hydrants to drop the static pressure at the test hydrant by at least 10 percent. Thus, if you initially record a static pressure of 90 psi on the test hydrant, the pressure at the test hydrant must drop at least

TABLE 5.1
Discharge Table for Circular Outlets* (U. S.)
Outlet Pressure Measured by Pitot Gauge

Outlet Pressure in lbs. per sq. inch	Outlet Diameter in Inches											
	$2\frac{3}{8}$	$2\frac{1}{2}$	$2\frac{5}{8}$	$2\frac{3}{4}$	$2\frac{7}{8}$	3	$3\frac{1}{8}$	$3\frac{7}{8}$	4	$4\frac{3}{8}$	$4\frac{1}{2}$	$4\frac{5}{8}$
	U.S. Gallons per Minute											
1	150	170	180	200	220	240	260	400	430	510	540	580
2	210	240	260	290	310	340	370	570	610	720	770	810
3	260	290	320	350	380	420	450	700	740	890	940	990
4	300	340	370	410	440	480	530	810	860	1030	1090	1150
5	340	380	410	450	500	540	590	900	960	1150	1220	1290
6	370	410	450	500	540	590	640	990	1050	1260	1340	1410
7	400	440	490	540	590	640	690	1070	1140	1360	1440	1520
8	430	480	520	570	630	680	740	1140	1220	1450	1540	1620
9	450	500	550	610	670	730	790	1210	1290	1540	1640	1720
10	480	530	580	640	700	760	830	1280	1360	1630	1730	1820
11	500	560	610	670	730	800	870	1340	1430	1710	1810	1910
12	520	580	640	700	770	840	910	1400	1490	1780	1890	1990
13	550	610	670	730	800	870	950	1450	1550	1850	1960	2070
14	570	630	690	760	830	900	980	1510	1610	1920	2040	2150
15	590	650	720	790	860	940	1020	1560	1660	1990	2110	2220
16	610	670	740	810	890	970	1050	1620	1720	2060	2180	2300
17	620	690	760	840	910	1000	1080	1660	1770	2120	2240	2370
18	640	710	780	860	940	1030	1110	1710	1820	2180	2310	2440
19	660	730	810	890	960	1050	1140	1760	1870	2240	2370	2510
20	680	750	830	910	990	1080	1170	1800	1920	2290	2430	2570
22	710	790	870	950	1040	1130	1230	1890	2020	2400	2550	2700
24	740	820	910	1000	1090	1180	1290	1970	2110	2510	2660	2810
26	770	860	940	1040	1130	1230	1340	2050	2190	2620	2770	2930
28	800	890	980	1070	1170	1280	1390	2130	2280	2720	2880	3040
30	830	920	1010	1110	1210	1320	1430	2210	2350	2820	2980	3150
32	860	950	1050	1150	1260	1370	1480	2280	2430	2910	3080	3250
34	880	980	1080	1180	1290	1410	1530	2350	2510	3000	3170	3350
36	910	1010	1110	1220	1330	1450	1580	2420	2580	3080	3260	3440
38	930	1040	1140	1250	1370	1490	1620	2480	2650	3170	3350	3540
40	960	1060	1170	1290	1400	1530	1660	2550	2720	3250	3440	3630

*Computed with Coefficient C = 0.90, to nearest 10 gallons per minute.

9 psi (to 81 psi) once the flow hydrants are opened. If more accurate results are required, the pressure drop should be as close as possible to 25 percent (in the above example that would mean a drop of at least 22.5 psi).

Fire protection professionals who have conducted water supply testing can often cite examples where the initial appearance of the capabilities of the system does not accurately portray the real capabilities of the system or the number of flow hydrants that will be needed to conduct a test. In one case at an international airport, the initial static pressure on the test hydrant measured 115 psi. This indicated a strong water supply system that would require opening several flow hydrants for the test. However, once the first hydrant was opened, the residual pressure at the test hydrant dropped to less than 15 psi. Thus, the volume of water available in the system was actually quite low; it just happened to be maintained under a high static pressure. Such results generally indicate a closed valve or other blockage in the system.

In another case, insurance authorities tested a designated high-flow, fire protection water supply system in a large U.S. city. The static pressure from the test hydrant measured 28 psi. Although, the test personnel had immediate concerns about the system's volume capabilities, they proceeded with the test. At a static pressure of 28 psi, they would have to drop the residual pressure by only 2.8, or about 3, psi in order to achieve the 10 percent pressure drop. What they found was that in order to get the 3 psi drop, they needed to open all of the discharge openings on 10 fire hydrants in the vicinity of the test hydrant. Thus, at the minimum desirable residual pressure of 20 psi for fire department pumper use, they calculated that over 18,000 gallons of water per minute were available.

Figure 5.15 Place a nozzle on the hydrant to boost the velocity pressure.

Another situation that sometimes arises during water flow tests is water mains with pressures so low that no flow pressure registers on the pitot gauge. To compensate for this, place a straight-stream nozzle with orifices smaller than 2 ½-inch on the hydrant outlet to increase the flow velocity to a point where the velocity pressure is measurable (**Figure 5.15**). When using this technique, you will have to adjust the water flow calculation (described later in the chapter) to include the smaller diameter and its respective coefficient of friction.

Flow tests conducted in areas very close to the base of an elevated water storage tank or standpipe can result in flows that are quite large in gallons per minute. Remember that such large flows can be sustained only as long as sufficient water remains in the elevated tank or standpipe. To determine how much water will be available when the storage has been depleted, you should conduct an additional flow test with the storage tank shut off.

Computing Test Results

Once the physical portion of the water flow test is complete, those results are used to calculate the amount of water available fire fighting. Modern computer equipment and programs allow us to simply plug in the results and get an answer. However, fire protection professionals should understand the methods used to calculate this information, should the need to do so arise.

Two basic types of calculations must be performed. The first involves determining the flow volume from the flow hydrant(s). The second involves plotting all of the given information on a chart to determine the available flow at the test hydrant. Both of these are described below.

DETERMINING VOLUME OF FLOW FROM FLOW HYDRANTS

Before the actual available flow at the test hydrant can be determined, it is necessary to calculate the volume of water that was flowing from the test hydrant(s) during the test. Of course, if flowmeters were used during the test, this step is not necessary.

Calculating Flow from Test Hydrants

The easiest way to determine how much water is flowing from the hydrant outlet(s) is to use prepared tables for nozzle/outlet discharge, such as was shown in Table 5.1. Individual jurisdictions may develop specific tables for their use based on common system discharge pressures and the style of hydrants/discharge openings used in their locales.

If flow charts or computer programs are not available, the flow from each individual discharge may be determined using the following formula:

Equation 5.1

$$\text{Flow in gpm} = (29.83)(C_d)(d^2)(\sqrt{P})$$

Given:

29.83 = A constant derived from the physical laws relating water velocity, pressure, and conversion factors

C_d = The coefficient of discharge

d = The actual diameter of the hydrant or nozzle orifice in inches

P = The pressure in psi as read at the orifice

One of the basic principles of this formula is that if a hydrant had an ideal frictionless discharge orifice the coefficient of discharge (C_d) would be 1.0. However, in real life a hydrant orifice or nozzle will have a lower coefficient of discharge, reflecting friction factors that slow the velocity of flow. The coefficient will vary with the type of hydrant outlet or nozzle used. When using a hydrant orifice, the operator will have to feel the inside contour of the hydrant to determine which of the three types of hydrant outlets is present **(Figure 5.16)**. When a nozzle is used, the coefficient of discharge depends on the type of nozzle. To determine the coefficient of discharge for a specific nozzle, refer to the manufacturer's specifications.

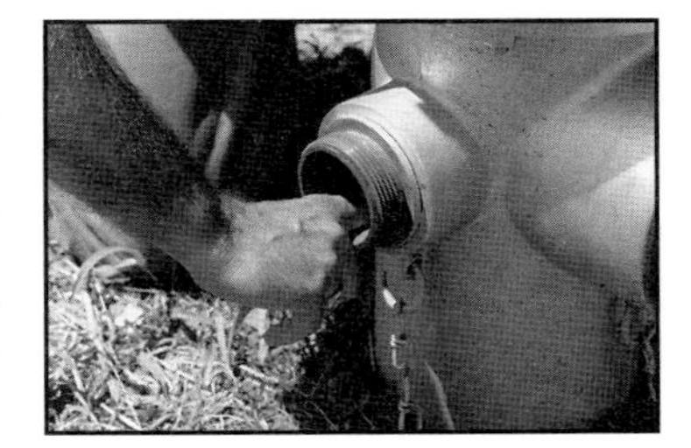

Figure 5.16 Determine the discharge coefficient by feeling the inside of the outlet.

Rarely does the actual internal diameter of the hydrant discharge equal exactly its stated size. In other words, a 2½-inch discharge is rarely exactly 2½-inches. It typically will vary by a couple of sixteenths of an inch one way or the other. This difference will have a significant impact on your calculations.

Let's look at a couple of sample calculations:

Example 5.1

During a flow test a single 2½-inch hydrant outlet is flowed. The actual diameter of discharge measures 2⁷⁄₁₆ inches (2.44 inches) with a C factor of 0.80. The flow pressure measured by the pitot gauge is 45 psi. What is the volume of water flowing from this discharge?

Solution:

Flow in gpm = $(29.83)(C_d)(d^2)(\sqrt{P})$

gpm = $(29.83)(0.80)(2.44)^2\ (\sqrt{45})$

gpm = $(29.83)(0.80)(5.9536)(6.708)$

gpm = 953.1 gpm

The test procedure stipulates that 2½-inch outlets should be used to conduct hydrant flow tests whenever possible because streams from large hydrant outlets (4 to 4½ inches) contain voids. In other words, their entire stream of water is not solid. Because of this, the flow formula will not give accurate results. If you must use the large outlets, using a correction factor can give more accurate results. The flow should be multiplied by the factor in **Table 5.2** that corresponds to the velocity pressure measured by the pitot tube and gauge.

TABLE 5.2
Conversion Factors for Large Diameter Outlets

Velocity Pressure	Factor
2 psi (13.8 kPa)	0.97
3 psi (20.7 kPa)	0.92
4 psi (27.6 kPa)	0.89
5 psi (34.5 kPa)	0.86
6 psi (41.4 kPa)	0.84
7 psi (48.3 kPa) or over	0.83

Example 5.2

The only discharge on the hydrant is 4½-inches. The actual diameter of discharge measures 4⅝ inches (4.625 inches), with a C factor of 0.90. The flow pressure measured by the pitot gauge is 60 psi. What is the volume of water flowing from this discharge?

Solution:

Flow in gpm = $(29.83)(Cd)(d^2)(\sqrt{P})$ x large diameter outlet correction factor from Table 5.2

$$gpm = (29.83)(0.90)(4.6252)^2(\sqrt{60})(0.83)$$

$$gpm = (29.83)(0.90)(21.39)(7.746)(0.83)$$

gpm = 3,692 gpm

Remember that if more than one discharge is used on any hydrant or if multiple hydrants are used, the flow from each individual discharge must be calculated before calculating the available fire flow.

REQUIRED RESIDUAL PRESSURE

From experience and water system analysis, fire protection engineers have established 20 psi as the minimum required residual pressure when computing the available water for area flow test results. This residual pressure is considered enough to overcome friction loss in a short 6-inch branch, in the hydrant itself, and in the intake hose, while allowing a safety factor to compensate for gauge error. Many state health departments require water supply systems to maintain this 20 psi minimum to prevent the possibility of external water being drawn into the system at main connections. Pressure differentials can cause water main collapse or cavitation, which is the implosion of air pockets drawn into pumps. More commonly, pumpers working at these low system pressures may be pumping near the water main's capacity. Shutting a valve on the pumper too quickly will create a water hammer. If this sudden surge in pressure is transferred to the water main, it can damage or break the main or connections. For the same reason, most fire department pumper operator training manuals likewise recommend that intake pressures on the pump during hydrant operations not be allowed to drop below 20 psi.

Determining Available Fire Flow

In addition to computer programs, available fire flow test results may be determined manually by graphical analysis or by mathematical computation. The following section details each of these methods.

Determining Available Water by Graphical Analysis

Perhaps the manual calculation most commonly used to determine available water flow is the graphical method. The graphical method relies on a logarithmic water flow chart that simplifies the process. The chart is accurate to an acceptable degree if you use a fine-point pencil or pen when plotting results. The figures on the vertical and/or horizontal scales may be multiplied or divided by a constant, as necessary to fit any problem.

The procedure for graphical analysis is as follows:

Step 1: Determine which of the three listed gpm scales should be used. This is based on the flows and pressures determined in the test and on the feasibility of plotting a line that will show the available flow at a residual pressure of 20 psi.

Step 2: Locate and plot the static pressure on the vertical scale at 0 gpm.

Step 3: Locate the total water flow measured during the test on the chart.

Step 4: Locate the residual pressure noted during the test on the chart.

Step 5: Plot the residual pressure above the total water flow measured.

Step 6: Draw a straight line from the static pressure point through the residual pressure point on the water flow scale.

Step 7: Read the gpm available at 20 psi and record the figure. This reading represents the total available water that can be relied upon.

The following examples of graphical analysis are for water flow tests using one and two flow hydrants.

Example 5.3: Two flow hydrants

Test hydrant = 60 psi static and 30 psi residual

Flow hydrant #1, using one 2½-inch outlet, with C = 0.80: pitot reading = 18 psi; actual discharge diameter = 2.56 inches.

$(29.83)(0.80)(2.56)^2(\sqrt{18}) = 664$ gpm

Flow hydrant #2, using one 2½-inch outlet, with C = 0.80: pitot reading = 9 psi; actual discharge diameter = 2.44 inches.

$(29.83)(0.80)(2.44)^2(\sqrt{9}) = 426$ gpm

Total Water Flow = 664 + 426 = 1,090 gpm

Solution:

Approximately 1,270 gpm would be available at 20 psi. This figure represents the minimum desired intake pressure **(Figure 5.17)**.

Figure 5.17 shows the test results plotted for graphical analysis of the water supply. The static pressure of 60 psi is plotted at 0 gpm. The residual pressure of 30 psi is above the total measured flow of 1,090 gpm on Scale B. The line drawn through the static and residual pressure points represents the water supply at the test location.

NOTE: Pitot pressures are never plotted on the graph; only the flow that corresponds to the pitot pressures is used.

Example 5.4: One flow hydrant

Test hydrant = 85 psi static and 45 psi residual

Flow hydrant, flowing two 2½-inch outlets (actual diameter 2.56 inches), each with C = 0.90; pitot reading for each = 17 psi.

$(29.83)(0.90)(2.56)^2(\sqrt{17}) = 725$ gpm x two outlets = 1,450 gpm

Solution:

The available water rate at 20 psi in Example 5.4 would be approximately 1,895 gpm (**Figure 5.18, p. 84**).

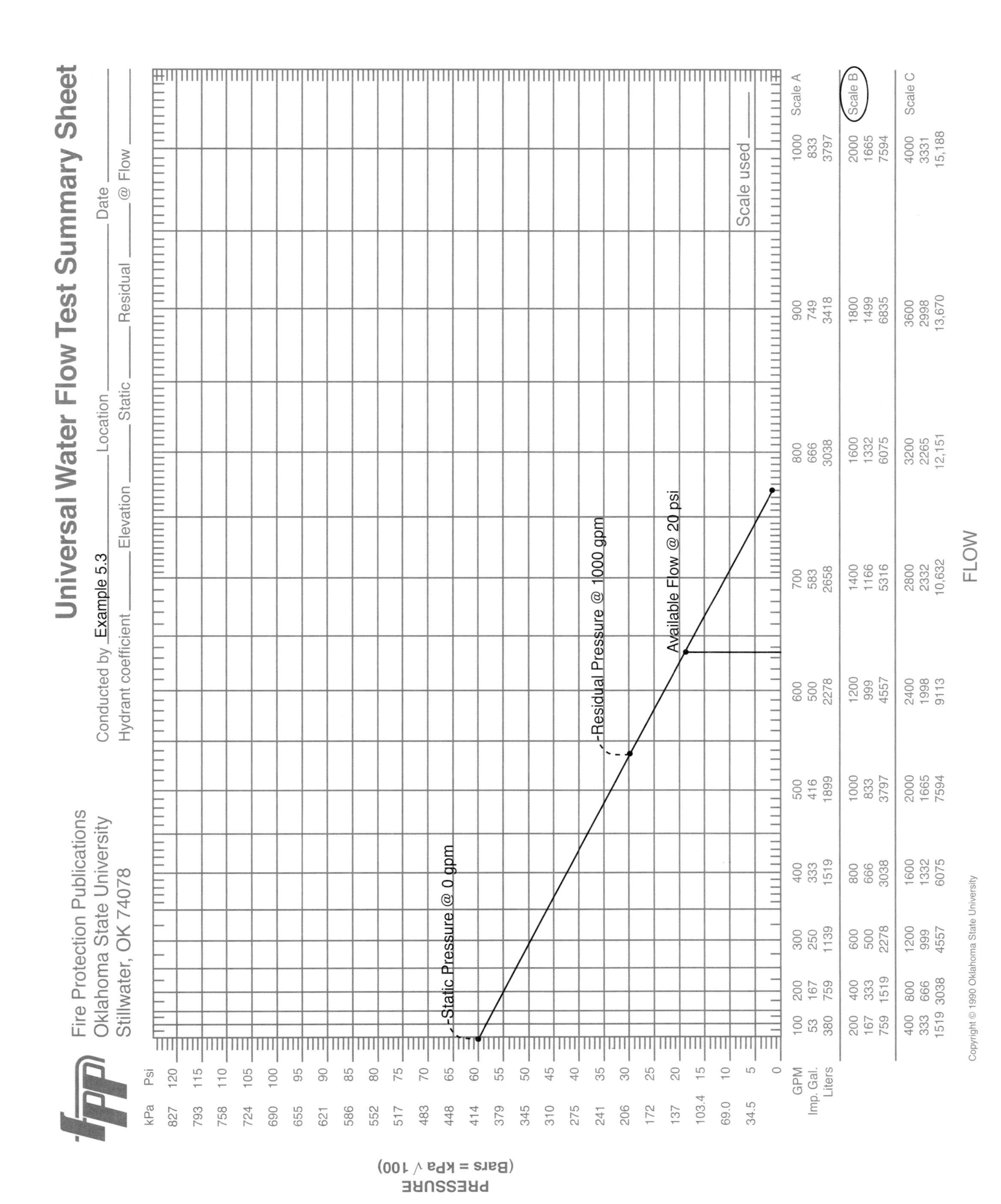

Figure 5.17 Example 5.3.

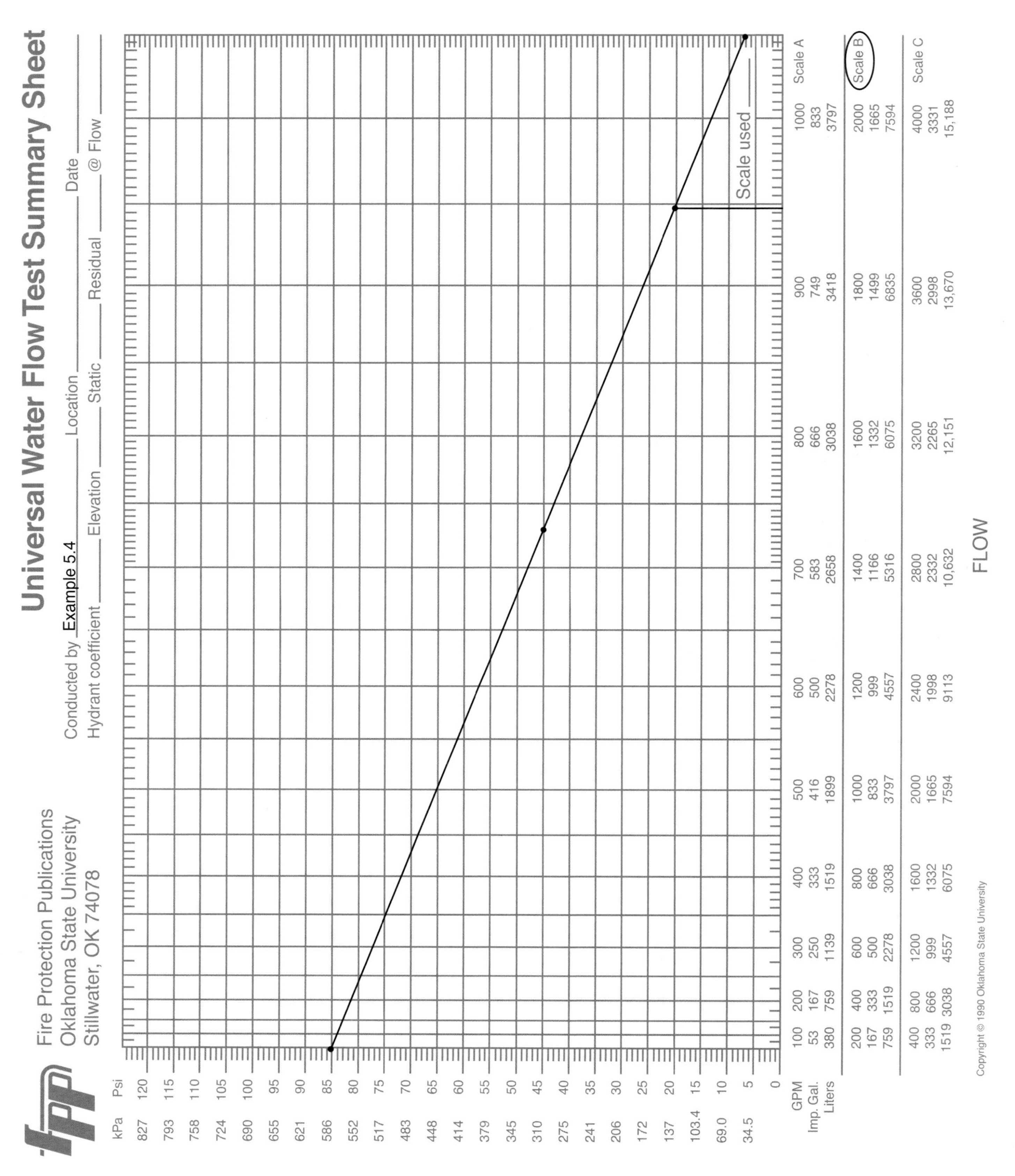

Figure 5.18 Example 5.4.

Determining Available Water by Mathematical Computation

While graphical analysis provides a fairly accurate picture of the available fire flow, some professionals prefer the precise results of mathematical calculation. This method uses a variation of the Hazen-Williams formula (see Chapter 3), as follows:

Equation 5.2

$$Q_r = \frac{(Q_f)(h_r^{0.54})}{h_f^{0.54}}$$

Given:

Q_r = Flow available at desired residual pressure

Q_f = Flow during test

h_r = Pressure drop to residual pressure (normal operating pressure minus required residual pressure)

h_f = Pressure drop during test (normal operating pressure minus residual pressure during flow test)

The values for h_r or h_f to the 0.54 power may be determined on a scientific calculator or found in **Table 5.3**.

Example 5.5

Using the conditions reported in Example 5.3, determine the available fire flow mathematically.

$$Q_r = \frac{(Q_f)(h_r^{0.54})}{h_f^{0.54}}$$

In this case:

Q_f = 1,090 gpm

h_r = 60 psi – 20 psi = 40 psi

h_f = 60 psi – 30 psi = 30 psi

So:

$$Q_r = \frac{(1{,}090)(40)^{0.54}}{30^{0.54}}$$

$$Q_r = \frac{(1{,}090)(7.33)}{6.28}$$

Q_r = 1,273.2 gpm

Note that the difference between the answers derived from the graphical method in Example 5.3 and from the mathematical method in Example 5.5 is only 3.2 gpm, or 0.2% of the final answer. This shows that either method is reliable when performed properly.

For more information on water supply analysis, see Pat D. Brock's *Fire Protection Hydraulics and Water Supply Analysis* (Fire Protection Publications).

TABLE 5.3
Values for Computing Fire Flow Test

h	$h^{0.54}$	h	$h^{0.54}$	h	$h^{0.54}$	h	$h^{0.54}$	h	$h^{0.54}$	h	$h^{0.54}$	h	$h^{0.54}$
1	1.00	26	5.81	51	8.36	76	10.37	101	12.09	126	13.62	151	15.02
2	1.45	27	5.93	52	8.44	77	10.44	102	12.15	127	13.68	152	15.07
3	1.81	28	6.05	53	8.53	78	10.51	103	12.22	128	13.74	153	15.13
4	2.11	29	6.16	54	8.62	79	10.59	104	12.28	129	13.80	154	15.18
5	2.39	30	6.28	55	8.71	80	10.66	105	12.34	130	13.85	155	15.23
6	2.63	31	6.39	56	8.79	81	10.73	106	12.41	131	13.91	156	15.29
7	2.86	32	6.50	57	8.88	82	10.80	107	12.47	132	13.97	157	15.34
8	3.07	33	6.61	58	8.96	83	10.87	108	12.53	133	14.02	158	15.39
9	3.28	34	6.71	59	9.04	84	10.94	109	12.60	134	14.08	159	15.44
10	3.47	35	6.82	60	9.12	85	11.01	110	12.66	135	14.14	160	15.50
11	3.65	36	6.93	61	9.21	86	11.08	111	12.72	136	14.19	161	15.55
12	3.83	37	7.03	62	9.29	87	11.15	112	12.78	137	14.25	162	15.60
13	4.00	38	7.13	63	9.37	88	11.22	113	12.84	138	14.31	163	15.65
14	4.16	39	7.23	64	9.45	89	11.29	114	12.90	139	14.36	164	15.70
15	4.32	40	7.33	65	9.53	90	11.36	115	12.96	140	14.42	165	15.76
16	4.47	41	7.43	66	9.61	91	11.43	116	13.03	141	14.47	166	15.81
17	4.62	42	7.53	67	9.69	92	11.49	117	13.09	142	14.53	167	15.86
18	4.76	43	7.62	68	9.76	93	11.56	118	13.15	143	14.58	168	15.91
19	4.90	44	7.72	69	9.84	94	11.63	119	13.21	144	14.64	169	15.96
20	5.04	45	7.81	70	9.92	95	11.69	120	13.27	145	14.69	170	16.01
21	5.18	46	7.91	71	9.99	96	11.76	121	13.33	146	14.75	171	16.06
22	5.31	47	8.00	72	10.07	97	11.83	122	13.39	147	14.80	172	16.11
23	5.44	48	8.09	73	10.14	98	11.89	123	13.44	148	14.86	173	16.16
24	5.56	49	8.18	74	10.22	99	11.96	124	13.50	149	14.91	174	16.21
25	5.69	50	8.27	75	10.29	100	12.02	125	13.56	150	14.97	175	16.26

SUMMARY

In this chapter we have taken a detailed look at water supply analysis as it relates to fire protection. We cannot state often enough that knowledge of the capabilities of the water supply system is crucial to fire protection professionals' ability to perform their job optimally. We perform water supply testing to determine whether a water supply system is operating as designed and to determine whether a properly functioning water supply system can provide an adequate flow of water for the target hazards in the area of the test.

To accomplish accurate testing, personnel must have reliable test equipment and know how to use it. The hydrants involved in the test should be tested visually and operationally before the flow test begins. The tried and true testing method described is this chapter should always be used for this process. The final results can be determined electronically, graphically, or through mathematical calculation.

Using the information in this chapter will provide a picture of the amount of water available for fire fighting operations. In the next chapter we will determine how much of that water or how much additional water will be needed to fight fires in a particular target hazard.

CHAPTER FIVE REVIEW QUESTIONS

1. List at two reasons for performing water flow analysis.

2. What agency is responsible for performing water flow testing on municipal water supply systems in most jurisdictions?

3. List at least five pieces of equipment needed to perform a water flow test.

4. List at least five safety precautions or visual inspections that should be made before performing the water flow test.

5. What is the ideal location in the water supply system for the test hydrant?

6. Which of the hydrant discharges are generally preferred for hydrant flow testing?

7. By what percentage should the residual pressure drop during testing in order to ensure accurate results?

8. During a flow test, a single 2½-inch hydrant outlet is flowed. The actual diameter of the discharge measures 2⅜ inches (2.375 inches) with a C factor of 0.70. The flow pressure measured by the pitot gauge is 83 psi. What is the volume of water flowing from this discharge?

9. During a flow test a single 4½-inch hydrant outlet is flowed. The actual diameter of the discharge measures 4⅜ inches (4.375 inches) with a C factor of 0.80. The flow pressure measured by the pitot gauge is 6 psi. What is the volume of water flowing from this discharge?

____

____

10. Using either the graphical or mathematical methods of calculation, determine the available volume of water at 20 psi residual pressure given the following test results:

    Test hydrant: 90 psi static pressure and 50 psi residual pressure

    Flow hydrant: Using one 2½-inch outlet, with C = 0.80, pitot reading = 42 psi, and actual discharge diameter = 2.56 inches.

____

____

11. Using either the graphical or mathematical methods of calculation, determine the available volume of water at 20 psi residual pressure given the following test results:

    Test hydrant: 70 psi static pressure and 55 psi residual pressure

    Flow hydrant: Using two 2½-inch outlets, with C = 0.90, pitot reading = 48 psi, and actual discharge diameter = 2.625 inches.

____

____

Answers found on page 426.

FESHE Course Objectives

The information in this chapter is intended to meet some of the objectives outlined for the Fire Protection Hydraulics and Water Supply course by the United States Fire Administration's National Model Core Curriculum as developed by the Fire and Emergency Services Higher Education (FESHE) initiative. The information in this chapter also allows the student to meet the requirements of NFPA® 1002, *Standard for Fire Apparatus Driver/Operator Professional Qualifications* (2009 edition) listed below.

FESHE Course Objectives

1. Explain and utilize the three common formulas used to calculate required fire flow rates for manual fire fighting operations.

2. List the fire flow requirements for automatic sprinkler and standpipe systems.

NFPA® 1002 Requirements

5.2.1 (A) and 5.2.2 (A) Requisite Knowledge. Hydraulic calculations for friction loss and flow using both written formulas and estimation methods.

Chapter 6

Calculating Required Fire Flows

INTRODUCTION

In the previous chapter we examined the methods used to determine the available water flow in a municipal or private water distribution system. We explained the importance of knowing how much water is available from the system so that fire protection professionals can perform adequate preincident planning. However, the amount of water available from the water distribution system is only one-half of the equation in terms of water for fire fighting. Equally important is the amount of water that must be applied to the fire in order to control and/or extinguish it in a reasonable amount of time. This often is referred to as the required fire flow (RFF) or as the needed or necessary fire flow (NFF). In this text we prefer to use the term required fire flow (RFF) for this purpose.

Determining the RFF has a number of benefits. First, it is important to compare the available fire flow to the RFF in order to determine if the water distribution system can supply the necessary amount of water should a major fire occur in the target hazard **(Figure 6.1;** *Courtesy of Chris Mickal, New Orleans F.D.*). Simply stated, if a water system can flow 800 gpm and a building will require an RFF of 1,200 gpm, firefighters have a problem. They will have to supplement the water from the system with an auxiliary source such as a nearby static water supply or a water shuttle operation. Determining the RFF is also helpful in deciding which apparatus and personnel need to respond to a fire at the target hazard. Suppose that the building will require an RFF of 1,800 gpm and the nearest hydrant is 500 feet away. A response that includes two 1,000 gpm pumpers will likely not be sufficient to control a well-advanced fire in this occupancy.

Figure 6.1 Major fires require large volumes of water to supply master stream nozzles. *Courtesy of Chris Mickal, New Orleans F.D.*

Required flow rates are important not only for manual fire suppression operations. Engineers and designers of fixed fire suppression systems, such as automatic sprinkler systems and standpipe systems must also know the flow requirements for their systems. As with manual fire suppression, if the water distribution system supplying the occupancy does not have sufficient volume or pressure to supply these systems adequately, it will also be necessary to find a way to supplement the water supply.

In this chapter we will examine three of the common methods to determine the RFF for manual fire fighting operations. Given the same occupancy and conditions, the results determined by these three methods may vary widely. No method for calculating this information has ever been proven to be totally accurate. Each jurisdiction or agency should determine which formula best suits its needs and then use it appropriately. Remember that when water for fire fighting is concerned, it is always better to plan on discharging too much water than too little. Thus, many jurisdictions choose the method that provides the highest flow figure.

The latter portion of this chapter will take a condensed look at the water supply requirements for automatic sprinkler and standpipe systems. In practice, some of these calculations can be quite comprehensive and difficult. This chapter's information on these topics is meant only to summarize the basic concepts and requirements.

REQUIRED FIRE FLOW FOR MANUAL FIRE FIGHTING OPERATIONS

Any fire service personnel who will be required to perform preincident planning or size-up and action plan development at working fires should be acquainted with at least one method to determine the required fire flow (RFF) for a given occupancy. The three methods, or formulas, commonly used by the fire service in the last fifty years are:

- The Iowa State Formula
- The National Fire Academy Formula
- The Insurance Services Office (ISO) Formula

The following section of this chapter details each of these calculation methods.

The Iowa State Formula

Before explaining the Iowa State Formula, we must discuss some important earlier fire service advances that led to its development. In the first part of the twentieth century, manual fire fighting efforts were limited mostly to exterior fire attacks using straight- or solid-stream nozzles. The limited number of interior fire attacks being conducted also relied on solid streams for applying water to the fire.

The Development of the Fog Nozzle

Though many credit the United States Navy or Coast Guard with pioneering fog-stream fire fighting technology, the initial foray into this topic actually can be traced to the oil fields of Santa Fe Springs, California, in the mid-1920s. It was there that Captain Glenn Griswold (later to become battalion chief) of the Los Angeles County Fire Department looked for a better way to extinguish the oil fires that plagued his response district. This was before foam concentrate for attacking Class B fires was available to the fire service. The traditional solid-stream nozzles in use for standard fire fighting operations were nearly useless in controlling oil fires. In fact, improper use of these nozzles often led to spreading the fire. Griswold correctly theorized that if water could be broken into little drops and discharged across the fire, it would absorb the heat and dilute the burning gases. As long as the drops of water were not large enough to agitate the fuel, the fire would be extinguished in place.

Griswold experimented with several designs of fog nozzles over 5 years or so before settling on an impinging-stream design that created optimally small droplets of water at pressures between 50 and 275 psi. Griswold secured numerous patents and founded the Fog Nozzle Company to produce the nozzles, which he called the Griswold Fognozl (also known as the California Fognozl). He ultimately produced several models of the nozzles as well as applicator pipes up to 30 feet long that could be used to insert the stream into otherwise unreachable areas.

The U.S. Coast Guard and Navy Experiments

Not until nearly 20 years after Griswold developed the fog nozzle did the Navy and Coast Guard begin experiments using fog-stream technology for interior fire attacks within confined spaces. In both cases, their initial research concerned containing oil fires within the lower levels of ships. Lloyd Layman conducted the Coast Guard research during his wartime stint as a fire instructor at the Coast Guard fire school in Baltimore, Maryland. In civilian life Layman was the fire chief in Parkersburg, West Virginia. His Coast Guard research involved inserting fog streams via applicator pipes into completely closed rooms (except for a small exhaust opening) containing oil fires **(Figure 6.2)**. Layman noted that immediately after the flow of water started, a massive amount of smoke exited the exhaust opening, followed by a mixture of smoke and steam. This was soon followed by condensed steam, and within minutes there was no discharge whatsoever. He called this the indirect attack method of fighting fires, and he widely published the results of his tests.

Layman's indirect attack theory and Griswold's direct attack theory differed in concept. Layman's theory was based on the atmospheric displacement of the heat and products of combustion by the expanding steam within the space. Griswold's direct attack theory was based on the dilution of the heat and combustion products. In reality, both theories were correct based on the conditions of their tests.

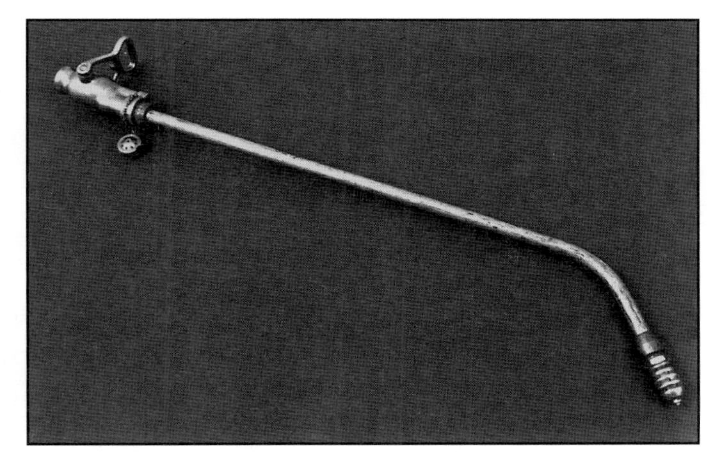

Figure 6.2 Early Navy fog nozzles were equipped with an applicator pipe that could be inserted through an opening into a fire area.

Layman's tenure with the Coast Guard lasted from 1943 through 1947, when he returned to his regular chief's job in Parkersburg. He then spent much of the rest of his career applying his shipboard oil fire fighting techniques to standard structural fires. By then, numerous manufacturers were marketing fog nozzles, and Layman worked with all of them. Over time he refined his fundamental principles of attacking and extinguishing interior fires to include the following:

- The indirect method will work only in confined spaces containing significant involvement of fire. It does not work well on incipient fires.
- The fog stream should be played on the upper portions of the involved space.
- Firefighters should not be in the space in which the fog will be introduced and furthermore should be protected from any steam that is developed.
- The stream should be discharged into the fire area without interruption until the condensed steam that emerges has decreased considerably.

About the same time that Layman was conducting his experiments for the Coast Guard, Chief Edward McAniff was performing similar research at the Navy Fire Fighting School at Pearl Harbor, Hawaii. McAniff, who later would become chief of the Fire Department of New York (FDNY), conducted tests under virtually the same conditions as Layman and generally got the same results. However, McAniff felt that the practical application of this method of fire attack would be limited to rare instances when perfect conditions for its use existed. He saw little or no application for the technique in standard structural fire fighting.

In 1950, Chief Layman was invited to present his research at the Fire Department Instructor's Conference (FDIC) in Memphis, Tennessee. His address, titled "Little Drops of Water," developed so much interest that the International Fire Service Training Association (IFSTA), in conjunction with the insurance industry, formed an exploratory committee to examine the issue. The committee's work involved controlled test burns on numerous houses in Miami, Florida, and Kansas City, Missouri. Ultimately, the committee validated Layman's theories, included the information in the IFSTA Fire Stream Practices manual, issued a report titled "An Analysis of the Extinguishment of Confined Structural Fires," and produced four films that showed the testing and principles of indirect fire attack.

While Layman and the others did an extensive amount of work to show how this method of fire attack worked, they never developed a formula for determining how much water must be injected into the fire area to achieve extinguishment. However, that research and information would soon follow.

The Iowa State Research

About the same time that Layman's principles were receiving wide attention in the fire service, Iowa State University's (ISU) state fire training program hired two gentlemen who would eventually take Layman's research to the next level. In 1951, Floyd W. (Bill) Nelson came to ISU from Atlantic, Iowa, where he was first assistant chief of the Atlantic Fire Department. Nelson started at Iowa State University as a field instructor in firemanship training. He later became chief instructor and remained in that position until his retirement in 1979. In 1951, Keith Royer also started work with Iowa State University as a field instructor. Royer had begun his career as a member of the Wichita (Kansas) Fire

Department. In 1949, he became the head of the University of Kansas's new fire training program. Within days of his hiring he traveled to Miami, Florida, where the first of a series of test fires sponsored by the Exploratory Committee on the Application of Water were being held. Royer eventually became the director of Iowa State Fire Training, a position he held until his retirement in 1988. In addition to the research that will be discussed in this section, Royer and Nelson made many other significant contributions to the fire service. Perhaps most notable was their role in the founding of the International Society of Fire Service Instructors (ISFSI).

By 1953 Nelson and Royer were engaged in a continual series of experimental fires designed to explore all aspects of the fire problem. These tests and some creative thinking on Nelson's and Royer's part ultimately would result in the Iowa State rate-of-flow formula. They felt this formula was highly practical because it enabled firefighters to calculate the RFF for any structure before a fire occurred.

The formula was based on to two fundamental principles:

1. One gallon of water will produce, with a margin of safety, 200 cubic feet of steam.
2 One gallon of water will absorb, with a margin of safety, all the heat that can be produced with the oxygen available in 200 cubic feet of normal air.

Following these two principles, cubic area in feet divided by 200 equals the gallons of water required for control of a specific area involved in fire:

Equation 6.1

$$\text{gallons} = \frac{\text{Volume of the room (in ft}^3\text{)}}{200\ \text{ft}^3/\text{gallon}}$$

Equation 6.1 is *not* a rate-of-flow formula. It gives only the total number of gallons of water needed to control a fire in a given area.

Example 6.1

Using Equation 6.1, determine how many gallons of water would be required to extinguish a fire in a room that is 20 feet wide, 30 feet long, and 10 feet high.

$$\text{gallons} = \frac{(20)(30)(10)}{200}$$

$$\text{gallons} = \frac{6{,}000}{200}$$

gallons = 30

Nelson and Royer derived the 200 cubic feet of steam figure from the expansion ratio of water to steam at 212° F, which is 1 to 1,700. Because 1 cubic foot of water contains 7.48 gallons:

$$\frac{1{,}700\ \text{ft}^3}{7.48\ \text{gal/ft}^3} = 227\ \text{ft}^3 \text{ of stream per gallon of water}$$

In their formula, they rounded down the 227 ft^3 to 200 ft^3; this allows for 90 percent efficiency in the conversion of water to steam and eases calculations.

Nelson and Royer based the number 200 in Principle 2 upon 1955 Factory Mutual Laboratories tests that determined 1 cubic foot of pure oxygen combined with ordinary fuels produced 535 BTUs. Air contains 21 percent oxygen, and flame production stops when the oxygen level falls below 15 percent. Therefore:

21 percent oxygen – 14 percent oxygen level = 7 percent of air

Thus, only 7 percent of air is available for heat production. Multiplying this number by the BTUs produced by a cubic foot of pure oxygen gives:

535 BTUs x .07 percent of air = 37 BTUs

This is the number of BTUs produced by 1 cubic foot of air. Because 1 gallon of water expands to more than 200 ft^3 of steam:

37 BTUs x 200 ft3 = 7,400 BTUs

One gallon of water converted to 200 ft^3 of steam absorbs 9,330 BTUs. Because 7,400 < 9,330, Royer and Nelson concluded that 1 gallon of water can absorb all the heat produced by 200 ft^3 of air.

Nelson and Royer spent considerable time and experimentation before they could use this information to determine a rate-of-flow formula. During the course of this research, they determined that 30 seconds was the maximum time required for effective use of the adjustable fog nozzle. Using Example 6.1, if 30 gallons of water are needed but the nozzle flows for only 30 seconds, then the rate-of-flow must be 2 times 30, or 60 gpm. Thus, to change their original formula into a rate-of-flow formula, they had to change the denominator from 200 to 100. This in turn resulted in what is now called the Iowa Rate-of-Flow Formula, or the Iowa State Formula:

Equation 6.2

$$\text{RFF} = \frac{\text{volume of the room (in ft}^3\text{)}}{100}$$

Given:

RFF = required fire flow in gallons per minute

100 = the cubic feet of steam created by 1/2 gallon of water

Example 6.2

Using Equation 6.2, determine the RFF for a detached garage that measures 30 feet wide, 60 feet long, and 14 feet high.

$$\text{RFF} = \frac{(30)(60)(14)}{100}$$

$$\text{RFF} = \frac{25{,}200}{100}$$

RFF = 252

Nelson's and Royer's experiments revealed that discharge times longer than 30 seconds were practical only if the space in which the fire was located was not significantly vented. In most real-life fire fighting situations, this would not be practical. Nelson and Royer also noted that their formula did not account for the protection of internal or external exposures. These duties would require additional water.

While the Iowa State Formula is now more than 50 years old, it remains widely used throughout the fire service and there is no scientific evidence to suggest it is not still relevant. Some professionals theorize that the higher fuel loads in today's structures may make this formula obsolete, but that has yet to be scientifically proven.

The National Fire Academy Formula

The Iowa State Formula remained the fire service's primary fire flow formula for nearly 30 years. For most of the period from the mid-1950s through the mid-1980s, most of the fire service used it exclusively (a few folks did use the ISO formula described in the next section). However, by the mid-1980s, the National Fire Academy was promoting a newer, simpler formula that quickly gained wide acceptance.

The National Fire Academy (NFA) is located at the National Emergency Training Center (NETC) in Emmitsburg, Maryland. The NFA provides advanced fire service managerial and administrative educational programs for career and volunteer firefighters and fire officers from all over the United States. It offers both courses-in-residence at Emmitsburg and local delivery courses throughout the U.S. The NFA is a federal government agency whose location within the governmental structure has changed several times since its inception in the mid-1970s. The NFA is a unit within the United States Fire Administration (USFA), which in turn is an agency within the Department of Homeland Security (DHS).

The NFA Fire Flow Formula was developed as a "quick-calculation" formula that can be used as a tactical tool to determine fire flow requirements at the scene of an incident. This formula was intended to provide a starting point for deciding the amount of water required, apparatus needed to deliver the water, and the number of companies that should be used to apply it.

The NFA Fire Flow Formula was derived through a study of fire flows that successfully controlled a large number of working fires. This research was combined with interviews of numerous experienced fire officers from throughout the country regarding the fire flows they had found to be effective in various fire situations.

This research indicated that the relationship between the area involved in fire and the approximate amount of water required to effectively extinguish the fire can be established by dividing the square footage of one floor of the structure by 3. This quick-calculation formula is expressed as:

Equation 6.3

$$\frac{L \times W}{3} = \text{fire flow in gpm for one floor at 100\% involvement}$$

Note that this formula utilizes the area or square footage of the structure involved, unlike the Iowa State Formula, which is based on the volume (in cubic feet) of the structure or room.

Example 6.3

Using Equation 6.3, determine the fire flow necessary to control a fire that fully involves one floor of a four-story structure that measures 50 feet long by 30 feet wide.

$$\frac{L \times W}{3} = \text{fire flow in gpm for one floor at 100\% involvement}$$

$$\frac{50' \times 30'}{3} = \text{fire flow in gpm for one floor at 100\% involvement}$$

$$\frac{1{,}500\ ft^2}{3} = \text{fire flow in gpm for one floor at 100\% involvement}$$

500 gpm = fire flow for one floor at 100% involvement

If more than one floor is involved in fire, the fire flow is simply adjusted by multiplying the number of floors involved by the required fire flow for each floor. In Example 6.3, if 3 of the 4 floors were on fire, then the required fire flow would be 1,500 gpm (3 floors x 500 gpm/floor). The NFA advises that the formula is considered reliable only if four floors or fewer are on fire.

The NFA formula also may be used to determine the required fire flow if only a portion of a floor is involved in fire. In these circumstances the fire officer must approximate not only the square footage of the floor but also the percentage of the floor that is involved in fire.

Example 6.4

Determine the required fire flows if a single-story structure measuring 60 feet by 40 feet is 75%, 50%, and 25% involved in fire.

$$\frac{L \times W}{3} = \text{fire flow in gpm for one floor at 100\% involvement}$$

$$\frac{60' \times 40'}{3} = \text{fire flow in gpm for one floor at 100\% involvement}$$

$$\frac{2{,}400\ ft^2}{3} = \text{fire flow in gpm for one floor at 100\% involvement}$$

800 gpm = fire flow in gpm for one floor at 100% involvement

75% involvement = 0.75 x 800 gpm = 600 gpm

50% involvement = 0.50 x 800 gpm = 400 gpm

25% involvement = 0.25 x 800 gpm = 200 gpm

The NFA formula indicates that if this structure were fully involved, effective control the fire would require approximately 800 gpm. If only half of the building were burning, 400 gpm should suffice, and if one-fourth of the building were burning, 200 gpm should be sufficient.

If the original fire building is exposing adjacent structures to fire, a 25 percent exposure charge of the required fire flow should be added for each side of the fire building with exposures. Should any exposure actually become involved with fire, that exposure then should be treated as a separate fire.

Example 6.5

Suppose that the structure in Example 6.4 has exposures on two sides. Calculate the total fire flow necessary to extinguish the fire and protect the exposures.

Required fire flow from Example 6.4 = 800 gpm

Exposure protection = 800 gpm x (25% x 2 exposures) = 400 gpm

Total fire flow required = 800 gpm + 400 gpm = 1,200 gpm

In buildings of fire-resistive construction, if other floors are not yet involved but are threatened by the possible extension of the fire, those floors should be considered an exposure, and 25% of the required fire flow for the fire floor should be added for exposure protection for each floor above the fire (up to four floors).

When using the NFA formula to determine required fire flows, it is important to remember that the answers provided are approximations of the amount of water needed to control the fire. They are based on an estimation of the area of the building and the amount of fire involvement within the building. Furthermore, unlike the Iowa State Formula, the NFA formula has never been validated by definitive scientific or practical testing. While most fire service instructors agree that the NFA formula provides firefighters and fire officers with a starting point to determine how much water they may need for an effective fire attack in normal situations, common sense and good judgment also are required to evaluate the water's effect once it is being applied. If unforeseen factors, such as barriers that prevent the water from being applied properly or building contents that cause unexpected fire behavior, prevent control from being achieved within a reasonable time, the amount of water being applied may have to be increased.

The NFA formula provides *approximations* of the amount of water needed to control the fire based on *estimations* of the area of the building and the amount of fire involvement within the building.

The ISO Formula

The third fire flow formula that has seen significant use throughout the fire service for many years is the ISO Required Fire Flow Formula. ISO is an abbreviation for the Insurance Services Office, an insurance industry trade organization whose mission is to provide useful and reliable risk evaluation information to property insurance companies. ISO evaluates many facets of risk including individual occupancies or facilities, community fire protection, and the community as a whole.

Of the three fire service required fire flow formulas covered in this text, fire departments use the ISO formula least, for several reasons. First, both the Iowa State and NFA formulas were designed for use on the fire ground. Both rely on a simple estimate of size of the structure and the amount of fire involvement, and both generally can be computed mentally with relative ease. These formulas can be quickly applied to almost any structure, with or without a preincident plan. Furthermore, neither the Iowa State nor the NFA formula requires the user to determine the construction type of the involved structure(s).

The Basic ISO Formula

The ISO formula is applicable to all structures other than one- and two-family dwellings not exceeding two stories. Because it is considerably more involved than the Iowa State or NFA formulas, it is highly impractical on the emergency scene, and its use generally is limited to preincident planning functions. However, when calculated and applied properly, the ISO formula is generally considered the most reliable of the three. The basic ISO Required Fire Flow Formula is:

Equation 6.4

$$F = 18\ C\ (A)^{0.5}$$

F is the required fire flow for the given occupancy in gallons per minute; 18 is a constant extrapolated during the research to develop the formula; C is a coefficient related to the type of construction for the occupancy in question; and A is the floor area.

The construction types for the coefficient C follow the construction classifications found in most model building codes. In general, buildings that provide a higher fire load receive a higher coefficient. This ultimately will result in a higher calculated fire flow than for a building of the same size but with a more fire resistant construction type. The coefficients for the formula are:

Wood frame construction: C = 1.5

Ordinary construction: C = 1.0

Heavy timber construction: C = 0.9

Noncombustible construction: C = 0.8

Fire-resistive construction: C = 0.6

For types of construction and/or materials that do not fall squarely within one of these categories, planners can interpolate a coefficient that more accurately reflects conditions. Such interpolation must be limited to occupancies that are listed consecutively in the above list. In other words, you can interpolate a coefficient of 1.3 for a part-wood-frame, part-ordinary-construction building. However you cannot interpolate a coefficient for a part-wood-frame, part-noncombustible hybrid. If the construction types are not listed consecutively in the above list, use the one with the highest coefficient. In no circumstance, however, should the coefficient be greater than 1.5 or less than 0.6.

In the ISO Formula, area (A) is the building's total floor area including all stories but excluding basements. If the building is of fire-resistive construction and its vertical openings are unprotected, consider the six largest successive floor areas as its area. If the vertical openings are properly protected, consider only the three largest successive floor areas.

ISO recommends that the calculated fire flow be rounded to the nearest 250 gpm if the flow is less than 2,500 gpm and to the nearest 500 gpm if the flow is greater than 2,500 gpm.

Example 6.6

Using the ISO formula, determine the required fire flow for a three-story, heavy timber construction mill building that measures 75 feet by 200 feet.

$F = 18\ C\ (A)^{0.5}$

$F = 18\ (0.9)\ (75 \times 200 \times 3\ \text{stories})^{0.5}$

$F = 18\ (0.9)\ (45{,}000)^{0.5}$

$F = 3{,}437$ gpm, or 3,500 gpm

The ISO formula may be applied to any type or size of structure, with the following stipulations. The minimum fire flow planned for any structure should be 500 gpm. The maximum fire flow, depending on the type of structure, should be:

- 8,000 gpm for wood frame construction
- 8,000 gpm for ordinary and heavy timber construction
- 6,000 gpm for noncombustible construction
- 6,000 gpm for fire-resistive construction
- 6,000 gpm for a normal one-story building of any type of construction

Adjusting the Final Results

Depending on the situation, the fire flow figure calculated using the basic ISO formula may have to be adjusted. Factors that require adjusting the fire flow include the level of fire hazard, the presence of sprinkler protection, and exposures. Depending on which of these applies to a given structure, adjustments for one or more may all be applied. Again, the fire flow should never drop below 500 gpm regardless of the calculation. However, if these factors are applied, the maximum fire flow may be increased to 12,000 gpm.

Fire Hazard The basic fire flow may need to be adjusted based on the level of hazard presented by the type of occupancy or its contents. The basic fire flow may be reduced up to 25 percent for low-fire-hazard occupancies or increased up to 25 percent for high-fire-hazard occupancies.

There are two ways to determine whether a structure falls into the low- or high-hazard categories. The first is to use the simple list provided by ISO as shown in **Table 6.1, p. 102**. The ISO gives the evaluator latitude in determining a percentage (+ or –) to apply to these occupancies, as long as it does not exceed +/– 25%.

The second method for adjusting the fire flow based on the level of content fire hazards is to apply one of the occupancy combustibility factors listed in **Table 6.2, p. 103**. Again, the evaluator has some latitude in applying one of the occupancy combustibility classes to the contents of the structure being evaluated. This typically will be based on the evaluator's experience with the products in the structure.

Table 6.1
ISO Occupancy Classifications

Low Hazard Occupancies	High Hazard Occupancies
Apartments	Aircraft hangers
Asylums	Cereal, feed, flour, and grist mills
Churches	Chemical works
Clubs	Cotton picking and opening operations
Colleges and universities	Explosives and pyrotechnic manufacturing
Dormitories	High-piled combustible storage in excess of 21 feet tall
Dwellings	Linoleum and oilcloth manufacturing
Hospitals	Linseed oil mills
Hotels	Match manufacturing
Institutions	Oil refineries
Libraries, except large stack room areas	Paint shops
Museums	Pyroxylin plastic manufacturing and processing
Nursing, convalescent, and care homes	Shade cloth manufacturing
Office buildings	Solvent extracting
Prisons	Varnish and paint works
Public buildings	Wood working and flammable finishing
Rooming houses	Any occupancies involving processing, mixing, storage, and dispensing flammable and/or combustible liquids
Schools	
Tenements	

Example 6.7

In evaluating the structure described in Example 6.6, the inspector determines that the structure is used to store bundled newspapers awaiting recycling. That is determined to be a free burning combustibility class. Adjust the required fire flow appropriately.

$F_{adjusted}$ = 3,500 gpm x occupancy factor

$F_{adjusted}$ = 3,500 gpm x 1.15

$F_{adjusted}$ = 4,025 gpm, or 4,000 gpm

Automatic Sprinkler Protection Reductions in the fire flow to account for automatic sprinkler protection are most commonly applied fire flows that have been determined by the basic formula and then adjusted for the fire hazard of

Table 6.2
ISO Occupancy Factor

Occupancy Combustibility	Occupancy Factor
C-1 Noncombustible	0.75
C-2 Limited combustible	0.85
C-3 Combustible	1.00
C-4 Free burning	1.15
C-5 Rapid burning	1.25

the contents. The adjusted fire flow may be reduced up to 50 percent for complete automatic sprinkler protection. Where buildings are either fire-resistive or non-combustible construction and have a low fire hazard, the reduction may be up to 75 percent. The reduction will depend upon the extent to which the automatic sprinkler system is judged to reduce the possibility of fires spreading within and beyond the fire area. Normally this reduction will not be the maximum allowed without proper system supervision including water flow and valves.

Exposures Reductions in the fire flow to give credit for exposure separation are also most commonly applied fire flows determined by the basic formula and then adjusted for the fire hazard of the contents. A percentage should be added to the adjusted fire flow for structures exposed within 150 feet of the fire area under consideration. This percentage shall depend upon the height, area, and construction of the exposed building(s), the separation, openings in the exposed building(s), the length of exposure, the provision of automatic sprinklers and/or outside sprinklers in the exposed building(s), the occupancy of the exposed building(s), and the effect of hillside location on the possible spread of fire. The percentage for any one side generally should not exceed the limits for separations shown in **Table 6.3, p. 104**. The total percentage shall be the sum of the percentages for all sides, but shall not exceed 75 percent of the fire flow.

Example 6.8

The structure described in Example 6.7 has exposures on two sides. Both are approximately 50 feet away. Determine the additional fire flow required to protect the exposures.

From Table 6.3, an exposure 50 feet from the fire building requires an additional 15 percent per side.

Table 6.3 ISO Exposure Protection Factor	
Separation	**Percentage**
0-10 feet	25
11-30 feet	20
31-60 feet	15
61-100 feet	10
101-150 feet	5

$F_{exposures} = 4{,}000 \text{ gpm} + (2)(0.15)(4{,}000)$

$F_{exposures} = 4{,}000 \text{ gpm} + 1{,}200 \text{ gpm}$

$\mathbf{F_{exposures} = 5{,}200 \text{ gpm}}$

Other Adjustments/Considerations

The ISO lists other types of adjustments or considerations that may be applied to the calculated fire flow. These adjustments are summarized as follows:

- The formula is not expected necessarily to provide an adequate value for lumber yards, petroleum storage, refineries, grain elevators, and large chemical plants but may indicate a minimum value for these hazards.
- Judgment must be used for business, industrial, and other occupancies not specifically mentioned in the tables.
- The configuration of the building(s) and its (their) accessibility to the fire department should be considered.
- Wood frame structures separated by less than 10 feet shall be considered one fire area.
- Normally an unpierced party (common) wall may warrant up to a 10 percent exposure charge.
- If a building is exposed within 150 feet, normally some percentage increase for exposure will be made.
- Where wood shingle roofs could contribute to spreading fires, add 500 gpm.
- Any noncombustible building is considered to warrant a 0.8 coefficient.

Dwellings

Rather than using the basic fire flow formula for groupings of one-family and small two-family dwellings not exceeding two stories in height, ISO recommends using the fire flows from **Table 6.4**. For other residential buildings, the basic formula and appropriate adjustments should be used. If the residences are continuous, as with row houses or townhouses, a minimum fire flow of 2,500 gpm is recommended.

Table 6.4
ISO Residential Fire Flows:

One-Family and Small Two-Family Dwellings, Not Exceeding Two Stories in Height

Exposure Distance	Suggested Required Fire Flow
Over 100 feet	500 gpm
31—100 feet	750—1,000 gpm
11—30 feet	1,000—1,500 gpm
10 feet or less	1,500—2,000* gpm

REQUIRED FIRE FLOW FOR AUTOMATIC SPRINKLER SYSTEMS

To this point in the chapter, we have focused on determining required fire flows for manual fire fighting operations. As the previous examples in this chapter show, manual fire fighting operations require large volumes of water. This is because the fires are often in advanced stages and consuming major portions of the structure by the time firefighters arrive. Comparatively, manual fire fighting is much less efficient than automatic sprinkler systems, which in essence detect a fire in its early stages and extinguish it with less water.

Records kept by the National Fire Protection Association (NFPA®) vary slightly from year to year, but in general they indicate that five or fewer activated sprinklers extinguish more than 90 percent of all fires in sprinklered buildings. With each sprinkler flowing about 8 to 15 gpm, this is significantly more efficient (and less damaging) than pouring hundreds or thousands of gpm on a well-advanced fire.

Most fire service personnel will never get involved with determining the required flow rates for an automatic sprinkler system. The system's designers determine those rates during its planning. The required flow rate will vary for each system based on a variety of factors, including the types of sprinkler system, the building's construction type and occupancy hazards, the size of the protected area, and numerous others.

Required flow rates for automatic sprinkler systems include only the amount of water the system is designed to flow on its own. They do not include the amount of water that a fire department pumper might pump into the system during a fire. In fact, there are no set rules for the volume of water that a fire department pumper must pump into a sprinkler system. (For more information on proper support of activated automatic sprinkler systems see Chapter 15, "Supporting Sprinkler and Standpipe Systems.")

The design of automatic sprinkler systems is regulated by NFPA® 13, Standard for the Installation of Sprinkler Systems. NFPA® 13 recognizes two methods of designing automatic sprinkler systems: the pipe schedule method and the hydraulic calculation method. Determining the required flow rate for a sprinkler system will depend on which of these two methods designers use to plan the system.

Pipe Schedule Systems

Of the two design methods, the pipe schedule design technique is the older, more traditional approach. Systems designed using this method have existed for over 100 years, and their record of performance is very good. Designing an automatic sprinkler system using the pipe schedule method involves sizing the pipe based upon the number of sprinklers supplied, as dictated by tables in NFPA® 13. When applied to proper occupancy considerations, systems designed by pipe schedule design provide excellent protection.

With a few exceptions, NFPA® 13 limits new sprinkler systems being designed by the pipe schedule method to occupancies of less than 5,000 square feet. It also may be used to modify or expand existing pipe schedule systems. The primary advantage of using the pipe schedule method for designing an automatic sprinkler system is ease of calculation, particularly with smaller systems. On the other hand, pipe schedule design systems tend to use larger pipe than hydraulically designed systems, therefore making them less economical in many instances.

From a required flow standpoint, NFPA® 13 does not give significant direction on system performance. The water supply requirements for pipe schedule systems are left substantially to the judgment of designers or authorities having jurisdiction. At best, NFPA® 13 provides only a few guidelines. **Table 6.5** outlines the basic requirements stated in NFPA® 13, Chapter 7, "Design Approaches." The requirements shown are for residual pressures at the location of the highest (multistory) or most remote (single-story) sprinkler in the building. To establish ground level pressure requirements, pressure must be added to compensate for elevation losses. The pipe schedule design method does not evaluate friction losses. Note that the standard requires extra hazard occupancies to have hydraulically calculated systems.

NFPA® 13 allows designers to exercise some judgment based upon building size, ceiling height, construction type, and other factors affecting the quantity of fuel available. For example, one of the exceptions to the 5,000 square foot

Table 6.5
Water Supply Requirements for Pipe Schedule Sprinkler Systems

Occupancy classification	Minimum residual pressure required	Acceptable flow at base of riser (Including hose stream allowance)	Duration (minutes)
Light hazard	15 psi	500—750 gpm	30—60
Ordinary hazard	20 psi	850—1,500 gpm	60—90

rule says that buildings over 5,000 square feet in size may have a pipe schedule system if they have a residual pressure of 50 psi available at the top line of sprinklers for both occupancy classes. Other NFPA® 13 requirements relative to fire flow and water supply for pipe schedule systems include:

The lower end of the duration requirements in Table 6.5 may be used only if the system is equipped with a remote station or central station water flow monitoring system.

The lower flow figures in Table 6.5 may be used only in buildings of noncombustible construction or in potential fire areas that do not exceed 3,000 square feet for light hazards and 4,000 square feet for ordinary hazards.

Hydraulically Designed Systems

Of the two methods for designing modern automatic sprinkler systems, the hydraulic calculation method is by far the more widely used. This method involves using mathematical calculations to determine the required piping and other design factors. Systems designed by this method are referred to as hydraulically calculated systems. This is the only acceptable method of designing protection for special and extra hazards, such as high-piled or rack storage of commodities. The numerous benefits of hydraulically calculated automatic sprinkler systems include:

The hydraulic calculation method is the only acceptable method of designing protection for special and extra hazards.

- The hydraulic calculations typically show that system performance will still be acceptable despite using smaller pipes sizes than required by pipe schedule systems.
- Smaller pipes and their associated fittings and equipment reduce material and labor costs on systems for light and ordinary hazard occupancies when compared to pipe schedule systems.
- Pipe schedule systems are limited to heavy wall steel piping. Hydraulically calculated systems may use other types of pipes, such as thin wall piping, copper tubing, and plastic piping. These pipes all cause less friction loss than heavy wall steel piping, thus allowing for the use of smaller sizes.

In the earliest days of hydraulically calculated automatic sprinkler systems, designers and engineers did all of the work manually using slide rulers or calculators. While this remains a viable technique, most hydraulically calculated systems today are designed using computer programs. The designers simply input the required information, such as occupancy class, size of the protected area, and other variables, and the computer program essentially designs the entire system.

The process of determining the required fire flow for a hydraulically calculated system is not simple. The required fire flow will be determined during the design process. It revolves around a number of variables, including the occupancy class, size of the sprinkler system, the sprinkler discharge design density, the size of the design area, and other factors. (For additional information on fire department support of all sprinkler systems, see Chapter 15. More detailed information on the procedures for designing hydraulically calculated automatic sprinkler systems can be found in Chapter 13 of Fire Protection Hydraulics and Water Supply Analysis by Pat D. Brock.)

REQUIRED FIRE FLOW FOR STANDPIPE SYSTEMS

Standpipe systems are fixed piping systems within tall or otherwise large buildings. During manual fire fighting operations, firefighters can connect attack hoses to these systems, eliminating the need to stretch long lays of supply hoses to remote parts of the structure. Most modern standpipe systems are connected to a water supply system. During large-scale fire fighting operations, the fire department may have to supplement that water supply by pumping water from a fire hydrant or other source into a fire department inlet connection on the standpipe system. This will boost the available pressure and volume of water at the attack location(s).

The design and operation of standpipe systems is dictated by NFPA® 14, Standard for the Installation of Standpipe and Hose Systems. Unlike the standard for automatic sprinkler systems, NFPA® 14 clearly defines the water supply requirements for standpipe systems. It identifies three classes of standpipe systems, each with different requirements for operation and water supply.

Class I Standpipe Systems

Class I standpipe systems are designed specifically for fire department use. These standpipes are equipped with 2 1/2-inch discharges, to which attack lines may be connected. In high-rise structures, the standpipe and its connections are almost always located in stairwells. In older structures, the discharge connections are located at each floor level. However, changes in NFPA® 14 allow discharge connections in some newer structures to be located on landings between floor levels.

If a structure is equipped with a single Class I standpipe, that system is required to have a minimum available flow rate of 500 gpm without being boosted by a fire department pumper. If the structure is equipped with more than one Class I standpipe system, the first system must have an available flow of 500 gpm, and each additional standpipe must have available 250 gpm, up to a maximum of 1,250 gpm. The system must be capable of supplying a residual pressure of 100 psi at its most hydraulically demanding discharge. The most hydraulically demanding discharge is defined as the one with the greatest loss of pressure to friction and elevation when delivering the 500 gpm minimum flow. In most high rise structures it will be the highest discharge in the system.

For Class I standpipes in buildings other than high rises, NFPA® 14 permits the system's flow and pressure requirements to be supplied by a fire department pumper, as opposed to requiring that the system be capable of supplying them on its own. This is more realistic, as the fire department pumper does not have to overcome the large elevation pressure losses of pumping into a high-rise system.

Class II Standpipe Systems

Class II standpipe systems are designed for use by building occupants to attack incipient fires. These systems have hose cabinets or reels to which 1 1/2-inch hose is preconnected. Older Class II standpipe systems typically contain 75 feet of unlined cotton or linen fire hose. Modern systems are more likely to have 100 feet of lined or synthetic hose. One-inch noncollapsible hose (similar to booster hose found on fire apparatus) may be found in some light hazard installations. Many jurisdictions no longer allow Class II standpipe systems in new structures as they tend to encourage the occupants to fight the fire themselves rather than notify the fire department and evacuate the structure. As well, OSHA requires occupants to be trained if they are going to use the hand lines to attack a fire. Most occupants do not have this type of training.

Class II standpipe systems require a flow of only 100 gpm at a minimum of 65 psi, with no increased requirement for multiple standpipes. Since the Class II service is for the building occupants' use, the required pressure and flow must be available independent of fire department pumping apparatus. Most buildings with Class II service will require a fixed fire pump to meet this demand.

Class III Standpipe Systems

Class III standpipe systems combine the elements of both Class I and Class II systems. Because Class III standpipe systems may be used to support fire department operations, their water supply requirements are identical to those for Class I systems.

SUMMARY

Being able to calculate the required fire flow for any given structure is an important tool both during preincident planning and during emergency operations. By being able to reliably estimate the flow of water that will be needed to control a fire, the incident commander is better able to request the resources necessary to control the incident. The three common methods for calculating the required fire flow for manual fire fighting operations are:

- The Iowa State Formula
- The National Fire Academy Formula
- The ISO Formula

Each of these methods will produce slightly different results. Each jurisdiction must decide which method works best for them.

Fire service personnel also must understand the water supply requirements for automatic sprinkler and building standpipe systems. This will help fire departments to support these systems effectively during emergency operations.

REVIEW QUESTIONS

1. Name the three most common formulas for determining required fire flows for manual fire fighting operations.

2. Describe the two basic principles upon which the Iowa State rate-of-flow formula was based.

3. Using Equation 6.1, how many gallons of water would be required to extinguish a fire in a room that is 30 feet wide, 45 feet long, and 20 feet high.

4. Using Equation 6.2, determine the RFF for a one-story fast food restaurant that measures 60 feet wide, 90 feet long, and 12 feet high.

5. Using the National Fire Academy formula (Equation 6.3), determine the fire flow necessary to control a fire that fully involves two floors of a six-story structure measuring 120 feet long by 100 feet wide.

6. Using the National Fire Academy formula (Equation 6.3), determine the fire flow necessary to control a fire and protect exposures on three sides of an 80-foot-by-50-foot, single-story structure.

7. Using the ISO formula, determine the required fire flow for a one-story, ordinary construction hard ware store that measures 80 feet by 150 feet.

8. In evaluating the structure described in Review Question 7, the inspector determines that the structure is a free-burning combustibility-class occupancy. Adjust the required fire flow appropriately.

9. Name the two methods for designing automatic sprinkler systems.

10. Which type of standpipe system is designed solely for fire department use?

11. According to NFPA® 14, what are the minimum and maximum required flows for a Class I standpipe system?

    Minimum: ______

    Maximum: ______

Answers found on page 427.

Photo courtesty Ron Jeffers

Part 3 Fire Apparatus and Pumps

Virtually every profession has its "tools of the trade." The fire service is certainly no exception to this rule. Every operational position in the fire service had a wide array of tools and equipment that firefighters must use in order to perform their tasks. For driver/operators, the primary tools are the fire apparatus itself, the pump and other mechanical systems on the fire apparatus, and the fire hose, appliances, and nozzles necessary to provide water for fire fighting operations.

The fire department pumper is the backbone of the fire service. Our primary mission remains to provide adequate water to the scene so that the fire may be extinguished. This is what pumpers do. In today's multidisciplinary fire departments, pumpers also serve as EMS first-response vehicles, rescue vehicles, and perform numerous other functions. As well, most other types of fire apparatus, including rescue vehicles, aerial apparatus, wildland fire fighting apparatus, and others are equipped with fire pumps.

Any driver/operator who will be expected to drive an apparatus equipped with a fire pump must be fully trained in the principles of operating that pump. A fire apparatus is of little more use than a taxi if the driver does not know how to operate its pump and equipment once he or she has delivered to the scene and positioned it. All fire departments must have a thorough training program that adequately prepares firefighters to drive the apparatus safely to and from emergency scenes and to operate the fire pump and other mechanical systems effectively. These functions are important from a tactical standpoint and absolutely critical from firefighter life safety standpoint. Each year approximately 25% of all firefighter fatalities result from being involved in collisions while responding to or returning from emergencies. The number of these deaths can be greatly reduced if proper driving procedures are followed. As well, failure to properly operate the pump or other systems on the apparatus can have tragic effects on firefighters in hazardous locations on the emergency scene.

The following part of this manual describes the driver/operator's "tools of the trade."

Chapter 7: Surveys the basic characteristics of fire apparatus equipped with fire pumps.

Chapter 8: Continues by providing a detailed look into the design and function of various types of fire service pumps.

Chapter 9: Concludes this part of the manual by providing detailed information on how to ensure that pumps are tested properly, both before they are placed in service and after.

Because this text is dedicated solely to fire service hydraulics, it does not address driving the fire apparatus. Fire detailed information on driving and positioning fire apparatus equipped with pumps, see the IFSTA *Pumping Apparatus Driver/Operator Handbook.*

FESHE Course Objectives

The information in this chapter is intended to meet some of the objectives outlined for the Fire Protection Hydraulics and Water Supply course by the United States Fire Administration's National Model Core Curriculum as developed by the Fire and Emergency Services Higher Education (FESHE) initiative.

FESHE Course Objectives

1. List and describe the characteristics of the various types of fire apparatus equipped with fire pumps, including:
   - Pumpers
   - Attack Pumpers
   - Wildland Apparatus
   - Tankers/Tenders
   - Aerial Apparatus
   - Rescue Vehicles
2. Explain the methods used for typing pumpers, wildland apparatus, and water tenders in the Incident Command System (ICS).

Chapter 7

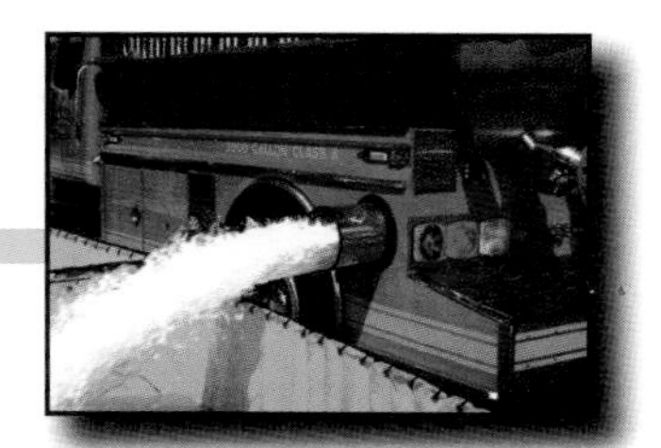

Apparatus Equipped with a Fire Pump

INTRODUCTION

Seldom, if ever, will municipal firefighters deliver water to fight a fire without the aid of some type of fire apparatus. After the firefighters themselves, the fire apparatus is perhaps the most important piece of the manual fire fighting puzzle. It is extremely important that firefighters, fire apparatus driver/operators, and fire officers have an intimate working knowledge of the fire apparatus to which they are assigned. This chapter examines the basic characteristics of the most common types of fire apparatus to be equipped with fire pumps.

Firefighters, fire apparatus driver/operators, and fire officers must have an intimate working knowledge of the fire apparatus to which they are assigned.

The minimum design criteria for most types of fire apparatus are specified in NFPA® 1901, *Standard for Automotive Fire Apparatus.* This standard covers all facets of fire apparatus design including, chassis and body construction, fire pump arrangements, water tank design, aerial device construction and mounting, and portable equipment considerations. Two major types of fire apparatus equipped with pumps are not covered in NFPA® 1901. Apparatus designed primarily for wildland fire fighting have their own standard in NFPA® 1906, *Standard for Wildland Fire Apparatus.* Likewise, apparatus designed to fill fire protection and suppression duties at airports must meet the specific requirements set forth in NFPA® 414, *Standard for Aircraft Rescue and Fire Fighting Vehicles.*

The descriptions of the basic types of fire apparatus in this chapter are very general for a number of reasons. First, the Fire Apparatus Manufacturer's Association (FAMA), the fire apparatus industry trade group, has identified well over 100 different manufacturers of fire apparatus in the United States. Over time, a certain percentage of these manufacturers go out of business and, correspondingly, a limited number of new manufacturers enter the market. Though most apparatus are constructed to meet the requirements of the appropriate NFPA® standard, each manufacturer tends to have its own design principles and nuances. Individual fire departments also tend to have specific requirements for the design of their apparatus, and this will result in apparatus differences from jurisdiction to jurisdiction.

A true fire-protection professional will have a working knowledge of the design differences and options available on various makes and types of fire apparatus. This knowledge will allow driver/operators to easily understand a different piece of apparatus and to operate it if necessary. It also will allow officers who may need to develop specifications for the purchase of new apparatus to make choices that benefit their department and the firefighters who will be expected to operate the apparatus.

FIRE DEPARTMENT PUMPERS

The most basic and common type of fire apparatus in use in North America is the fire department pumper. The vast majority of all emergency calls to which fire departments respond are handled using the pumper **(Figure 7.1)**. Depending on the jurisdiction, a number of other names or nicknames may be applied to the pumper, including:

- *Engine or engine company.*
- *Wagon* – Commonly applied to the second piece of a two-piece engine company whose specific duty it is to lay supply lines to the attack pumper.
- *Triple* – A West Coast term, short for triple-combination pumper (an apparatus that carries hose, water, and a fire pump).
- *Pipeline* – A mid-Atlantic term for a pumper equipped with large diameter supply hose.
- *Squad* – A primarily East Coast term for a pumper that carries additional rescue, forcible entry, or salvage equipment above and beyond that which a standard pumper would carry. Sometimes also referred to as a rescue-pumper.

Figure 7.1 The vast majority of all emergency calls to which fire departments respond, such as this vehicle fire, are handled using the pumper. *Courtesy of Bob Esposito*

The main purpose of the fire department pumper is to provide water at an adequate pressure to produce an effective fire stream from a nozzle. Pumpers are needed because most municipal water supply systems do not supply water from a hydrant at a pressure high enough to produce an effective stream once it travels through a hose to the nozzle. Furthermore, many locations are void of a municipal water supply system, and the pumper must impart pressure on water drawn from a static water supply source, such as a lake, pond, or portable tank or from an apparatus water tank.

NFPA® 1901 specifies that the minimum pump capacity for a vehicle to be considered a fire department pumper is 750 gpm. Standard pump capacities larger than 750 gpm typically increase in increments of 250 gpm. Municipal fire department pumpers rarely have pump capacities exceeding 2,000 gpm. However, it is common for industrial fire pumpers to have pump capacities in excess of 3,000 gpm. In addition to the fire pump itself, a fire department pumper must also have intake and discharge pump connections, pump and engine controls, gauges, and other instruments to allow the driver/operator to use the pump. (Pumps and their accessories are highlighted in Chapter 8.)

NFPA® 1901 also requires the vehicle to have an on-board water tank capacity of at least 300 gallons in order to be considered a fire department pumper. Most fire department pumpers have water tank capacities ranging between 500 and 1,000 gallons. Some agencies, particularly those in rural jurisdictions, may operate pumpers with water tanks in excess of 1,000 gallons. Technically these apparatus fall into the category of pumper-tankers, which are covered in more detail later in this chapter.

Fire department pumpers may be built on custom or commercial truck chassis. A custom chassis is one that is designed specifically for application as a fire apparatus (**Figure 7.2**). It may be built by the actual builder of the whole fire apparatus (as is the case with a Seagrave or American LaFrance), or it may be built by a fire chassis manufacturer, such as Spartan or HME and then provided to a fire apparatus manufacturer for completion of the fire body. The primary advantage of using a custom chassis for a pumper is that the chassis is designed for the harsh conditions under which emergency vehicles operate. Furthermore, custom chassis typically are equipped with larger cabs that accommodate more firefighters and equipment than do commercial cabs. Their primary disadvantage is that they tend to be considerably more expensive than commercial chassis at the initial purchase.

Figure 7.2 A custom chassis is one that is designed specifically for application as a fire apparatus.

A commercial chassis is not designed specifically for application as a fire apparatus (**Figure 7.3, p. 118;** *Courtesy of Steve Loftin*). In addition to fire apparatus, these chassis can be used for an endless number of other applications, such as garbage trucks, dump trucks, cement mixers, or delivery trucks. Some departments prefer to place pumpers on commercial chassis because of their

Figure 7.3 A commercial chassis is not designed specifically for application as a fire apparatus. *Courtesy of Steve Loftin*

lower purchase cost and the ready availability of replacement parts at the local level. The downside of commercial chassis is that unless they are specified properly and modified appropriately, they may have insufficient power, braking ability, or load carrying capacity for safe use as fire apparatus. Each jurisdiction must evaluate its individual needs and resources when determining which type of chassis best meets its requirements.

NFPA® 1901 contains a list of the minimum portable equipment that all fire department pumpers must carry. However, local preferences generally dictate exactly what equipment any particular pumper carries. The list of equipment that may be carried is endless, but some of the more common items include:

- Ground ladders, including roof, extension, and attic ladders
- Various sizes and lengths of attack and supply fire hose
- Various sizes and types of nozzles, adapters, and fittings
- Self-contained breathing apparatus (SCBA) and spare cylinders
- Rescue/extrication tools
- Forcible entry equipment
- Ventilation fans
- Salvage equipment
- Portable fire extinguishers
- First aid/medical equipment

Since the early 1990s, the fire service increasingly has combined the functions of a rescue vehicle with those of a pumper. These apparatus, commonly called *rescue pumpers*, feature all the standard engine company equipment but also carry a larger than standard amount of rescue and extrication equipment. They typically are designed with more compartment or storage space than a standard fire department pumper. Depending on the nature of the call, the personnel on the apparatus can function as either an engine company or a rescue company. In addition to standard engine company equipment, rescue pumpers commonly contain items such as:

- Hydraulic extrication tools, hose reels, and power units
- Pneumatic lifting and cutting equipment
- SCBA cascade refilling systems
- Moderate to large quantities of cribbing
- Battery and electrically operated cutting, impact, and drilling tools

INITIAL ATTACK FIRE APPARATUS

Initial attack fire apparatus perform most of the same functions as full-sized fire department pumpers. The primary differences between the two are their size and capabilities. Initial attack fire apparatus are essentially scaled-down versions of fire department pumpers. Typically they have smaller pumps and water tanks than regular pumpers, and they carry considerably less equipment than their larger counterparts. All initial attack apparatus are constructed on commercial truck chassis. Many are equipped with all-wheel drive.

Initial attack apparatus are used by departments of all sizes and in a variety of different roles. Small, rural fire departments often use this type of apparatus to make a quick response to a fire or other emergency with limited personnel until additional personnel and larger apparatus can respond and arrive. Suburban and urban fire departments often use them to supplement larger apparatus on minor responses or to approach heard-to-reach places; these departments also might couple initial attack apparatus in a task force arrangement with larger apparatus. One common practice is to couple a mini-pumper with a small quint (combination pumper/aerial apparatus). This often is referred to as a "mini-maxi" group, or concept.

Though all initial attack fire apparatus are smaller than full-sized pumpers, they vary in size. NFPA® 1901 does not differentiate between different sizes, but manufacturers and fire departments typically classify initial attack fire apparatus as minipumpers and midipumpers.

Minipumpers

The smaller sizes of initial attack fire apparatus are commonly referred to as *minipumpers* **(Figure 7.4)**. Minipumpers most often are mounted on 1-ton or 1½-ton chassis. NFPA® 1901 requires all initial attack fire apparatus to have a minimum pump capacity of 250 gpm. Most minipumpers have fire pumps with a capacity no larger than 500 gpm, although some one-ton vehicles have pumps rated up to 1,000 gpm. These larger

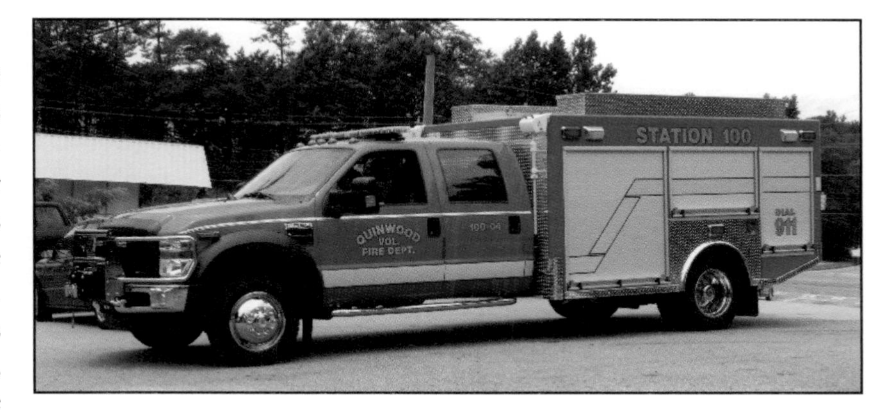

Figure 7.4 The smaller sizes of initial attack fire apparatus are commonly referred to as minipumpers.

pumps are made feasible in recent years because of more powerful engines in the smaller chassis vehicle. The standard also requires them to carry at least 200 gallons of water on board.

In addition to smaller quantities of standard engine company equipment, some minipumpers also carry basic medical and extrication equipment, enabling them to serve as a rescue unit as well as a fire fighting unit. Some minipumpers are equipped with a turret gun that can be supplied directly from another pumper. The small size and maneuverability of the minipumper allow it to get into small spaces and set up a master stream.

Minipumpers that have all-wheel-drive capabilities are commonly used to provide fire fighting capabilities or emergency medical services during inclement weather such as ice or snow storms. Also they often are used in rural areas, where unimproved, unpaved roads can be challenging or dangerous for larger, heavier apparatus.

Midipumpers

Midipumpers are used in much the same manner as are minipumpers. The main differences between a midipumper and a minipumper are size, pump capacity, and amount of equipment carried **(Figure 7.5).** *Midipumpers* are built on a chassis usually over 12,000 pounds gross vehicle weight (gvw). They routinely are equipped with pumps as large as 1,000 gpm. These units typically carry the same types of equipment as minipumpers, but they carry more of it because of their greater size.

Figure 7.5 The main differences between a midipumper and a minipumper are size, pump capacity, and amount of equipment carried.

WILDLAND FIRE APPARATUS

The requirements for fire apparatus designed primarily to attack wildland fires differ from those designed to attack structural fires. Wildland fire apparatus need to be highly maneuverable and able to operate on poor roads or off the road. They also must be able to pump water while moving, commonly referred to as pump and roll or as module pumping. Fire apparatus specifically adapted for fighting wildfires are designed to fulfill these requirements. As noted earlier, NFPA® 1906, *Standard for Wildland Fire Apparatus,* stipulates the design requirements for wildland fire apparatus. Wildland fire apparatus are known by a wide variety of synonyms throughout the fire service, including:

- Brushers
- Brush pumpers
- Brush breakers
- Field units
- Tac units
- Booster apparatus

Wildland fire apparatus vary greatly in size and capabilities. The smallest of these vehicles are mounted on four- or six-wheel or tracked all-terrain vehicles (ATVs) or Jeeps **(Figure 7.6;** *Courtesy of 1st Attack Engineering, Inc.* **).** These apparatus typically carry less than 100 gallons of water and have pump capacities

Figure 7.6 The smallest of these vehicles (wildland fire apparatus) are mounted on all-terrain vehicles (ATVs) or Jeeps. *Courtesy of 1st Attack Engineering, Inc.*

less than 100 gpm. More wildland fire apparatus are constructed on pickup-truck style 1-ton or 1 1/2-ton commercial chassis **(Figure 7.7;** ***Courtesy of Steve Loftin*****)**. These vehicles commonly carry from 200 to 400 gallons of water and have pump capacities ranging from 100 gpm to 500 gpm. Many jurisdictions operate larger wildland apparatus that are similar in size and design to midipumpers **(Figure 7.8)**. These vehicles may carry 500 or more gallons of water and have pump capacities of 1,000 gpm or greater. Regardless of their size, many of these vehicles have all-wheel-drive capabilities.

Figure 7.7 More wildland fire apparatus are constructed on pickup-truck style 1-ton or 1 1/2-ton commercial chassis. *Courtesy of Steve Loftin*

A growing practice is to equip wildland fire apparatus with Class A foam systems. Class A foam is extremely effective in attacking wildland fires and protecting exposures. Both high-energy and low-energy foam systems are used on these apparatus. (For more on foam systems, see Chapter 9.)

Figure 7.8 Many jurisdictions operate larger wildland apparatus that are similar in size and design to midipumpers.

For agencies that operate within the NIMS or FIRESCOPE Incident Command System (ICS), wildland fire apparatus are categorized by capability **(Table 7.1)**. This *apparatus typing*, as it is called, makes it easier for incident commanders to call for exactly the types of resources they need to handle an incident. Individual states or jurisdictions may have their own method of typing apparatus. The driver/operator should know the method used in his jurisdiction.

Table 7.1
ICS Pumper Typing

Pumper Type	Minimum Pump Capacity in GPM	Minimum Water Tank Capacity in Gallons
Type 1	1,000	400
Type 2	500	400
Type 3	120	300
Type 4	50	200
Type 5	70	750
Type 6	50	500
Type 7	50	300
Type 8	20	125

Figure 7.9 There are two proper methods for making a moving fire attack. The first is to have firefighters using short sections of attack hose walk alongside the apparatus and extinguish fire as they go. *Courtesy of Monterey County Training Officers*

The ability to pump and roll is a tremendous advantage when combating wildland fires. Vehicles that can pump and roll use a separate motor or a power take-off (PTO) to power the fire pump. (For more detail on pump and roll capability, see Chapter 8.)

There are two proper methods for making a moving fire attack. The first is to have firefighters using short sections of attack hose walk alongside the apparatus and extinguish fire as they go **(Figure 7.9)**. The second is to use nozzles that are remotely controlled from inside the cab (**Figure 7.10;** *Courtesy of 1st Attack Engineering, Inc.*). Some jurisdictions design their apparatus so that firefighters may ride on the outside of the vehicle and discharge water as the vehicle is driven; however, NFPA® 1500, *Standard for Fire Department Occupational Safety and Health Program* strictly prohibits this practice.

Figure 7.10 The second is to use nozzles that are remotely controlled from inside the cab. *Courtesy of 1st Attack Engineering, Inc.*

In the primary means of attacking wildland fires, firefighters deploy booster (noncollapsible) hose, forestry hose, or small-diameter attack lines, typically 1½-inch diameter or less. Two types of fixed nozzles are also commonly incorporated into wildland fire apparatus. Some jurisdictions use remote control nozzles similar to but smaller than those used on airport fire apparatus. These allow a person riding within the cab of the apparatus to manipulate the fire stream while the vehicle is moving. This is much safer than having someone ride outside the apparatus to manually direct a nozzle attached to a fire hose. Fixed ground-sweep nozzles may be attached to the front, sides, or undercarriage of the apparatus to extinguish fire in close proximity to the apparatus. Ground-sweep nozzles are effective in protecting the front of the apparatus and extinguishing small fires in short vegetation as the apparatus advances.

In addition to hose and nozzles, wildland fire apparatus also carry a variety of other types of portable equipment, including:

- Fire rakes, Pulaskis, McLeods, brush hooks, and shovels
- Portable water fire extinguishers
- Chain saws, axes

- Bolt cutters adapted for fence cutting
- Drip torches, flares, or fusees for lighting backfires
- Portable pumps for refilling the apparatus's water tank from remote static water sources

MOBILE WATER SUPPLY APPARATUS

Most fire department pumpers carry enough water to initiate an initial attack but not enough to sustain an extended attack at a larger fire. Fire departments that operate in rural areas require *mobile water supply apparatus* to transport additional water for extended fire fighting operations to areas without an adequate water source. Mobile water supply apparatus have considerably larger water tanks than standard fire department pumpers.

NFPA® 1901 uses the term *mobile water supply apparatus* to describe these vehicles. In the field, the more common terms for these apparatus are *tanker* or *tender*. As *tanker* is the most commonly used term for these apparatus, this book will follow that practice.

Agencies that operate within the NIMS or FIRESCOPE Incident Command Systems call these apparatus tenders or water tenders, as they reserve the term *tanker* for fixed-wing aircraft that drop water or fire retardant on wildland fires. ICS classifies tenders by their capabilities, similar to the system for classifying wildland fire apparatus (**Table 7.2**).

Table 7.2
ICS Water Tender (Tanker) Typing

Tender/Tanker Type	Minimum Pump Capacity in GPM	Minimum Water Tank Capacity in Gallons
Type 1	300	5,000
Type 2	200	2,500
Type 3	200	1,000

Tankers are used to support fire fighting operations in two basic ways. The first method uses the tanker as a reservoir, or "nurse tanker." This involves parking the tanker close to the pumpers attacking the fire and supplying them directly. This typically is not an effective practice for long-term fire fighting operations, although it is preferred when tankers are supplying foam concentrate for extended foam fire fighting operations.

The second method uses tankers to shuttle water. In water shuttle operations, tankers dump their loads into a portable water tank or nurse tanker/tender and then go to a fill site to reload (**Figure 7.11, p. 124**). This is the preferred method for supplying water to extended fire fighting operations beyond the reach of a fixed water supply system, long hose lay, or a reliable static water supply source.

Figure 7.11 In water shuttle operations, tankers dump their loads into a portable water tank or nurse tanker/tender and then go to a fill site to reload.

Some fire departments operate tankers to carry commodities other than water. Agencies that engage in large-scale flammable and combustible-liquids operations often use tankers to carry bulk quantities of foam concentrate. These large quantities are required when fighting fires such as those in petroleum storage tanks, refining operations, or large transportation fires. Some larger municipal fire departments also use fuel tankers to refuel fire apparatus during long emergency operations.

NFPA® 1901, Chapter 5, "Mobile Water Supply Apparatus," states the basic design requirements for new fire department tankers. The standard's major requirements include:

- To be considered a fire department tanker, the apparatus must have a water tank with a capacity of at least 1,000 gallons.
- The tanker must have a minimum of 20 cubic feet of compartment storage space.
- All tankers must carry at least the following equipment:
  - — One 6-pound flat- or pick-head axe
  - — One 6-foot or longer pike pole
  - — At least 200 feet of 2½-inch or larger fire hose
  - — Two portable hand lights
  - — A portable dry chemical fire extinguisher with a rating of at least 80-B:C
  - — A 2½-gallon or larger portable water fire extinguisher
  - — At least two SCBA and one spare cylinder for each
  - — A first-aid kit

— Two spanner wrenches and a hydrant wrench

— One each, double-male and double-female hose adapters at least 2½ 2-inches in diameter

— Two wheel chocks that fit the vehicle's tires properly

If the tanker is equipped with a fire or transfer pump, the following minimum provisions also apply:

- The pump should meet the requirements of either NFPA® 1901, Chapter 14, "Fire Pumps," or Chapter 16, "Transfer Pumps," whichever applies.
- A minimum of 15 feet of soft intake or 20 feet of hard intake hose with an intake strainer, of the appropriate diameter for the pump, must be carried on the apparatus.
- At least 400 feet of 1½-, 1¾-, or 2-inch fire hose and two combination spray nozzles capable of flowing at least 95 gpm.
- One gated swivel intake connection
- One rubber mallet

Figure 7.12 Single-rear-axle chassis equipped with elliptical water tanks should be limited to about 2,000 gallons of water. *Courtesy of Dennis Wetherhold, Jr.*

Figure 7.13 Tankers with a tandem rear axle generally are limited to about 4,000 gallons of water.

The amount of water carried on a tanker depends on local preference and the vehicle's chassis capabilities. One of the most common weight-related design problems associated with fire department tankers is trying to carry too much water on a specific chassis. Pumper-tankers and square-side tankers equipped with T-shaped water tanks on a single-rear-axle chassis should carry no more than 1,500 gallons of water. Single-rear-axle chassis equipped with elliptical water tanks should be limited to about 2,000 gallons of water (**Figure 7.12;** *Courtesy of Dennis Wetherhold, Jr.*). Tankers with a tandem rear axle generally are limited to about 4,000 gallons of water (**Figure 7.13**). All of these figures vary depending on the gross vehicle weight rating of the chassis, the type of fire pump (if any) carried on the apparatus, and the amount of hose and other equipment expected to be carried on the apparatus. Departments that desire to transport more than 4,000 gallons of water on a single vehicle will need to use a tractor-trailer arrangement (**Figure 7.14;** *Courtesy of John Reith*).

One of the most common weight-related design problems associated with fire department tankers is trying to carry too much water on a specific chassis.

A number of factors must be considered when a fire department determines the type and size of tanker it wishes to operate. These include:

- *Terrain* – Larger tankers do not perform well in jurisdictions where they will be required to climb steep hills or to operate on winding roads.
- *Bridge weight limits* – Fire departments should determine the load carrying capacity of the bridges in their response district. They should make every effort to specify and design apparatus that will not exceed these

Figure 7.14 Departments that desire to transport more than 4,000 gallons of water on a single vehicle will need to use a tractor-trailer arrangement. *Courtesy of John Reith*

limits. Some bridges have weight restrictions so severe that no apparatus can be designed to traverse them. The fire department should identify alternative routes around these bridges.

- *Budgetary constraints* – Some fire departments that have a response district suitable for operating a larger tanker may not have the financial resources to purchase one. These departments will choose a smaller tanker that fits their budgetary constraints.
- *Compatibility with mutual aid tankers* – Water shuttles flow more easily when they use tankers of similar size. Departments that operate together on a regular basis should coordinate the size and design of their tankers.

Figure 7.15 Many jurisdictions use tankers equipped with large fire pumps as attack apparatus. These apparatus are commonly referred to as pumper-tankers. *Courtesy of Ron Jeffers*

Most of the information on tankers discussed to this point assumes that the apparatus will be used in a support role supplying water to other apparatus that attack a fire. Many jurisdictions, however, use tankers equipped with large fire pumps as attack apparatus. These apparatus are commonly referred to as pumper-tankers **(Figure 7.15;** *Courtesy of Ron Jeffers*). Pumper-tankers are commonly equipped like standard fire department pumpers and typically have fire pump capacities similar to pumpers. Their primary difference is the amount of water they carry. While standard pumpers typically carry 1,000 gallons of water or less, pumper-tankers may carry as much as 3,000 gallons of water. This increased water capacity allows the apparatus to sustain a longer initial fire attack independent of an external water supply. Disadvantages of these apparatus are that they are slower to respond and may have difficulty accessing some fire scenes because of their enormous size. (For more detailed information on operating tankers and water shuttles, consult IFSTA's *Pumping Apparatus Driver/Operator Handbook.*)

AERIAL APPARATUS EQUIPPED WITH FIRE PUMPS

The primary function of fire department aerial apparatus is to provide access for firefighters to upper levels of a structure and to deploy elevated master streams at large fires. While neither of these functions requires the aerial apparatus to be equipped with its own pump, some fire departments do choose to equip aerial apparatus with fire pumps. Aerial apparatus equipped with fire pumps are commonly referred to as quads or quints. Specifically, a quint is a fire apparatus that is equipped with an aerial device, ground ladders, fire pump, water tank, and fire hose **(Figure 7.16;** *Courtesy of Ron Jeffers*). Some of the more common reasons for equipping aerial apparatus with pumps include:

- The apparatus will be capable of supplying its own elevated master stream. This is important in jurisdictions with limited apparatus resources. In these situations, it may not be desirable to commit a pumper to supplying an elevated master stream device.

Figure 7.16 Specifically, a quint is a fire apparatus that is equipped with an aerial device, ground ladders, fire pump, water tank, and fire hose. *Courtesy of Ron Jeffers*

- Some fire departments operate on a "total quint concept." This involves the use of quints in all fire stations within the jurisdiction. Each quint is equipped with a full complement of both engine and ladder company equipment. The apparatus and the company assigned to it perform engine or ladder company functions depending on the needs of the particular incident and their assignment by the incident commander.
- The apparatus may be used to initiate an attack on a fire when an engine company is not present.
- The apparatus can protect itself in high radiant-heat situations.

Aerial apparatus may be equipped with a variety of different aerial devices. The standard aerial ladder is the oldest and most common type of aerial device. A variety of different articulating boom and telescoping platform aerial devices also are used widely throughout the fire service. Each type of aerial device has advantages and disadvantages that the local jurisdiction must evaluate before finally choosing one.

The capacity of the pump on aerial apparatus will vary depending on the local jurisdiction's preference. In order to be considered a true quint according to NFPA® 1901, the aerial apparatus must have at least a 1,000 gpm pump. Most departments that operate quints specify pumps considerably larger than the minimum. Pump capacities of 1,500 to 2,000 gpm are more common. A quint equipped with a large pump can supply its own elevated master stream and portable master streams or hand lines, given sufficient water supply. NFPA® 1901 also requires quints to have a water tank capacity of at least 300 gallons.

Departments that do not choose to use an apparatus as a full quint but still wish it to have some fire attack or self-preservation capabilities may specify a pump with a capacity between 250 and 750 gpm. NFPA® requires these smaller pumps to supply this limited volume of water at a pressure of 150 psi. This amount will supply one or two small hand lines to handle small fires or allow

Figure 7.17 Many fire departments equip full-blown rescue vehicles with small fire pumps, water tanks, and in some cases, foam systems to provide the rescue vehicle with limited fire fighting capabilities. *Courtesy of Ron Jeffers*

firefighters to cool the apparatus when it is exposed to high levels of radiant heat. Apparatus with these smaller pumps may also be equipped with small water tanks ranging from 100 to 400 gallons.

NFPA® 1901 contains an extensive list of fixed and portable equipment that an aerial apparatus must carry. This list varies depending on whether or not the apparatus will be operated as a quint. In general, the standard requires quints to carry fewer ground ladders and less ladder company equipment and more hose and engine company equipment than non-quint aerial apparatus. (For more information on aerial apparatus, quints, and operating aerial devices and their associated components, see IFSTA's *Aerial Apparatus Driver/Operator Handbook* and NFPA® 1901.)

RESCUE VEHICLES EQUIPPED WITH FIRE PUMPS

Rescue pumpers have become increasingly popular in the fire service during the last decade. Rescue pumpers are basically standard fire department pumpers altered to carry additional rescue and extrication equipment. Conversely, many fire departments equip dedicated rescue vehicles with small fire pumps, water tanks, and in some cases, foam systems to provide the rescue vehicle with limited fire fighting capabilities **(Figure 7.17)**.

Medium and heavy-duty rescue vehicles often are equipped with small pumps and water tanks similar to those on non-quint aerial apparatus. Most departments that add this equipment to rescue vehicles do so in order to allow the rescue vehicle to make a limited fire attack when a pumper is not immediately on the scene. Another common reason for adding this equipment is to allow the rescue vehicle to provide protection lines or extinguish fires during automobile extrications when another fire department vehicle may not be on the scene.

Figure 7.18 Fire fighting equipment in rescue vehicles takes up valuable compartment space.

Having some type of fire extinguishing capabilities is certainly an advantage for a rescue vehicle. On the other hand, it is also a drawback in that the pump, pump panel, and water tank will consume valuable compartment space **(Figure 7.18)**. In designing a rescue vehicle, the fire department must weigh the value of adding the pump and water tank versus the loss of compartment space based on the amount of rescue equipment needed and the number of personnel who typically will be riding the apparatus. The department also must weigh its own experience with needing fire attack capabilities when only a rescue vehicle was on the scene of an emergency.

AIRCRAFT RESCUE AND FIRE FIGHTING APPARATUS

Protecting aircraft and airport facilities requires specialized apparatus that can launch an expedient attack on flammable liquid spills and fires, using foam extinguishing agent independent of an external water supply. *Aircraft rescue and fire fighting (ARFF) apparatus*, formerly referred to as crash, fire, rescue (CFR) vehicles, are designed exactly for this purpose. ARFF apparatus range from small, rapid-intervention vehicles to extremely large, major fire fighting vehicles that are the largest land-based fire apparatus.

Figure 7.19 Major fire fighting (ARFF) vehicles are the largest land-based fire apparatus.

NFPA® 414, *Standard for Aircraft Rescue and Fire Fighting Vehicles,* contains the requirements for ARFF apparatus. Additional requirements for airports and ARFF vehicles based at airports in the United States are found in Federal Aviation Administration (FAA) regulations 14 CFR Part 139, *Certification and Operations: Land Airports Serving Certain Air Carriers.* In addition to Part 139, two other FAA documents apply specifically to ARFF apparatus:

Figure 7.20 RIVs are similar in design and function to major fire fighting vehicles, only smaller in size and capabilities. *Courtesy of Ron Jeffers*

- Federal Aviation Administration (FAA), AC 150/5220-10, *Guide Specifications for Water/Foam Type Aircraft Fire and Rescue Trucks*
- Federal Aviation Administration (FAA), AC 150/5220-14A, *Airport Fire and Rescue Vehicle Specification Guide*

Airports outside the United States follow the International Civil Aviation Organization (ICAO) Annex 14, *International Standards and Recommended Practices, Aerodromes.*

Figure 7.21 Combined agent vehicles. These small, initial attack vehicles are intended to arrive on a scene quickly and knock down or extinguish smaller fires. *Courtesy of Dennis Wetherhold, Jr.*

NFPA® 414 divides ARFF apparatus into three general classifications:

- *Major fire fighting vehicles.* These are the largest land-based fire apparatus **(Figure 7.19)**. They commonly have pumping capacities of up to 2,000 gpm and carry as much as 6,000 gallons of water and 600 gallons of foam concentrate.
- *Rapid intervention vehicles (RIV).* RIVs are similar in design and function to major fire fighting vehicles, only smaller in size and capabilities **(Figure 7.20;** *Courtesy of Ron Jeffers*). They typically have pump capacities of 1,250 gpm or less and carry no more than 1,500 gallons of water.
- *Combined agent vehicles.* These small, initial attack vehicles are intended to arrive on a scene quickly and knock down or extinguish smaller fires **(Figure 7.21)**. They may or may not be equipped with a small fire pump, foam system, and water tank. They all will have a variety of pressurized extinguishing agents including dry chemical, halogenated agents, and premixed foam/water agent.

SUMMARY

Fire apparatus are a crucial part of the overall municipal, and in some cases industrial, fire protection picture. Fire apparatus driver/operators must be knowledgeable of the different types of apparatus that they will be expected to drive. Fire officers must understand the capabilities of the different types of apparatus in making assignments during emergency operations. Other than meeting the minimum design requirements in the NFPA® standards, there is little consistency in the fire service in designing, constructing, and equipping fire apparatus. Each jurisdiction must determine the combination that best suits its particular needs.

REVIEW QUESTIONS

1. Name the NFPA® standard that sets forth the minimum design and performance requirements for all fire apparatus except wildland and aircraft rescue and fire fighting vehicles.

   ____

2. Name the fire apparatus industry trade group and its acronym.

   ____

3. What minimum pump capacity must a pumper have to meet the requirements of NFPA® 1901?

   ____

4. What is a triple-combination pumper?

   ____

5. What minimum water tank capacity must a pumper have to meet the requirements of NFPA® 1901?

   ____

6. Name the two basic types of initial attack fire apparatus.

   1. ______________________________
   2. ______________________________

7. What are the two basic methods of operation for using a tanker/tender?

   1. ______________________________
   2. ______________________________

8. List at least three factors a fire department must consider when designing a new tanker.

   1. ______________________________
   2. ______________________________
   3. ______________________________

9. List the five components necessary for an apparatus to be considered a quint.

   1. ______________________________
   2. ______________________________
   3. ______________________________
   4. ______________________________
   5. ______________________________

10. What is the primary drawback of equipping a rescue vehicle with a fire pump and water tank?

    ____

11. Name the three categories of ARFF vehicles identified in NFPA® 414.

    1. ______________________________
    2. ______________________________
    3. ______________________________

Answers found on page 427.

FESHE Course Objectives

The information in this chapter is intended to meet some of the objectives outlined for the Fire Protection Hydraulics and Water Supply course by the United States Fire Administration's National Model Core Curriculum as developed by the Fire and Emergency Services Higher Education (FESHE) initiative. The information in this chapter also allows the student to meet the requirements of NFPA® 1002, *Standard for Fire Apparatus Driver/Operator Professional Qualifications* (2009 edition) listed below.

FESHE Course Objectives

1. Describe the operation of positive displacement pumps and explain their use in the modern fire service.
2. Explain the design, components, and operating principles of single-stage and multistage centrifugal fire pumps.
3. List and describe the various pump driver arrangements used on modern fire apparatus.
4. Describe the various types of pressure regulating devices used on fire apparatus pumps.
5. Explain the operation and use of flowmeters with fire pumps.
6. List and describe the operation of the three basic types of priming devices used on modern fire pumps.

NFPA® 1002 Requirements

5.2.1 (B) and 5.2.2 (B) Requisite Skills. Power transfer from vehicle engine to pump, operate pumper pressure control systems, operate the pressure/volume transfer valve (multistage pumps only), operate auxiliary cooling systems.

Chapter 8

Fire Service Pump Design

INTRODUCTION

Most municipal water systems in service today cannot provide adequate pressure to the hydrant system for the reach and penetration of direct fire fighting application. In other words, those systems cannot supply attack hoses straight from a fire hydrant with high enough pressures for fire attack. Thus, to produce effective fire streams requires increasing the water-system pressure by using the fire pump on a fire apparatus as an intermediary between the water system and the attack hose(s). Fire pumps on fire apparatus also become necessary when attack lines must be supplied from the apparatus water tank or from an external static water supply source such as a portable tank, lake, stream, pond, or river.

The use of fire pumps on fire apparatus dates to the late eighteenth century. These early "pumpers" were operated manually in the truest sense. First, they had to be hand pulled to the fire scene, and then firefighters had to operate by a hand pump handle that drove a piston in a cylinder **(Figure 8.1)**. This forced the water out of the pump with enough velocity to push it through the hose or nozzle. Not far behind the early piston pumps was the development of rotary pumps. These early rotary pumps had a hand crank that rotated a gear, again forcing the water out of the pump at a workable pressure. Both of these pumps were early examples of *positive displacement pumps* because each positive action forced a specific amount of water and/or air from the pump body. Most modern apparatus still utilize some form of positive pressure pump as a secondary mechanism for drawing water from a static water source into the main fire pump.

Nearly all fire department pumpers in service today utilize a centrifugal-type pump for their main fire pump. The *centrifugal pump* uses centrifugal force to impart a velocity to the water. The pump then transforms that velocity into the pump discharge pressure necessary for effective fire stream operation. This chapter will cover the basic concepts and operational principles of both positive displacement and centrifugal fire pumps. It also will summarize the various components that make up the total apparatus pumping system. Today's fire apparatus increasingly have foam extinguishing systems as an integral part of the fire pump and apparatus; Chapter 9 covers foam systems.

Figure 8.1 Some of the earliest fire pumpers were hand-operated piston pumps.

POSITIVE DISPLACEMENT PUMPS

Since the decline of the Ahrens-Fox piston pumpers in the 1950s and of the John Bean high-pressure piston pumpers of the 1960s and 1970s, the centrifugal pump has replaced the positive displacement pump as the *main* fire pump on modern fire apparatus. However, that does not mean that positive displacement pumps have been eliminated from use on modern fire apparatus. Quite the contrary, virtually every modern pumping fire apparatus is still equipped with a positive displacement pump of some kind. Because centrifugal pumps cannot move air, positive displacement pumps are used as priming pumps to evacuate air from the centrifugal pump, thus allowing atmospheric pressure to force water from a static supply source through the intake hose and into the centrifugal pump. This process is based on the second principle of pressure: *Pressure applied on a confined liquid from an external source will be transmitted equally in all directions throughout the liquid without a reduction in magnitude* (Chapter 2).

Modern positive displacement pumps operate on the same basic principles as the earliest piston and rotary pumps. Obviously, though, the modern piston and rotary pumps are no longer manually operated. The following section provides an overview of the modern versions of these types of positive displacement pumps.

Piston Pumps

The operating principle of the piston-type positive displacement pump is rather simple: a piston moves back and forth inside a cylinder. As the piston moves forward, it compresses the air within the cylinder, creating a higher pressure inside the pump than the atmospheric pressure in the discharge manifold. This pressure opens a discharge valve and forces the air through the discharge lines **(Figure 8.2)**. Once the piston completes its travel on the forward stroke and stops, the pressures equalize and the discharge valve closes. Then, as the piston begins its return stroke, the area in the cylinder behind the piston increases and the pressure decreases, creating a partial vacuum. This opens an intake valve connected to the pump intake, allowing some of the air from the suction hose to enter the pump **(Figure 8.3)**.

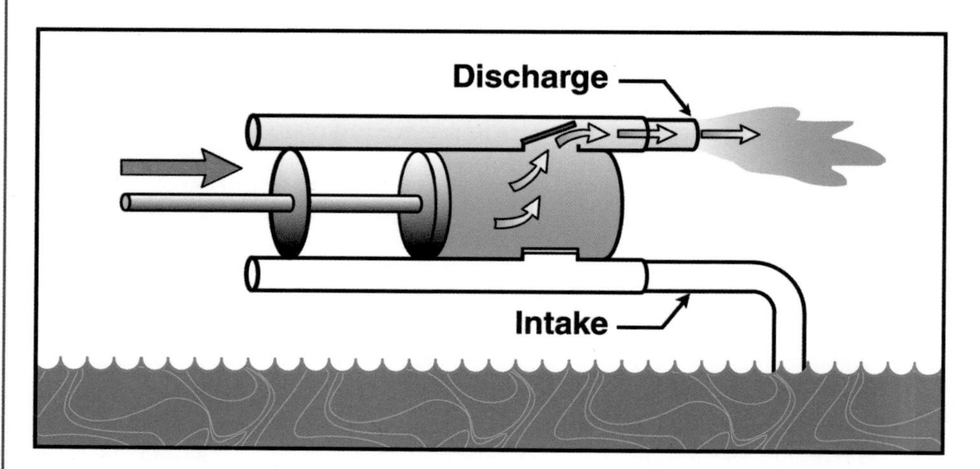

Figure 8.2 Piston pump forward stroke. The higher pressure inside the pump causes the discharge valve to open, allowing air to escape through the discharge lines.

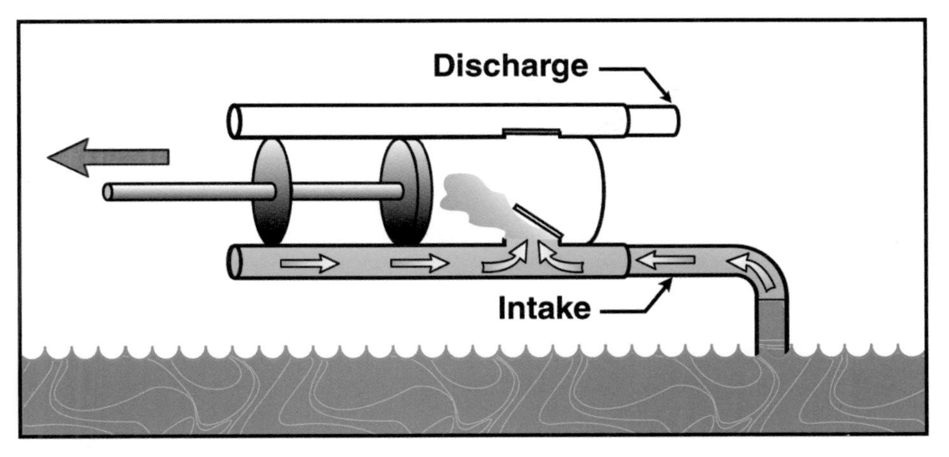

Figure 8.3 Piston pump return stroke. The partial vacuum created as the piston begins the return stroke causes the intake valve to open. This allows air from the suction hose to enter the pump.

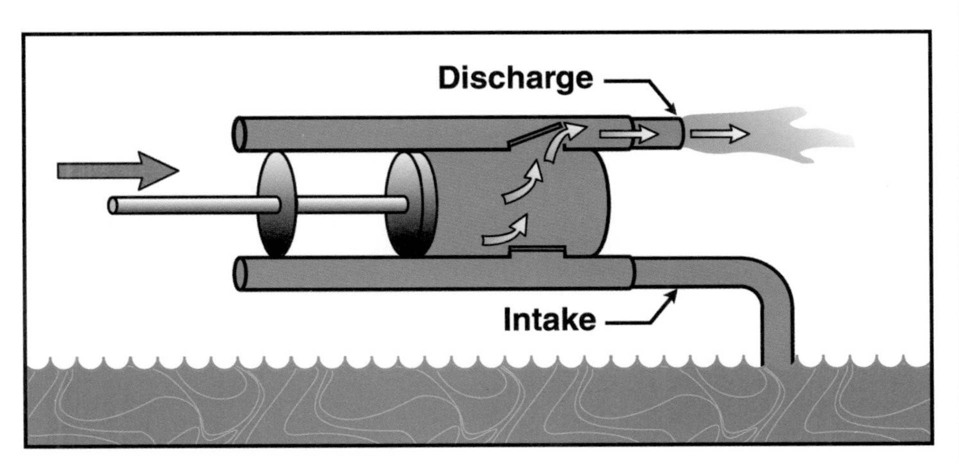

Figure 8.4 Primed piston pump. Once all the air has been evacuated, only water is pushed through the pump.

The pump repeats this process until it has removed a significant portion of the air from inside the suction hose and the intake area of the pump. This reduces pressure inside the hose to less than atmospheric pressure. Technically, this is called a partial vacuum. Now the higher atmospheric pressure forces the water to rise a small distance within the hose until the piston completes its travel and the intake valve closes. This action is repeated until all the air has been removed and the intake stroke introduces water into the cylinder itself. Once the pump has filled with water it is considered to be primed, and further strokes force water instead of air into the pump discharge **(Figure 8.4)**.

As this process continues, every forward stroke causes water to be discharged, and every return stroke causes the pump to fill with water again. Pumps that operate on this principle are known as *single-acting piston pumps.* They do not produce a usable fire stream because no water flows during the return stroke; the discharge is a series of surges of water, each followed by an equal time with no water.

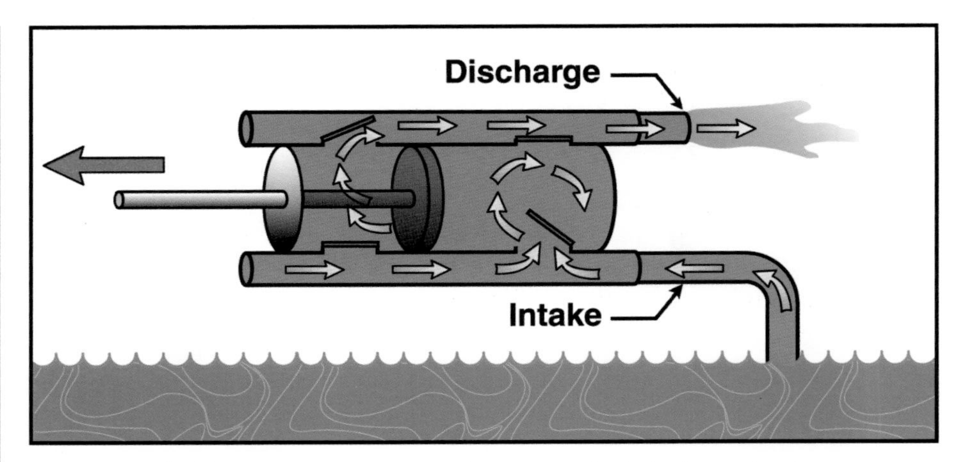

Figure 8.5 A double-acting piston pump pushes water on both the forward and return strokes.

The addition of two more valves enables the pump both to receive and to discharge water on each stroke of the piston, producing a more constant stream **(Figure 8.5)**. This is known as a *double-acting piston pump.* Although the double-acting pump still produces an output that is a series of pressure surges with periods of no flow when the piston ends its travel in either direction, these pulses are not as noticeable in fire stream operation as those of a single-acting piston pump.

Because the pump cylinder contains a defined volume, each stroke of the piston delivers just that amount. Thus, the output capacity of a piston pump is determined by the volume of the cylinder and the speed of the piston travel. Because there is a practical limit to the speed at which a pump can be operated, its capacity is usually determined by the size of the cylinder.

Most piston pumps used for fire service applications were double-acting, multicylinder pumps. Multicylinder pumps were more practical than large single-cylinder pumps because some of the cylinders could be disengaged when the pump's full capacity was not needed. Multicylinder pumps also provided a more uniform discharge because their cylinders were arranged to reach their peak flows at different points in the cycle.

Some large-capacity fire service piston pumps had a pressure dome or air chamber on the discharge to smooth the pulses. During peak pressures, the pump forced water into the air chamber. The water compressed the air trapped in the chamber until the pressure of the air equaled the pressure of the water. When the pressure in the pump discharge dropped, the air pressure in the chamber forced the water from the chamber into the line, thus counteracting the drop in pressure from the pump. These pressure domes were particularly common on Ahrens-Fox pumpers.

The piston pump has not been used as the primary fire pump in fire department pumpers since the 1950s. In addition to the problems of the pulsating fire stream, these pumps are very susceptible to wear and thus are maintenance intensive. Their efficiency depends upon a close tolerance between the piston and the cylinder walls. As the piston wears, water can escape from the discharge side of the pump to the intake, reducing the pump's capacity. Wear also occurs on the walls of the cylinder. Lastly, water contaminated by dirt, sand, and other debris can quickly damage the pump.

Although piston pumps were no longer used as high-volume pumps after the 1950s, later versions, such as the John Bean high-pressure pumps common in the 1960s and 1970s, were used for high-pressure, low-volume applications. These multicylinder, PTO-driven pumps provided pressures up to 1,000 psi for high-pressure fog lines that typically flowed less than 15 gpm. This application requires a dependable relief valve because pressures quickly build to dangerous levels if the discharge flow is interrupted. Some high-pressure, low-volume piston pumps remain in service today, most commonly in wildland fire fighting.

Rotary Pumps

Rotary-type positive displacement pumps have the simplest operational design of all fire pumps. A few types of older apparatus used them as their primary pumps. Some were also used for small booster-type pumps. Almost all modern pumping fire apparatus use some form of rotary-type positive displacement pump as a priming pump for their main centrifugal fire pump. Today's rotary-type pumps are of either rotary gear or rotary vane construction. They are driven either by a small electric motor or through a clutch that extends off the apparatus drive shaft.

Rotary Gear Pumps

The rotary gear pump consists of two tightly meshed gears that rotate inside a watertight case **(Figure 8.6)**. The gears are constructed so that their teeth contact each other and come in close proximity to the case. This allows the gears to form watertight and airtight pockets within the case as they rotate away from the intake and toward the outlet. As each gear tooth reaches the discharge chamber, it forces the air or water from that pocket out of the pump. As the tooth returns to the intake side of the pump, the gears mesh tightly enough to prevent the water or air that has been discharged from returning to the intake side.

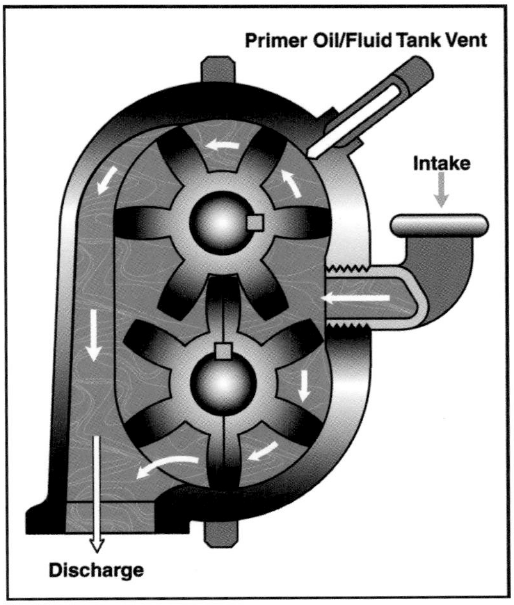

The rotary gear pump is a positive displacement pump because each pocket in the gears contains a fixed volume. As well, each time the gears turn they force water out of the pump with a positive action. A rotary gear pump's capacity depends upon the size of the pockets formed by the gears and the speed of the gear rotation. If the pump is trying to move more water than the discharge lines use can carry, pressure builds. Because of this, a pressure-relief device must be provided to discharge or divert any excess pressure.

Figure 8.6 The basic design of a rotary gear pump.

Like the piston pump, the rotary gear pump requires very tight operating tolerances between its moving parts and the pump case. Thus, the rotary gear pump also is very susceptible to damage from normal wear and from water containing sand and other debris. To minimize damage to the pump casing, most rotary gear pumps have gears made of bronze or another soft metal. These gears may be easily replaced if necessary. A stronger metal alloy, such as cast iron or steel, is used for the pump casing.

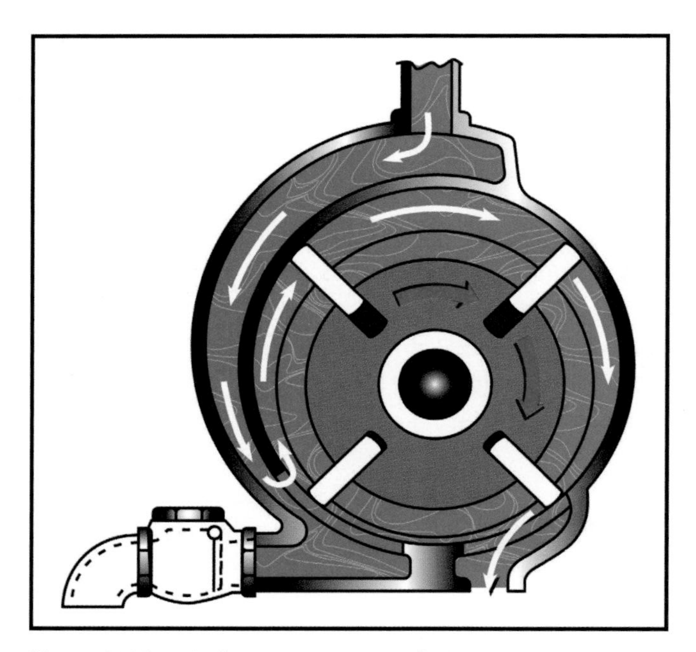

Figure 8.7 A typical rotary vane pump design.

Some rotary gear pumps have only one powered gear. The other gear is simply moved by the rotation of the powered gear. These pumps' drive gears are usually steel with bronze inserts; the steel gives the drive gears the strength needed to withstand the torque that will be developed. Another common design utilizes steel pilot gears mounted outside the pump casing. The pilot gears in turn are connected to a shaft that drives the pump gears.

Rotary Vane Pumps

The *rotary vane pump* contains a rotor that is mounted off-center inside the pump casing **(Figure 8.7)**. The distance between the rotor and the casing is much greater at the intake area than at the discharge area. The rotor is equipped with a series of vanes that are free to move within the slot where they are mounted. As the rotor turns, centrifugal force forces the vanes outward against the casing. When the surface of the vane that contacts the casing becomes worn, centrifugal force causes the vane to extend further, thus automatically maintaining a tight fit. This self-adjusting feature makes the rotary vane pump much more efficient at pumping air than a standard rotary gear pump.

As the rotor turns, air is trapped between the rotor and the casing in the pockets between adjacent vanes. As the vanes turn, these pockets become smaller, which compresses the air and builds pressure. When the vanes reach the pump discharge, the pressure reaches its maximum level, forcing the trapped air out of the pump. The close spacing of the rotor at that point prevents the air or water from returning to the intake. As with all positive displacement pumps, the air being evacuated from the intake side causes a reduced pressure (partial vacuum) and atmospheric pressure forces water into the pump. When the pump is filled with water, it is primed, and it forces water out the discharge in the same manner that it forced out air.

Driver/operators should be familiar with the type of rotary pump that their apparatus uses as a priming pump and with the manufacturer's instructions for operating it. These pumps must be serviced and tested regularly to ensure proper operation. Some manufacturers also recommend maximum operating times for their pumps. If the pump does not achieve a prime within that time, the operator should rest the pump and check hose connections, pump valves, and pump drains for signs of air leaks.

CENTRIFUGAL PUMPS

The primary fire pump on all modern fire apparatus is a *centrifugal pump*. The centrifugal pump is considered a nonpositive displacement pump because it does not pump a definite amount of water with each revolution. Instead, it imparts velocity to the water and then converts that velocity to pressure within the pump casing. Centrifugal pumps are extremely flexible and, depending on their particular design, allow fire personnel a variety of options to meet the needs of the incident. This section explores the design and operating concepts of the centrifugal pump and highlights the flexibility of its use.

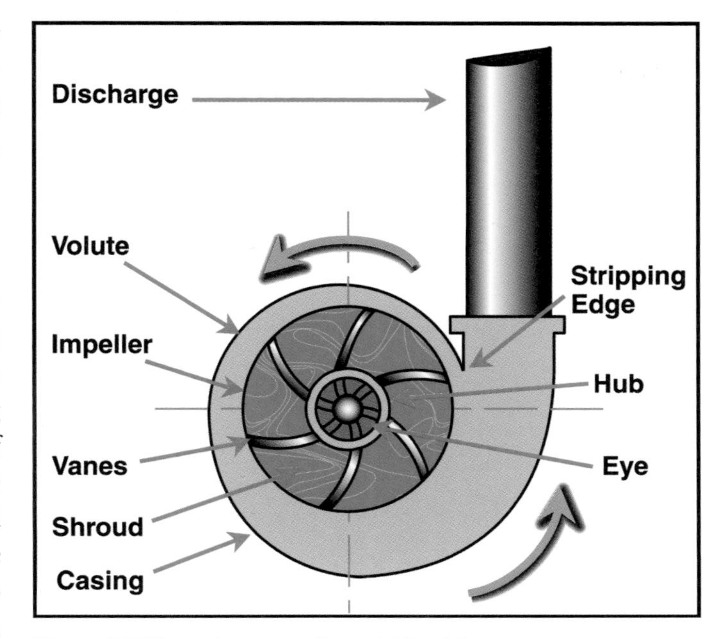

Figure 8.8 The major parts of a centrifugal fire pump.

Principles of Operation and Construction of Centrifugal Pumps

The operating principle of a centrifugal pump is rather simple. Inside the pump casing are one or more disks called impellers **(Figure 8.8)**. Water enters the pump at the center of the casing, and the rapidly revolving impellers throw the water toward the outer edge of the impeller. The faster the impeller turns, the more velocity it imparts to the water. If the water is contained at the edge of the disk, the water at the center of the container, in this case the pumping casing, begins to move outward. The pump casing confines the water, converting its velocity to pressure. With its movement limited by the pump casing, the water moves upward in the path of least resistance. The amount of pressure created depends upon the speed of impeller rotation; the faster the rotation, the greater the pressure.

The amount of pressure that a centrifugal pump creates depends on the speed of impeller rotation; the faster the rotation, the greater the pressure.

In summary, the centrifugal pump consists of two primary parts: an impeller and a casing. The *impeller* (disk) transmits energy to the water in the form of velocity. The *casing* (container) collects the water and confines it in order to convert the velocity to pressure. Then, the casing directs the water to the pump discharge.

Water is introduced into the pump through the eye of the impeller **(Figure 8.9)**. Because of this, the pump's capacity depends partially on the size of the eye. The greater the diameter of the eye, the greater the flow capacity.

The impeller rotates very rapidly within the casing. Speeds of 2,000 to 4,000 rpm are common depending on the design of the pump and the amount of pressure or volume desired. As the water contacts the vanes of the spinning impeller, centrifugal force throws it to the outside of the impeller **(Figure 8.10, p. 140)**. The shrouds of the impeller confine this water in its travel, which increases in velocity with the speed of rotation.

The impeller is mounted off-center in the casing. This creates a water passage that gradually increases in cross-sectional area as it nears the pump's discharge outlet. This passage is known as the *volute*. The increasing size of the volute is necessary because the amount of water passing through the volute increases as it approaches the discharge outlet. The gradually increasing size of

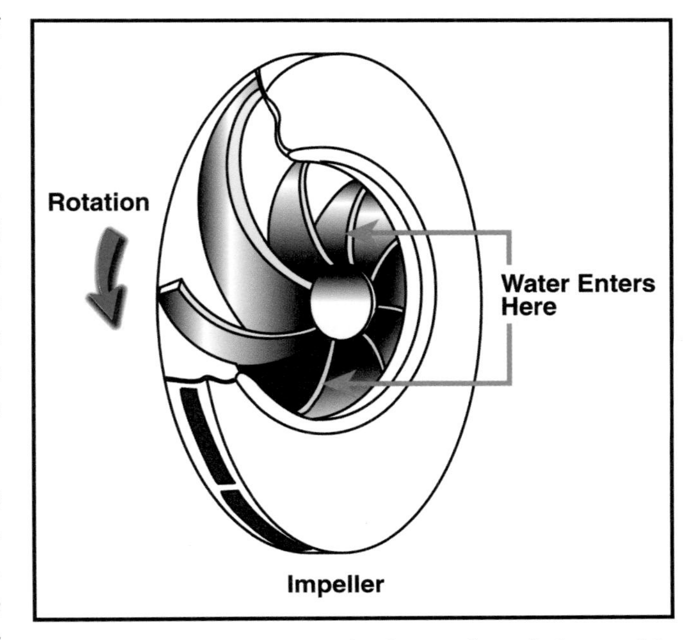

Figure 8.9 Water enters a centrifugal pump through the eye of the impeller.

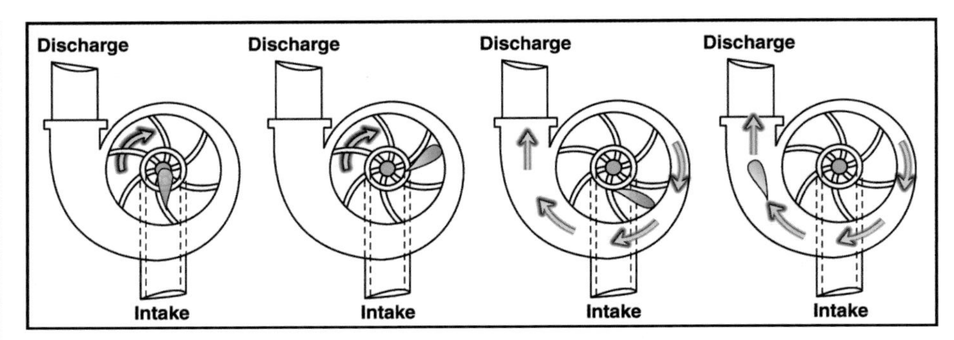

Figure 8.10 These schematics trace the path of water flow through the centrifugal fire pump.

the waterway reduces the velocity of the water, thus enabling the pressure to build proportionately. Three main factors influence a centrifugal fire pump's discharge pressure:

- Amount of water being discharged
- Speed at which the impeller is turning
- Pressure of water when it enters the pump from a pressurized source (hydrant, relay, etc.)

The greater the volume of water being flowed by a centrifugal pump, the lower the discharge pressure will be.

If all of the water entering the pump were immediately discharged as it was thrown from the impeller, the water would have little ability to build pressure. However, restricting the discharge of the water, as by partially or totally closing a discharge valve, develops an increasing amount of pressure on the water. Thus, if all other factors remain constant, the amount of discharge pressure that a pump can develop is directly dependent upon the volume of water it is discharging. The greater the volume of water being flowed, the lower the discharge pressure will be.

Because the discharge pressure in the centrifugal pump results from the velocity and amount of moving water, impeller speed is important in determining the pressure to be developed. The greater the impeller speed, the greater the pressure. This increase in pressure is approximately equal to the square of the change in impeller speed. If all other factors remain constant, doubling the speed of the impeller quadruples the discharge pressure.

The third factor that influences the discharge pressure is the intake pressure. Centrifugal pumps have no mechanical blockage between their intake and discharge sides. Because of this, water will flow through a centrifugal pump even if the impeller is not turning. When water is supplied to the eye of the impeller under pressure, any movement of the impeller increases both the velocity of the water and the corresponding pressure buildup in the volute. Conversely, if the intake pressure were to be decreased at any point in the operation, the discharge pressure also would drop accordingly.

The centrifugal pump's ability to create pressure on water also is due in part to water's being incompressible. When pressure is imparted on water, it forces the water to move. This is not the case with a compressible medium such as air. This is why the centrifugal pump cannot pump air and is not self-priming. For a centrifugal pump to draft water, some type of external priming pump must remove the air and allow atmospheric pressure to force the water into the pump. Either a rotary gear pump or a rotary vane pump typically is used for this purpose.

For a centrifugal pump to draft water, some type of external priming pump must remove the air and allow atmospheric pressure to force the water into the pump.

The fire service uses two basic types of centrifugal pumps: single-stage and two-stage. The following section details the operation of each of these.

Single-Stage Centrifugal Fire Pumps

A *single-stage centrifugal pump* contains only one impeller within its pump casing **(Figure 8.11)**. In the days when most fire apparatus were powered by gasoline engines, the volume capacities of single-stage pumps were limited to 1,250 gpm. The substantially higher horsepower provided by modern diesel engines has allowed the development of larger single-stage pumps. The added horsepower enables the pump to operate larger impellers. Modern pumpers equipped with single-stage pumps of up to 2,000 gpm are now common. In reality, single-stage pumpers are easier to operate and most fire department no longer have a need for two-stage pumps.

Figure 8.11 A single-stage pump impeller. *Courtesy of Hale Fire Pump Company.*

However, these more powerful engines, larger impellers, and greater amounts of moving water presented new design challenges. A well-known law of physics states that for every action there is an equal and opposite reaction. In the case of a pump, the action of the moving water creates a reaction of stress on the pump, its bearings, other moving parts, and the casing's mounting to the truck frame.

To minimize these stresses and the lateral thrust created by large quantities of water entering the eye of the impeller, engineers designed a double intake impeller. The double intake impeller takes in water equally from both sides. The resultant equal and opposite reaction cancels any lateral thrust. The dual intakes also allow greater volumes of water to enter the pump, further increasing the pump's capacity.

Because the impeller turns at a very high rate, a radial thrust is developed as the water is delivered to the discharge outlet. Stripping edges in the opposed discharge volutes divert the water 180 degrees apart. Removing the water at two places and directing it in opposite directions cancels the radial thrust. These combined design elements provide a hydraulically balanced pump. This lessens stress on the pump and chassis and helps to lengthen the serviceable life of the pump and the apparatus.

Figure 8.12 The impellers of a two-stage fire pump as they appear mounted on the impeller shaft. *Courtesy of Hale Fire Pump Company.*

Two-Stage Centrifugal Fire Pumps

As its name implies, the *two-stage centrifugal pump* has two impellers mounted within its pump casing **(Figure 8.12)**. Characteristics of these impellers include:

- They are usually mounted on a single shaft.
- They are driven by a single power source.
- They are usually identical in size and design and have the same capacity.

Two-stage centrifugal pumps were designed in the days when most fire apparatus had gasoline-powered engines. Because those early gas-powered engines did not develop enough power to turn large-capacity, single impeller pumps, pumps with capacities exceeding 1,250 gpm required more than one impeller. By operating parallel to each other, the two impellers could move larger volumes of water.

However, the two-stage pump does not only allow large *volumes* of water to be pumped. A valve that switches the water from entering both impellers at the same time to entering the impellers in series from one to the next enables the pump to develop extremely high pressures. The following sections examine the two-stage centrifugal pump's ability to operate in either of these modes.

Pumping in the Volume (Parallel) Position

When a two-stage centrifugal pump's changeover valve, sometimes also referred to as the transfer valve, is set to the volume position, both of the impellers take water from the intake and deliver it to the discharge simultaneously **(Figure 8.13)**. Each of the impellers can deliver the pump's rated pressure while flowing half of its rated capacity. Thus, if the pump is rated at 2,000 gpm at 150 psi, each of the impellers supplies 1,000 gpm to the pump discharge manifold at that pressure. The volume position is most commonly used when pumping multiple or large diameter supply lines or feeding master stream devices.

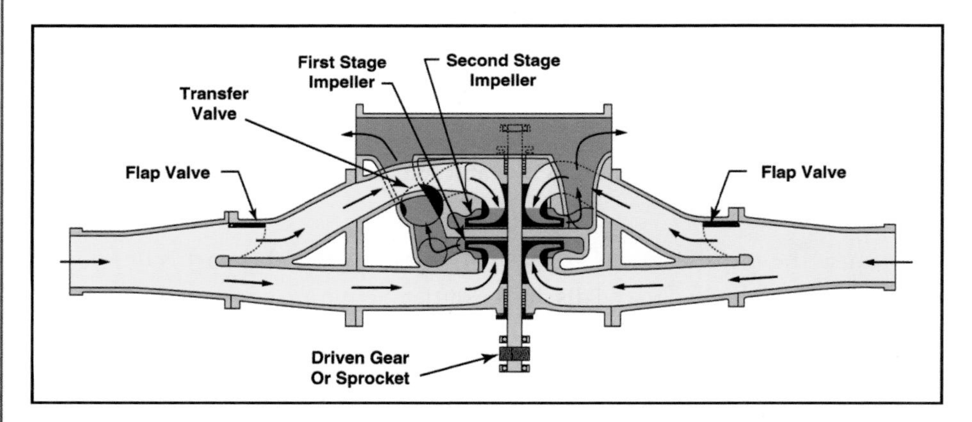

Figure 8.13 The route of water through a two-stage fire pump during *volume* pumping operations. *Courtesy of Waterous Company.*

Pumping in the Pressure (Series) Position

When the changeover valve is in the pressure position, it directs all of the water from the intake manifold into the eye of the first impeller **(Figure 8.14)**. Depending on the pump manufacturer, the first stage increases the pressure and

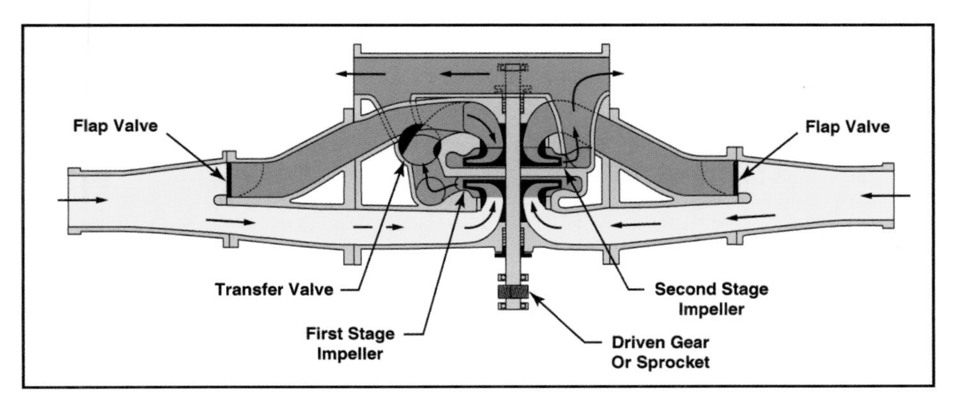

Figure 8.14 The route of water through a two-stage fire pump during *pressure* pumping operations. *Courtesy of Waterous Company.*

discharges from 50 to 70 percent of the volume capacity through the changeover valve and into the eye of the second impeller. The second impeller increases the pressure and delivers the water at the higher pressure into the pump discharge port. At this point, the pressure is much higher than in the parallel (volume) position because the same stream of water has passed through two impellers, each adding to the pressure. With only one impeller delivering water to the pump discharge port, the total volume of water is limited to the amount that one impeller can supply. The pressure position is most commonly used when supplying attack lines, especially when operating from the apparatus water tank.

Figure 8.15 The changeover, or transfer, valve is used to switch between the *volume* and *pressure* operations.

Operating Two-Stage Fire Pumps

Each fire pump manufacturer specifies when the changeover valve on their particular pump should be in the volume or pressure positions **(Figure 8.15)**. The process of switching between pressure and volume is sometimes referred to as *changeover*. Historically, the fire service has taught the rule of thumb that the changeover valve stays in the pressure position until it is necessary to supply more than one-half the rated volume capacity of the pump. However, advances in design and efficiency now allow most pump manufacturers to specify that the pump may remain in the pressure system until it is necessary to flow more than 70 percent of the rated volume capacity. At lower flow rates, operating in the series (pressure) position reduces the load and the required rpm of the engine. The driver/operator should consult the owner's manual for the specific pump being operated to determine when the transfer should occur.

Before beginning operations, the pump operator should attempt to anticipate which mode of operation will be required so that the pump will be in the proper position from the start. If the pump operator is not sure which setting to use, parallel (volume) is better than series (pressure). While the parallel (volume) position may make attaining the higher pressures difficult, at least it will be capable of supplying 100 percent of the rated volume at 150 psi. The series (pressure) mode of operation may make it impossible to supply the necessary volume of water to the attack lines.

Having a two-stage centrifugal pump in the correct mode from the start is important because it negates the need to change over in the middle of operation.

Having the pump in the correct mode from the start is important also because it negates the need to change over in the middle of operations. The maximum net pump discharge pressure at which the changeover valve should be operated varies depending on the pump's age and manufacturer. In most cases, the recommended maximum pressure would not exceed 75 psi. If the pump is already operating at a higher pressure, the operator will have to throttle down the engine and reduce pressures. This could endanger firefighters in forward positions if it interrupts their water supply. Coordinating the changeover with attack crews is imperative to prevent shutting down lines when crews are in precarious positions.

Another hazard of performing the transfer in the middle of an operation is that sudden changes in pressure occur as the water changes its direction of flow. Switching from volume to pressure immediately doubles the discharge pressure. This can damage hoselines and fire pumps as well as cause injuries to firefighters on the hoselines if it takes place suddenly with excessive pressure on the pump.

On most older two-stage pumps, changing over is a manual operation. Most of these pumps contain a built-in safeguard that makes it physically impossible to perform the transfer while the pump is operating at high pressures.

Newer pumps utilize a power-operated changeover valve. These changeover valves use electricity, air pressure, vacuum from the engine intake manifold (gasoline engines only), or even the water pressure itself to accomplish the transfer. The power control may have some type of manual override to allow transfers should the power equipment fail. Many power-operated changeover valves operate at pressures as high as 200 psi. Such high pressures can present extreme danger to personnel and equipment, so driver/operators must use special care with these controls.

Changeover at high pressure can present extreme danger to personnel and equipment.

Centrifugal pumps with more than two impellers have been and continue to be used in some portions of the fire service. They are most common in fire departments that protect tall high-rise structures. These multistage pumps have as many as four impellers connected in series and can develop pressures up to 1,000 psi. They may be referred to as three-stage or four-stage pumps. Another design involves a single-stage high-pressure centrifugal pump with a separate drive system connected to a conventional two-stage pump. When circumstances require pressures higher than the 250 psi that a conventional two-stage pump can create, engaging the separate third stage will greatly increase the pressure from the second stage.

Some of these multistage pumps were previously used for high-pressure fog fire fighting. The use of high-pressure fog declined rapidly in the last two decades of the twentieth century, although it is still somewhat common in wildland fire fighting. In recent years multistage pumps more commonly have been used in cities that have a substantial number of high-rise structures. Their increased pressure capabilities allow fire department pumpers to overcome the elevation pressure loss and supply sprinkler and standpipe systems in tall buildings. Pumpers designed to supply high pressures must be equipped with fire hose that is rated and tested for those pressures. In many cases the standard fire hose carried by most engine companies is not designed for such extreme pressures.

Pumpers designed to supply high pressures must be equipped with fire hose that is rated and tested for those pressures.

MOUNTING AND DRIVE ARRANGEMENTS

When specifying the design of a fire apparatus, a fire department has a number of choices for how the fire pump will be driven. Factors in the decision can include cost, appearance, space required, ease of maintenance, and tradition. However, the most important consideration may be the pump's expected use. Each pump driving system has certain characteristics that make it more or less adaptable to a particular fire department's needs. The most common pump drive arrangements on modern fire apparatus are:

- Auxiliary-engine driven pumps
- Power take-off (PTO) driven pumps
- Front-mount driven pumps
- Midship transfer driven pumps
- Rear mount pumps

Auxiliary-Engine Driven Pumps

Auxiliary-engine driven pumps receive their power from a separate engine independent of the engine used to propel the vehicle **(Figure 8.16)**. Depending on the wishes of the purchaser, the pump engine may use the same fuel type as the propulsion engine or it may use an entirely different fuel. Auxiliary engine-driven pumps occasionally are used for structural fire fighting apparatus. However, they are more likely to be found in special applications such as:

- Airport rescue and fire fighting (ARFF) vehicles
- Wildland fire apparatus
- Mobile water supply apparatus
- Trailer-mounted fire pumps
- Portable fire pumps

Figure 8.16 Auxiliary engine pumps are commonly used on small wildland fire apparatus.

Auxiliary engine-driven pumps are preferred when the department wants either the ability to pump and roll or the ability to place a fire pump remote from an apparatus. Pump and roll refers to pumping water while the apparatus is in motion. This is a common tactic in ARFF and wildland fire fighting. By having a separate engine to drive the pump, the pump's operation and discharge do not depend on the speed of the propulsion engine. This is important in situations that require large volumes of water or high pressures while the apparatus operates at a slow speed. Separate engine-driven pumps also provide flexibility in apparatus design in that they can be mounted anywhere on the apparatus.

Auxiliary-engine driven pumps used for wildland apparatus, mobile water supply apparatus, and portable pumps typically are powered by engines with less than 50 horsepower and a pumping capacity of 400 gpm or less. In many cases, these pumps are mounted on a skid assembly that includes a small water tank, booster reel, and hose trays for small-diameter attack lines. Fire depart-

Figure 8.17 Trailer pumps have auxiliary engines.

ments purchase these assemblies and mount them in the rear of a pickup truck to make a small attack or wildland fire apparatus. Modern auxiliary-engine driven pumps also may be equipped with foam proportioning and compressed air systems.

Auxiliary-engine driven pumps used on structural pumpers, ARFF apparatus, and trailer-mounted applications tend to be large-capacity pumps of up to 4,000 gpm. They are powered by full-sized diesel engines capable of up to 500 horsepower or more. Trailer-mounted pumps commonly are used at industrial and fire training facilities **(Figure 8.17)**.

While these pumps do provide flexibility in apparatus design and operation at emergencies, they have one inherent drawback. Having a second engine on the apparatus creates additional maintenance work for the driver and departmental mechanics. If the engine driving the pump fails, the apparatus may arrive at the fire scene but be unable to pump water to fight the fire.

Power Take-Off (PTO) Drive

Power take-off (PTO) driven pumps are powered by a driveshaft connected to a PTO on the chassis transmission. The optimum design for a PTO pump installation includes a minimum of angles in the driveshaft between the PTO connection and the pump casing. The driveshaft also should not extend so far below the chassis that it could be damaged easily as the truck travels on or off the road. If the driveshaft does extend below the chassis' frame rails, some type of skid plate should be used to provide protection.

The PTO pump permits pump and roll operation but not as effectively as the auxiliary-engine driven unit. The PTO connection receives its power from an idler gear in the chassis transmission **(Figure 8.18)**. When the pump is in use, the speed at which the PTO shaft turns is independent of the gear in which the chassis transmission is operating. However, if the chassis transmission is the manual shifting type, the clutch controls the spinning of the PTO driveshaft. Thus, when the driver/operator disengages the clutch (pushes in the clutch pedal) to stop momentarily or to change gears, the driveshaft to the pump also stops turning.

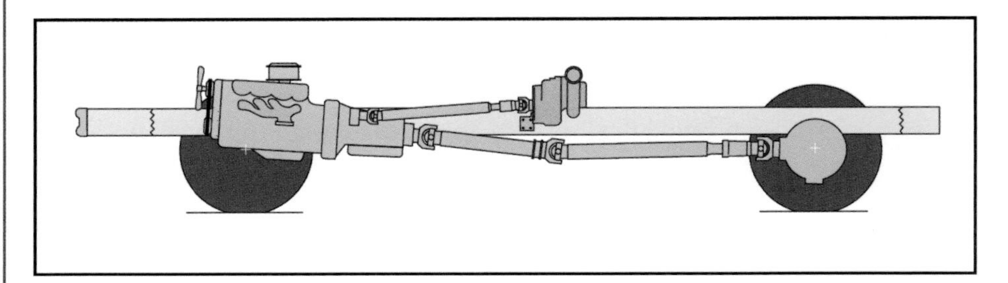
Figure 8.18 The basic layout of a PTO pumping system.

As with any type of pump, the amount of pressure developed depends on the speed of the engine turning the pump. Thus, if the driver increases the apparatus speed or revs the engine, the pressure in the pump will increase accordingly. If the truck is designed for pump and roll operation, a pressure gauge should be mounted inside the cab in full view, and the driver should operate the vehicle by the pressure gauge instead of the speedometer while hoselines are in operation. However, the driver/operator must still take caution not to drive the apparatus too fast for conditions.

In the past, conventional PTO-driven fire pumps were used most commonly on initial attack, wildland, rescue, and mobile water supply apparatus. The conventional PTO arrangement unit limited the capacity of the pump to about 500 gpm. This was because the PTO unit was mounted on the side of the transmission, and the maximum strain the transmission housing could withstand limited the connection to about 35 horsepower. Most pump manufacturers also limited the pump drive's rpms.

In recent years, some pump manufacturers have developed "full torque" power take-offs that permit the installation of pumps as large as 1,500 gpm. This has made PTO driven pumps available for structural and ARFF apparatus. This type of PTO is especially common on some of the modern automatic transmissions where the flywheel of the engine drives the PTO unit. This provides enough torque to drive larger pumps than traditional PTOs could power.

Modern apparatus use a PTO arrangement to power more than just fire pumps. Electric generators, hydraulic rescue tool power systems, and large-scale air compressors may also be powered by a PTO arrangement.

Front-Mount Pumps

Front-mount pumps are similar to PTO pumps in design and operation. The primary differences between the two are the location of the pump on the apparatus and the location of the connection for the driveshaft that powers the pump.

Figure 8.19 Some jurisdictions prefer front-mount pumps.

As their name implies, front-mount pumps are located on the very front of the apparatus. To accommodate the pump, extensions are added to the chassis frame rails and the front bumper is moved forward 3 to 5 feet from its normal position. The pump is mounted on these extensions between the bumper and the grill **(Figure 8.19)**. To power a front-mount pump, a chassis should have a front-mount PTO option that provides a coupling to the front of the engine's crankshaft. The chassis also should have an opening in the radiator for the driveshaft when required.

The front-mount pump's driveshaft is driven by a gear box and a clutch connected by a universal-joint shaft to a PTO connection on the front of the crankshaft **(Figure 8.20, p. 148)**. The gear box uses a step-up gear ratio. This causes the pump's impeller to turn faster than the engine's crankshaft. The gear ratio is set to match the torque curve of the engine to the rotation speed required

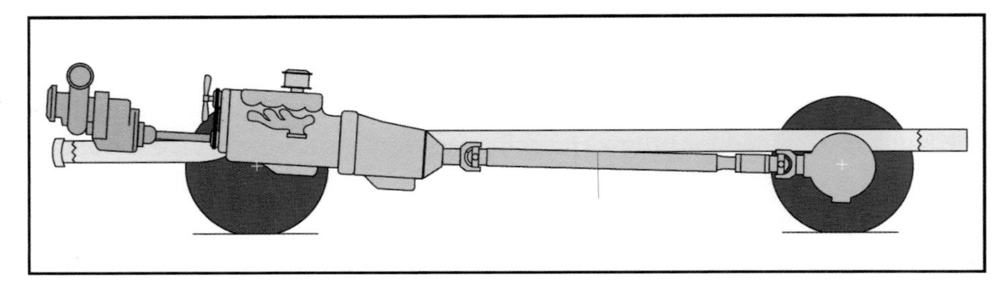

Figure 8.20 The drive arrangement for front-mount pumps.

for the impeller to deliver the pump's rated capacity. The most common gear ratios for front-mount pumps are between 1:1 and 2 1/2:1. The maximum capacity pump that can be used in front-mount applications depends on the capability of the engine driving it. Capacities of up to 1,500 gpm are possible without major modifications by the transmission, pump, or apparatus manufacturer.

The front-mount pump is engaged and controlled from the pump location itself. This puts the driver/operator in a vulnerable spot while operating the pump, namely, standing in front of the vehicle. Therefore, a lock is essential to prevent the road transmission from being engaged while the pump is operating. This applies to either a manual or an automatic transmission, and the lock should always be used when the pump is in service. Because the operating lever that engages the pump is located at the pump, a warning light in the cab should be provided to alert the driver that the pump is engaged. If the vehicle is driven with the pump turning and no water being discharged, damage to the pump results. Damage occurs either to the packing from running the pump dry or to the pump itself from overheating the water by churning it inside the pump with no water flowing.

Fire departments that frequently have to draft water from static water supply sources particularly prefer front-mount pumps. Having the pump on the front of the apparatus allows the driver/operator to drive the rig straight toward the static water source. The driver/operator can then connect the hard intake hose to the pump and place the hose into the water without having to bend it or maneuver it.

Of the major pump drive and mounting arrangements, the front-mount pump is the only one that places the pump outside of the apparatus body. This creates one potential disadvantage in that the pump and gauges are more susceptible to freezing in cold climates than pumps contained within the body of the apparatus. Using external lines that circulate radiator coolant through the pump body can overcome this problem. Modern front-mount pumps may also have some type of electric heating equipment to prevent freezing. A radiator warming system will not appreciably thaw the gauges and connecting pipes. If the apparatus will be operated in extremely cold weather it is desirable to enclose the gauges and protect them from the weather.

Another potential disadvantage of the front-mount pump is its vulnerable position in the event of a collision. Making the frame and bumper extension modifications correctly can lessen this risk. Properly installed front-mount pumps have withstood accidents that caused considerable damage to the apparatus.

Like PTO-driven pumps, front-mount pumps can be used for pump and roll operations. Again, if the unit is to be used for pump and roll, a pressure gauge is needed inside the cab so that the driver can refer to the gauge instead of the speedometer.

Midship-Transfer Drive

The midship-transfer-driven fire pump is the most common drive and mounting arrangement for modern pumping apparatus including pumpers, quints, and pumper-tankers. Midship-transfer-driven pumps are mounted laterally across the frame rails behind the transmission. A split-shaft gear case, called the transfer case, supplies power to the pump driver. It is located in the drive line between the transmission and the rear axle **(Figure 8.21)**. When the driver/operator activates the pump drive controls in the apparatus cab, a gear and collar arrangement inside the transfer case diverts the power from the driveshaft for the rear axle to the fire pump. The pump is then actually driven by either a series of gears or a drive chain.

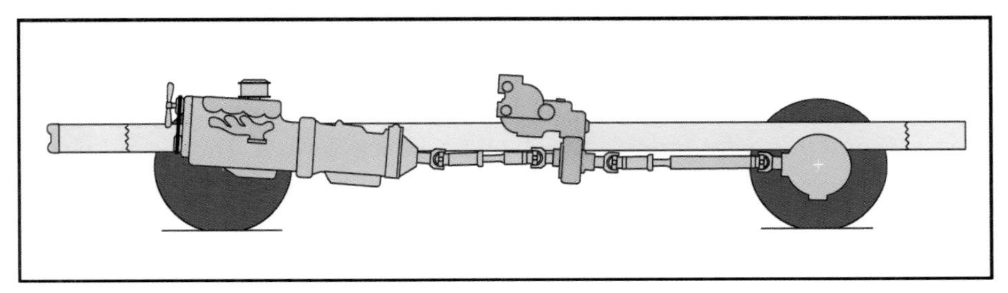

Figure 8.21 The drive arrangement for a midship transfer fire pump.

As with front-mount pumps, a gear ratio matches the engine torque curve to the speed of rotation required for the pump impeller(s) to deliver the pump's rated capacity. This ratio is calculated so that the impeller turns faster than the engine, usually one and a half to two and a half times as fast. This system's maximum capacity is limited only by engine horsepower and pump size. Most fire pumps can operate over a range of capacities that are determined by the piping arrangement and gear ratio. For example, one particular two-stage centrifugal pump popular in the fire service can be rated at anywhere from 500 to 1,500 gpm with no major changes to the pump itself. Midship transfer-driven pumps used in today's fire service commonly range in capacity from 500 to 3,500 gpm.

The midship transfer case controls are inside the cab of the apparatus. On older pumpers the control was often a manually-operated mechanical linkage lever at the panel. On modern apparatus, it is more likely to be electrically, hydraulically, or pneumatically-operated. To use the pump, the driver/operator must engage the pump and put the road transmission in the proper gear before leaving the cab. Each manufacturer will specify exactly which gear should be used when the pump is operated. This is generally, although not always, the highest gear available. If the road transmission is not placed in the correct gear, the pump will not turn at the needed rpm to operate the pump at its fullest capacity.

When the pump is engaged the transmission is turning, so most apparatus will register the engine speed in road miles per hour on the speedometer even though the apparatus is not actually moving. One way to ensure that the transmission is in the correct gear is by observing the speedometer reading after the pump is engaged. With the engine idling and the pump engaged, most speedometers read between 10 and 15 mph. Some newer apparatus may be designed so that the speedometer does not go above 0 mph when the pump is engaged. In that case it will not be possible to verify the correct gear by observing the speedometer.

When transferring power from the rear axle to the pump driver, the driver/operator must first disengage the clutch and then place the road transmission in neutral to avoid damaging the gears.

The transfer of power from the rear-axle to the pump driver involves the proper meshing of gears. To prevent damage to these gears, the driver/operator must first disengage the clutch and then place the road transmission in neutral while making the power transfer. The engine/transmission should be allowed to slow to idle speed after shifting to neutral before operating the pump transfer control (either in or out of pump gear). Some power transfer arrangements have a manual override in case of difficulty with the power unit.

One disadvantage of midship transfer-driven pumps is that they cannot be used for pump and roll operations. This is because engaging the pump transfers the power from the rear-axle coupling shaft to the pump drive gears; thus, transmitting power to the rear axle while the pump is engaged is impossible. An apparatus whose primary pump is midship transfer-driven may be equipped with a second pump for pump and roll operations if desired. The second pump may be either auxiliary-engine driven or PTO driven. This is particularly common on structural fire apparatus in jurisdictions prone to wildland-urban interface fire situations.

Figure 8.22 The automatic transmission locks in place when the pump is operating.

To prevent an automatic transmission gear selector from moving during a pumping operation or a manual transmission from slipping out of gear, a lock on the transmission or shift lever holds it in the proper gear for pumping **(Figure 8.22)**. It is also possible to operate the pump shift control and not have the gears complete their travel. If this happens, the vehicle may begin to move as the driver/operator increases engine rpm to build pressure. To protect against movement, most modern apparatus have an indicator light on the dash. Once the shift is made, the pumping operation should not begin until the green indicator light goes on. Later model automatic transmissions for fire apparatus are equipped with a gear lockup to prevent improper shifting of an automatic transmission while the pump is engaged. Hydraulic circuitry ensures that the transmission is shifted to the correct gear for direct drive of the fire pump.

Rear-Mount Pumps

In recent years, fire departments have increasingly specified pumpers with rear-mount pumps **(Figure 8.23)**. Having the pump on the rear of the apparatus offers a number of advantages. First, it helps to distribute weight more evenly on the apparatus chassis. As well, it typically allows the apparatus to have more compartment space for tools and equipment than a similar sized vehicle with a midship pump. A disadvantage of the rear-mount pump is that the driver/operator may be more directly exposed to oncoming traffic than in other pump positions. Placing the pump controls on one of the rear sides of the vehicle can offset this somewhat.

Figure 8.23 Rear-mount pumps have gained acceptance in recent years. *Courtesy of Ron Jeffers.*

Rear-mount pumps may be powered either by a split-shaft transmission or by a power take-off, depending on the manufacturer or fire department's preference **(Figure 8.24)**. In either case, a driveshaft of appropriate length and size is connected between the transmission and the pump. Other than the location being the rear of the apparatus, the operation of these pumps is the same as that of PTO and split-shaft (midship transfer) drive pumps.

FIRE PUMP COMPONENTS

To be useful for fire service applications, a fire pump must be equipped with a wide variety of ancillary devices. Without these devices to support fire fighting, operating the pump safely, consistently, and effectively would be virtually impossible. This section looks at some of the more important ancillary devices with which fire pumps may be equipped.

Automatic Pressure Control Devices

Because fire fighting is a very dynamic activity, changes occur constantly on the fire ground. Very commonly a single pumping apparatus will simultaneously supply a number of hose lines and/or master stream devices. Each of these may require different discharge pressures to operate efficiently and safely. This would be no problem if all of the lines were to be opened and closed at the same time; however, that is rarely, if ever, the case in fire fighting operations. The volume of water moving through the pump changes frequently as firefighters shut down or gate down nozzles or change the setting on a variable gallonage nozzle. When a pump is supplying multiple attack lines, any sudden flow change in one line

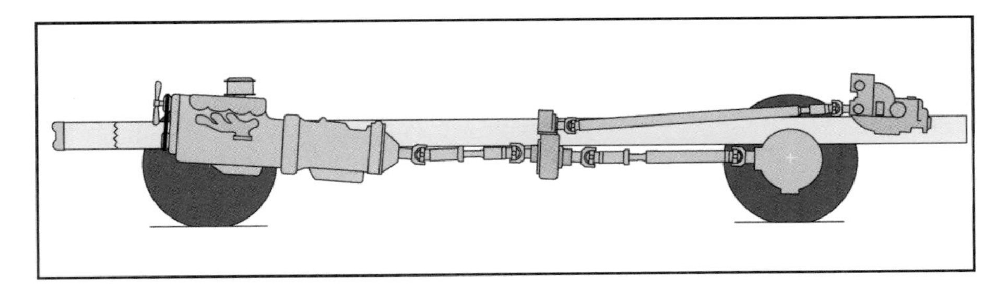

Figure 8.24 A split-shaft drive arrangement for a rear-mount pump.

can cause a pressure surge on the others. While modern nozzles can tolerate changing pressures and still maintain an effective fire stream, the firefighter on the nozzle often cannot handle any sudden changes in pressure and may lose control of the nozzle if the change is dramatic. Not even an alert operator can compensate for these sudden changes in time to protect the firefighters on the other lines.

Thus, some type of automatic pressure monitoring and regulation is essential to ensure the safety of personnel operating the hoselines and to prevent water hammer from damaging the apparatus.

NFPA® 1901, *Standard for Automotive Fire Apparatus*, requires any fire apparatus pumping system to have some type of pressure control device. The standard requires this system to operate within 3 to 10 seconds after the discharge pressure rises at least 30 psi above the set level. A yellow indicator light on the pump operator's panel must illuminate when the pressure control system is in control of pump pressure. Devices designed to discharge excess water to the atmosphere must do so without exposing the driver/operator or other personnel to a pressurized water stream. Modern fire apparatus use two basic types of pressure control devices: relief valves and pressure governors.

Pressure Relief Valves

Pressure relief valves are intended to reduce the possibility of water hammer damaging the pump and discharge hoselines when valves/nozzles are closed too quickly. Fire department pumps may be equipped with one or both of two basic types of pressure relief valves: those that relieve excess pressure on the discharge side of the pump and those that relieve excess pressure on the intake side of the pump.

NFPA® 1901 requires all fire pumps that are not equipped with a pressure governor to be equipped with a discharge pressure relief valve as an integral part of the pump. Discharge pressure relief valves sense pressure changes on the discharge side of the pump and relieve excess pressure when it occurs.

A number of designs of discharge pressure relief valves are on the market. The most common design uses an adjustable spring-controlled pilot valve. When the pump discharge pressure is lower than the pilot valve setting, water flows from the pump discharge to the pressure chamber of the pilot valve **(Figure 8.25)**. A diaphragm separates this pressure chamber from the pilot valve. Tension against the diaphragm is regulated by adjusting the handle of the pilot valve against the spring. As long as the hydraulic force in the pressure chamber is less than the force of the spring, the pilot valve stays closed. The water is then transmitted back to the main valve chamber, and the main valve stays closed.

When the pump discharge pressure exceeds the pilot valve setting, the spring in the pilot valve compresses **(Figure 8.26)**. This allows the needle valve to open, causing water to dump back into the intake side of the pump. This in turn reduces the pressure in the tube and behind the main valve. The hydraulic force on the smaller end of the main valve is now greater than that behind the main valve. When the discharge pressure drops below the set pressure in the pilot valve, the pilot valve reseats, pressure increases behind the larger end of the main valve, and it closes.

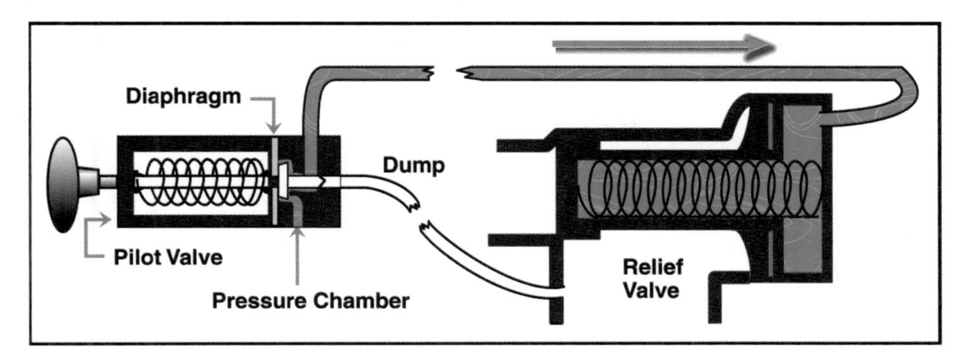

Figure 8.25 Schematic of water flow when the pump discharge pressure is lower than the pilot valve setting of the pressure control system. *Courtesy of Waterous Company.*

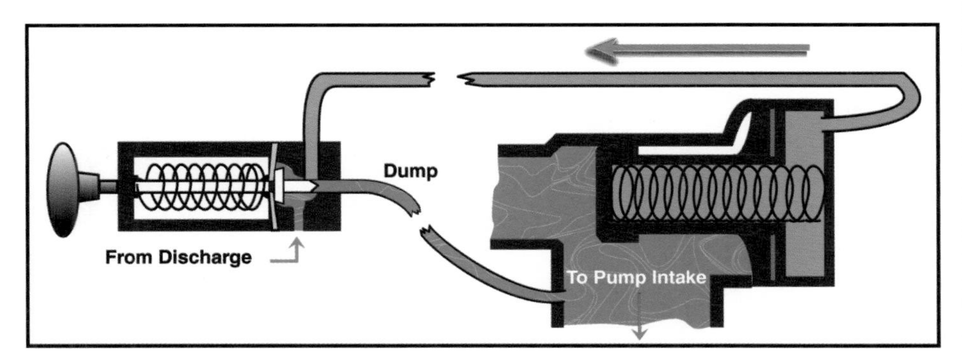

Figure 8.26 Schematic of water flow when the pump discharge pressure is higher than the pilot valve setting of the pressure control system. *Courtesy of Waterous Company.*

A second type of discharge pressure relief valve is also equipped with a pilot valve but operates in a slightly different manner **(Figure 8.27)**. When the pressure goes above the set pressure, the pilot valve moves, compressing its spring until an opening in the pilot valve housing is uncovered. Water flows through this opening and through the bleed line and into the pump intake relieving the excessive pressure. This flow reduces the pressure on the pilot valve side of the churn valve, allowing the higher pressure on the discharge side to force open the churn valve.

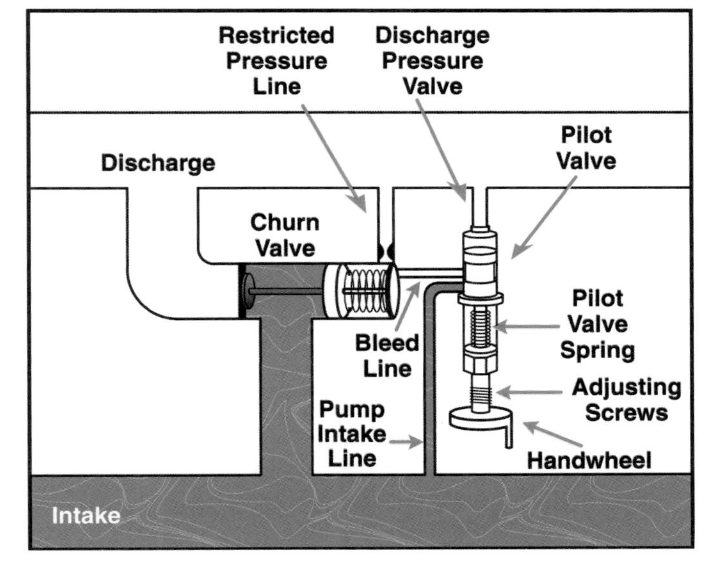

Figure 8.27 An alternative relief valve.

The driver/operator should remember that while discharge pressure relief valves react quickly to overpressure conditions, they are somewhat slower to reset to "all-closed" positions. Therefore, it takes a short time for the pump to return to normal operation.

The other basic type of pressure relief valve is the intake pressure relief valve. There are two common designs for intake pressure relief valves. One type is supplied by the pump manufacturer as an integral part of the pump intake manifold. The second type is an add-on device that is screwed onto the pump intake connection

(**Figure 8.28**). In either case, the driver/operator sets these devices to allow a maximum pressure into the fire pump. If the incoming pressure exceeds the set level, the valve activates and dumps the excess pressure/water until the incoming pressure drops to the preset level. It is generally recommended that intake relief valves be set to open when the intake pressure rises more than 10 psi above the desired operating pressure.

Most screw-on intake pressure relief valves are also equipped with a manual shut-off valve that allows the driver/operator to stop the water supply to the pump if desired. Bleeder valves on the intake pressure relief valve allow air to bleed off as the incoming supply hose is charged. This is particularly important when using these devices in conjunction with large-diameter supply hose. A large amount of air pushes through these hoses until a solid column of water reaches the valve.

Figure 8.28 External intake pressure relief valves are screwed onto the main pump intake.

Pressure Governor

The second manner in which pressure can be regulated on centrifugal fire pumps is through a mechanical or electronic governor that is pressure activated to adjust the engine throttle. The basic operational principle of the pressure governor is that it regulates the engine's power output to match the pump's discharge requirements. When the pressure in the pump's discharge piping exceeds the pressure necessary to maintain safe fire streams, the governor slows the engine, which in turn slows the pump impeller, thus reducing the pressure being developed by the pump.

When excessive pressure builds, a tube from the discharge side of the pump transmits the pressure rise to the governor, which then cuts back the throttle to slow the engine. The device's design varies with each manufacturer; it may be

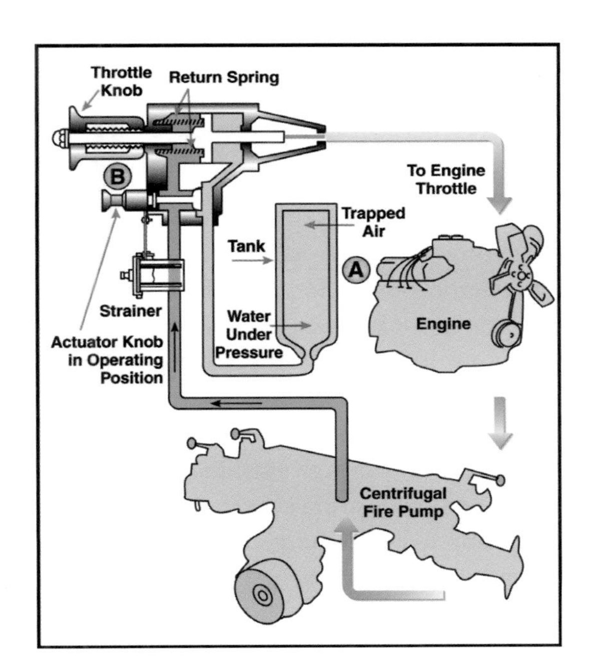

Figure 8.29 When activated, the pressure governor uses the trapped air and water in Area A as a reference pressure for the piston that controls the engine throttle. A change in pump pressure affecting Area B moves the piston, thus adjusting the engine speed. *Courtesy of Hale Fire Pump Company.*

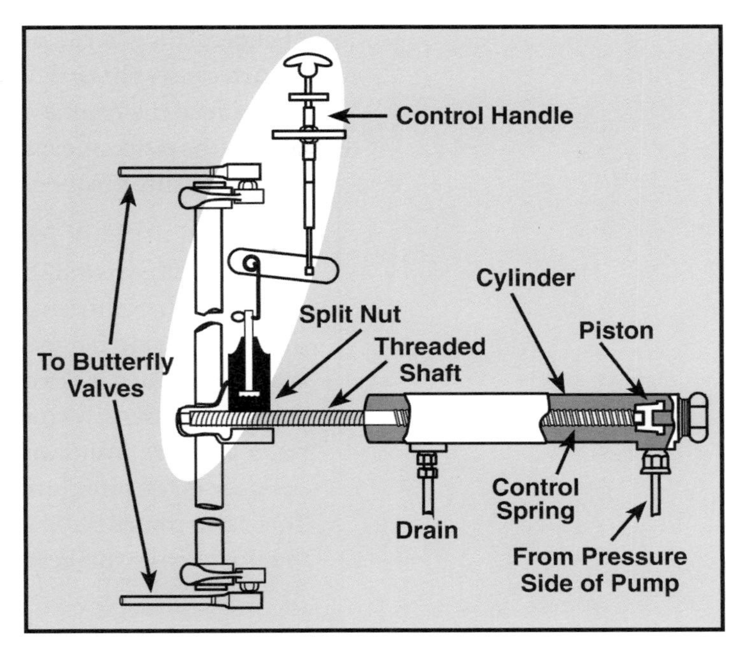

Figure 8.30 This type of mechanical governor's control handle is pulled out to disengage the split nut from the serrated shaft.

attached to either a regular or an auxiliary throttle. A pressure governor can be used in connection with a throttle control, engine throttle, and/or pump discharge **(Figure 8.29)**.

Older apparatus are equipped with mechanically operated governors. One style of mechanical governor utilizes a control handle. Pulling out the control handle lifts a split nut from a serrated piston shaft **(Figure 8.30)**. The driver/operator advances the engine throttle until the desired pump pressure is reached. Then the driver/operator opens the water control valve from the discharge side of the pump to permit water from the pump discharge to enter the governor cylinder. The hydraulic pressure in the cylinder causes the piston to compress the control spring and move the serrated shaft. This movement activates the carburetor linkage and butterfly valve to control engine speed.

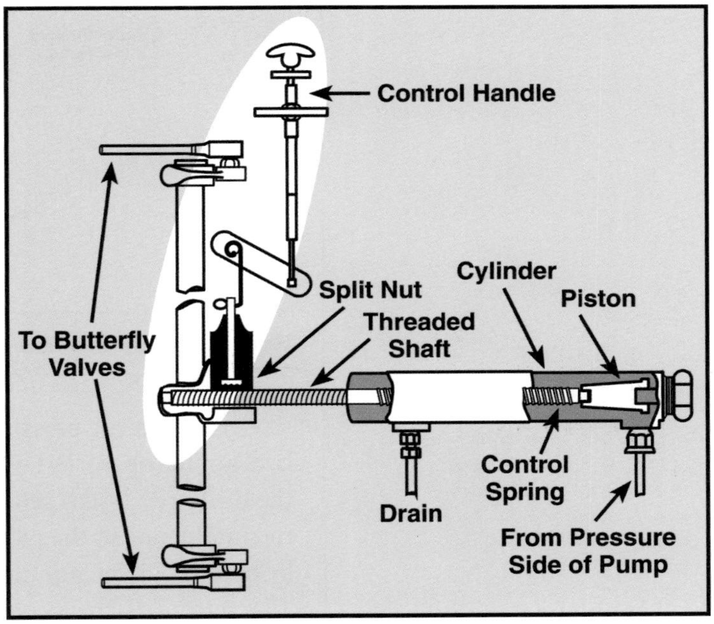

Figure 8.31 After the desired pressure is reached, the mechanical governor's control handle is pushed back in, setting the governor for that pressure.

When the control handle is pushed in to lower the split nut against the serrated piston shaft, the water pressure on the discharge side of the pump governs the engine speed **(Figure 8.31)**. If a hoseline is shut off, the increase in pressure travels through the pipe from the pump discharge side into the governor cylinder. This will cause the serrated piston to move from its set position, which in turn causes the carburetor (fuel) control linkage to adjust the butterfly valve and reduce the engine speed. When a discharge line is reopened, the reverse occurs: the piston moves back to increase engine speed and returns to the preset pump pressure position.

A second type of mechanical governor is known as the piston assembly governor **(Figure 8.32)**. The piston assembly governor is mounted onto the carburetor (gasoline engines) or throttle link (diesel engines) and reduces or increases the engine speed under the control of a rod connected to a piston in a water chamber. A flexible diaphragm in the water chamber isolates the piston and spring assembly from the water while exposing the face of the piston to the water pressure. Water enters the water chamber through a strainer on the pressure side of the pump and leaves through a two-position governor return valve. This valve routes the water from the pump to be discharged on the ground or to return to the intake side of the pump.

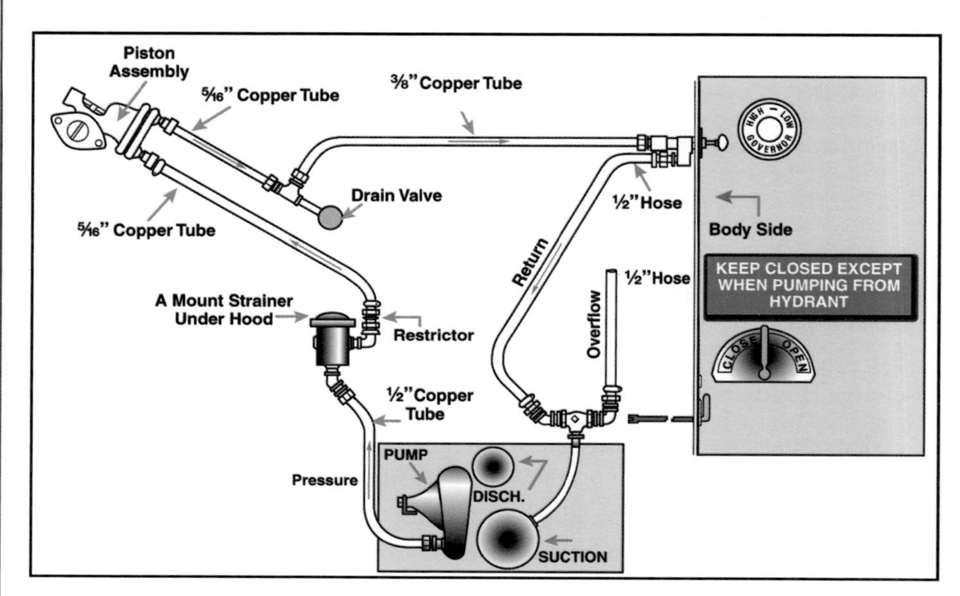

Figure 8.32 The piston assembly governor is mounted on the carburetor.

A mechanical pressure governor has two settings, draft (or closed) and hydrant (or open). When drafting or pumping from the apparatus water tank, the draft (or closed) setting must be used. This setting returns excess water to the intake side of the pump. When pumping from a hydrant or being supplied by hoselines from another pumper, the hydrant (or open) setting must be used. This setting discharges excess water onto the ground through a drain line. The drain line can be run through a check valve to the tank to prevent water from continuously running on the ground.

Most modern fire apparatus have electronic governors **(Figure 8.33)**. Electronic governors utilize a pressure sensing element connected to the pump discharge manifold. This element controls a computer that compares the pump discharge pressure to a set reference point. When the element senses excess pressure, the computer decreases the throttle by reducing the amount of fuel supplied

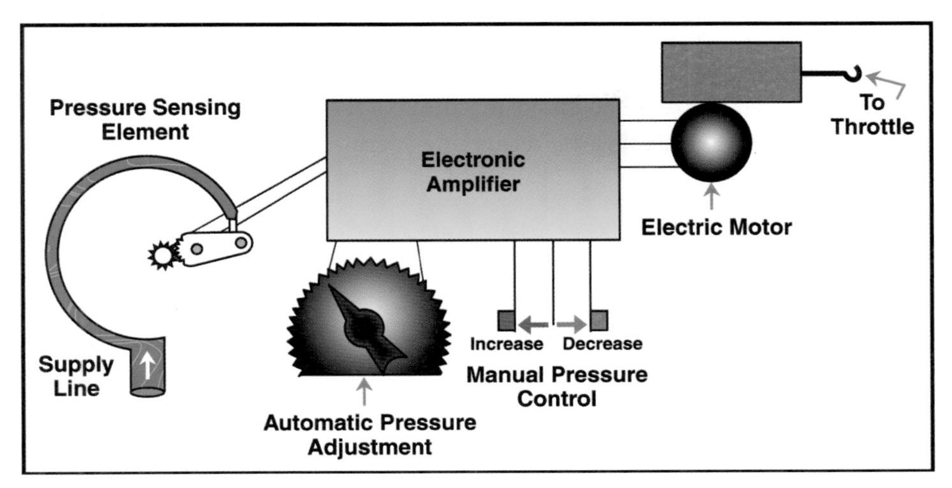

Figure 8.33 An electronic pressure governor has a pressure-sensing element that drives an electric motor to change the throttle setting. In this way, the desired operating pressure is achieved.

to the engine. The lower throttle returns the pump to the desired operating pressure. The electronic governor has a control knob that the operator can set to maintain any pressure above 50 psi. The governor returns the engine to idle speed anytime the pump *discharge* pressure drops below 50 psi. It also protects against pump cavitation (running the pump dry) by returning the engine to idle when the pump *intake* pressure drops below 30 psi. Electronic governors are so accurate and quick to respond that they virtually eliminate the need to equip the pump with a relief valve; however, many jurisdictions equip their apparatus with both for added safety.

While both mechanical and electronic governors operate reliably, electronic governors have one major advantage over their mechanical counterparts. Mechanical governors reduce engine rpm in order to lower the discharge pressure when a pump discharge is closed. This in turn briefly lowers all discharge line working pressures. Electronic governors, on the other hand, adjust the engine rpm but maintain a steady discharge pressure on the lines that are still flowing.

Pumping Priming Devices

To draft water from a static water supply source, a fire pump must be able to create a lower pressure within the pump and intake hose than exists in the atmosphere. This is referred to as a partial vacuum. When a partial vacuum is created in the pump and intake hose, the greater atmospheric pressure forces water from the static source up into the intake hose and fire pump. Because today's centrifugal fire pumps cannot create a partial vacuum by themselves, some other device must do so. These devices are called priming devices, or simply *primers*. Primers fall into three categories: vacuum, exhaust, and positive displacement pumps.

Vacuum Primers

The vacuum primer is the oldest and simplest type of primer used by the fire service. Vacuum primers were common on earlier, gasoline-powered fire apparatus. They are not commonly used on modern apparatus; however, some jurisdictions may still have older apparatus that are equipped with them.

Vacuum primers make use of the vacuum already present in the intake manifold of any gasoline engine. These devices utilize a small hose or pipeline that is connected between the intake manifold of the engine and the intake side of the fire pump. A control valve in that line operates the primer when desired.

In addition to the control valve, the line between the engine manifold and pump intake has a check valve that protects against two potentially dangerous conditions. First, it prevents explosive gases from the intake manifold from being drawn into the fire pump and damaging it. Second, it prevents water from being drawn through the pump and into the intake manifold and damaging the engine **(Figure 8.34)**. When the check valve is under pressure, as occurs during a backfire, it closes, isolating the pump from the engine. When the check valve is under a vacuum, it allows the pump to prime normally. As the air comes through the line from the intake side of the pump, the pump is primed and fills with water. When the water enters the wet chamber of the primer, a float closes a ball valve. This opens a vent valve and allows the vacuum line to draw air from outside the pump. If the float does not operate, the water continues up into the dry chamber, where a cork float rises and closes the safety valve. When the priming valve is closed, the drain opens and allows the water from the primer to drain, and the primer is ready to operate again.

Because engine vacuum is at the maximum near idle speed, the primer works best at low engine rpm. The driver/operator has to increase engine speed only enough to keep it from stalling when the vacuum line is opened to prime the pump.

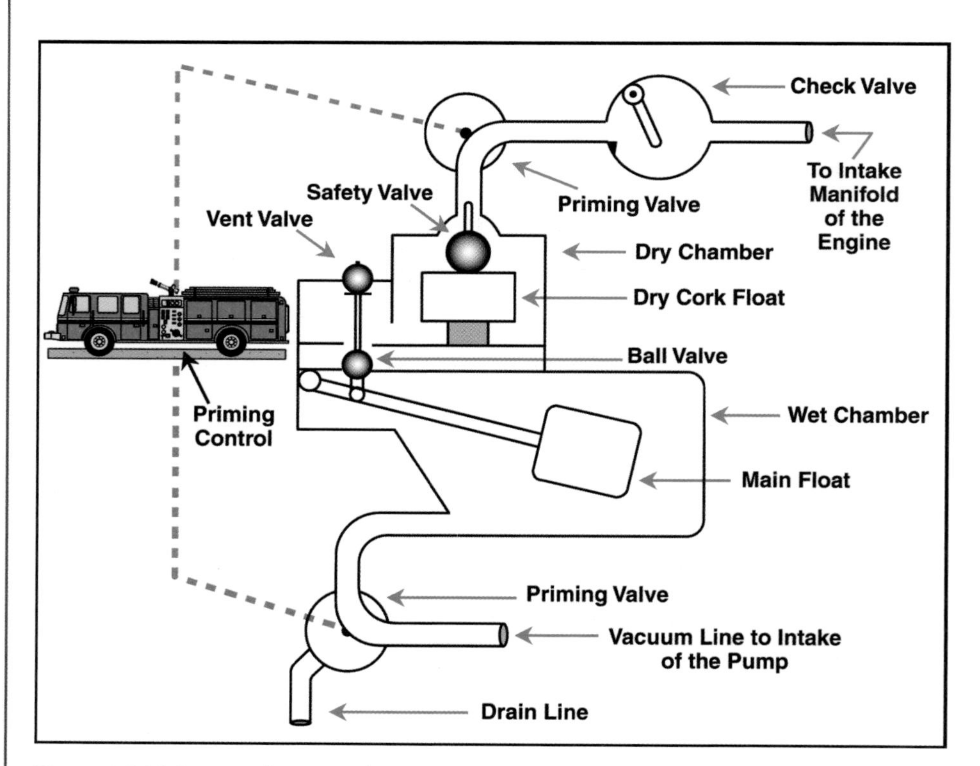

Figure 8.34 Schematic diagram of a vacuum primer system.

Exhaust Primers

Like vacuum primers, exhaust primers typically are found only on older apparatus and portable pumps. The exhaust primer operates on the same principle as a foam eductor. An exhaust deflector prevents engine exhaust gases from escaping to the atmosphere. It diverts the gases to a chamber where their changing velocity through a venturi creates a vacuum. This chamber is connected through a line and a priming valve to the pump intake **(Figure 8.35)**. There the air is evacuated into the venturi chamber and then discharged into the atmosphere along with the exhaust gases.

Unlike vacuum primers, exhaust primers require high engine rpm to generate the exhaust gases' maximum velocity. At best, they are not very efficient. As well, exhaust primers require a great deal of maintenance because the passing exhaust gases tend to leave carbon deposits that further decrease the device's efficiency. Also, exhaust primers require keeping any air leaks in the pump to an absolute minimum and keeping the suction hose and gaskets in good condition.

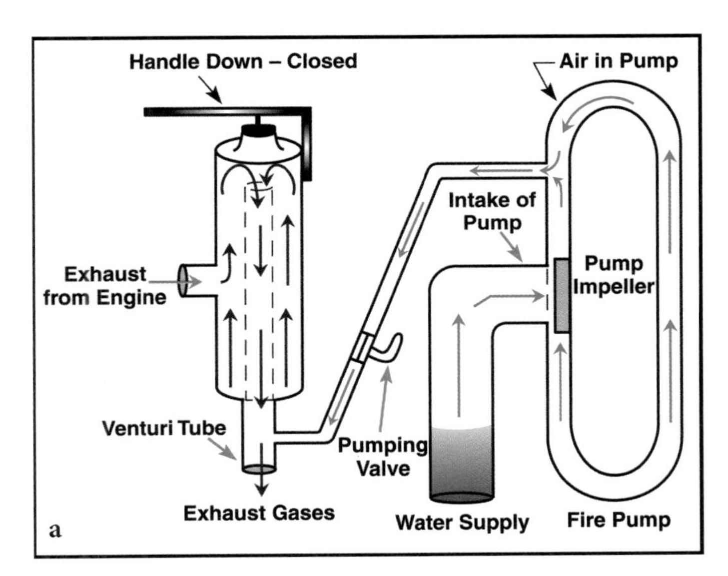

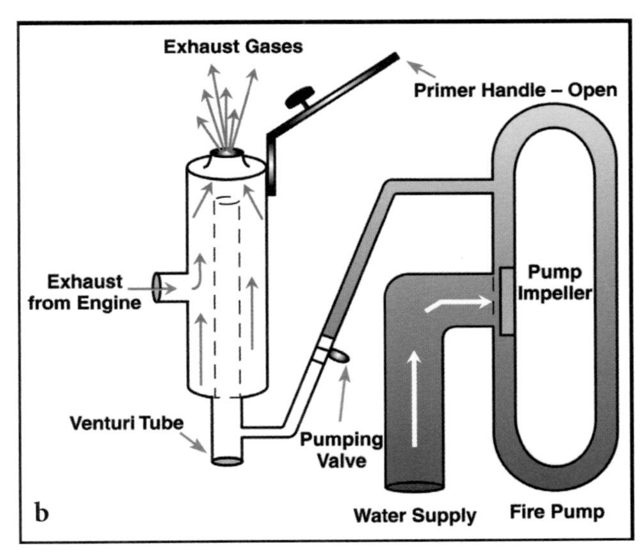

Figure 8.35 An exhaust primer with the primer handle in the *closed* position **(a)** and in the *open* position **(b)**. *Courtesy of Bennie Spaulding.*

Positive Displacement Pump Primers

Most modern fire apparatus use positive displacement pump primers. Small-capacity sizes of both the rotary vane and rotary gear type positive displacement pumps are most common. The rotary vane primer requires a relatively high rpm compared to a rotary gear primer.

Both types of positive displacement pump primers can be driven either through the pump transfer gear case or by an electric motor. Nearly all modern apparatus have electric motor driven primers. Electric motor drivers allow maximum flexibility in mounting and application of common designs of primers to different chassis and pumps. Another advantage of electric primers is that they can operate effectively regardless of engine speed.

The primer pump inlet is connected to a primer control valve that is in turn connected to the fire pump. If there are high spots in the way the pump is constructed or mounted, the line from the priming pump may be connected to the pump in several different locations to aid in evacuating air from those difficult-to-reach places. On electrically driven primer pumps, the priming valve usually incorporates a switch so that only one operation is necessary to prime the pump.

Most positive displacement primers use an oil supply or some other fluid to serve two purposes **(Figure 8.36)**. As the pump wears, the clearances between the gears and the pump casing increase, resulting in a loss of pump efficiency. A thin film of oil/fluid is drawn into the pump and seals the gaps between the gears and the case. The oil/fluid fills any irregularities in the housing caused by pumping contaminated water, and it improves the efficiency of the primer. The oil/fluid also acts as a preservative and minimizes deterioration of the metal parts by inhibiting corrosion when the pump is not in use. To derive the maximum benefit from the oil/fluid, operating the primer periodically is necessary so that a coating of oil/fluid will form on all the metal parts.

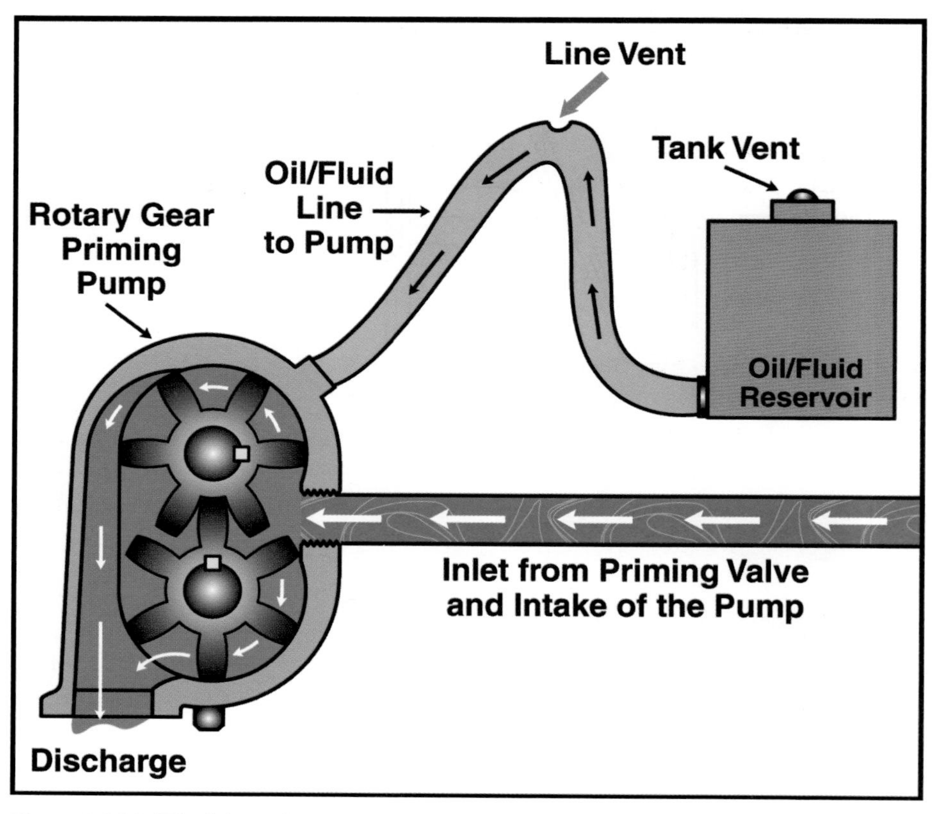

Figure 8.36 Oil/fluid from the reservoir is siphoned into the positive displacement primer.

Because the tank is frequently mounted higher than the priming pump, a siphon action can drain the oil/fluid tank after the primer has operated. A vent in the oil/fluid line from the reservoir to the priming pump breaks the siphon.

The vent must be large enough to serve its purpose but small enough to allow the priming pump to draft oil from the tank when the primer is in use. This vent should be checked frequently to ensure that it is free of dirt.

If the apparatus has a transfer case driven primer, the desired engine rpm for primer operation depends on the construction of the primer, the gear ratio of the transfer case, and other features unique to the particular installation. The operating manual for the fire pump or apparatus should specify the desirable engine speed (rpm) for priming. On most apparatus this will be in the neighborhood of 1,000 to 1,200 rpm. The driver/operator should activate these primers with the engine at idle speed and then increase the throttle to the indicated rpms. This minimizes the wear on the mechanical clutch pack.

Auxiliary Cooling Systems

NFPA® 1901 requires all pumping apparatus to be equipped with a supplementary heat exchanger cooling system, more commonly called an auxiliary cooling system, for the engine that drives the fire pump. This system must be capable of maintaining the temperature of the coolant in the pump drive engine at or below the engine manufacturer's rated maximum temperature under all pumping conditions. These systems are especially important on gasoline-powered engines, which tend to run very hot when operated for extended periods. However, they may also be needed to cool diesel engines during extended use in warm weather.

All pumping apparatus must be equipped with an auxiliary cooling system for the engine that drives the fire pump.

Two types of auxiliary coolers commonly found on fire apparatus are the marine type and the immersion type. The marine type is inserted into one of the engine's cooling system hoses so that the circulating engine coolant must travel through it **(Figure 8.37)**. The cooler itself contains a number of small tubes similar to the flues in a steam boiler. A water jacket that surrounds the tubes is connected to the fire pump discharge. When the fire pump is operating, water from the pump can circulate through this water jacket. As the radiator coolant passes through the tubes in the cooler, the colder water from the fire pump contacts the outside of the metal tubes and conducts heat away from them, thus reducing the temperature of the engine coolant. A valve on the pump panel allows the driver/operator to control the amount of water supplied to the auxiliary cooler.

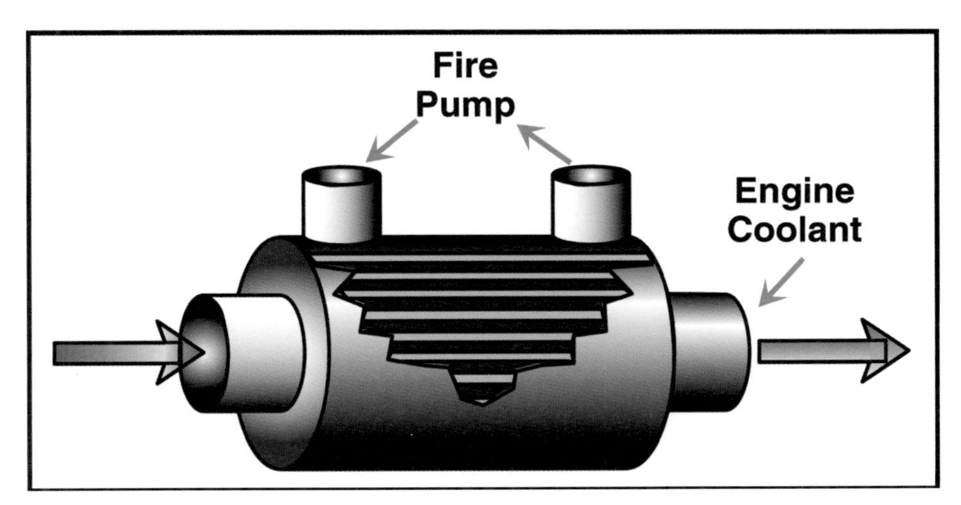

Figure 8.37 A marine-type auxiliary cooler.

The immersion-type auxiliary cooler is mounted in a similar manner to the marine type, with the radiator coolant passing through the body of the cooler **(Figure 8.38).** In the immersion-type cooler, however, the water from the fire pump passes through a coil or other type of tubing mounted inside the cooler so that it is immersed in the coolant. As the cooler water from the fire pump passes through the tubing, the tubing absorbs some of the heat from the coolant and dissipates it into the water from the fire pump. As with marine-type coolers, a valve on the pump panel allows the driver/operator to regulate the degree of cooling desired.

Both types of auxiliary coolers are constructed so that the coolant in the radiator does not contact the cooling water from the fire pump. Thus, the auxiliary cooler can be used without contaminating the engine coolant.

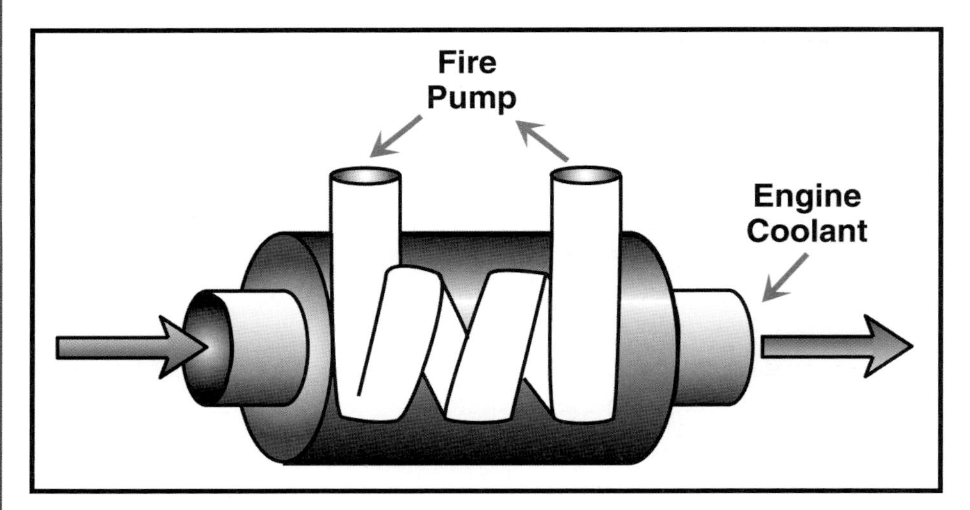

Figure 8.38 An immersion-type auxiliary cooler.

Pump Wear Rings

Centrifugal fire pumps have no positive mechanical isolation between the discharge side of the pump and the intake in the impeller. Generally, the velocity of the water moving through the intake into the impeller prevents most of the water in the discharge side of the pump from escaping back into the intake. Because the pressure in the volute is much higher than that in the intake side of the pump at the eye of the impeller, a very close tolerance must be maintained between the pump casing and the hub of the impeller. This opening is usually limited to 1/100 of an inch or less. Any increase in this opening lessens the pump's effectiveness.

Whether it is being pumped from a municipal water supply system or a static water supply source, all water contains impurities, sediment, and dirt. As these materials pass through the pump and contact the impeller, they cause wear. This process accelerates greatly when pumping water from a static supply source that has a high sand or dirt content. The sand and dirt particles pass between the impeller and the pump casing and act like a sand blaster, pitting and wearing down the metal surfaces. This increases the gap between the pump casing and the impeller hub. As the gap increases, it allows greater amounts of water to slip back into the intake instead of being available at the discharge. In severe situations this may prevent the pump from reaching its rated capacity. One indication that this wear may be becoming a problem can be noted during pump tests. If running the pump at its rated capacity requires an engine rpm substantially higher than listed on the UL pump plate, wear in the pump may be a problem.

Figure 8.39 Replaceable wear rings are vital in maintaining the proper spacing between the hub of the impeller and the casing.

When wear becomes a problem, it is possible to restore the capacity of the pump without replacing the whole pump. Most manufacturers supply replaceable wear rings, also called clearance rings, that can be inserted into the pump casing to maintain the desired spacing between the hub of the impeller and the casing **(Figure 8.39)**. If the impeller hub also is worn down, it is possible to install smaller wear rings to compensate for the smaller size and maintain the proper clearance.

The centrifugal pump differs from the positive displacement pump in that little or no harm results from briefly shutting off all discharges. However, when the discharges are shut off and the water in the pump is allowed to churn, the energy being supplied to the impeller dissipates to the water in the form of heat. If this is allowed to continue for extended periods, the water in the pump can become boiling hot and the metal pump parts may expand. The faster the pump turns, the faster this heating will occur. If the wear rings and the impeller expand too much, they may contact each other, and the friction between their surfaces may cause even more heat. In extreme cases, the wear rings may seize, causing serious pump damage. The best protection against this is to ensure that some water is moving through the pump at all times by using either a circulating line or by cracking open a discharge.

The best protection against overheating a centrifugal pump is to ensure that some water is moving through the pump at all times.

The driver/operator should not put the pump in a position where it might overheat. He or she can check the pump temperature by placing a hand on the direct pump intake pipe (main barrel) of the pump. If it is warm to the touch, he or she should open a discharge or circulator valve. If no water is expected to be discharged for an extended period, the pump should be disengaged until needed.

Pump Packing

A centrifugal pump's impellers are fastened to a shaft that connects to a gear box. The gear box transfers the energy necessary to spin the impellers at a very high speed. The seal where the shaft passes through the pump casing must remain semi-tight to prevent air leaks that could interfere with drafting. In most fire pumps packing rings make this seal **(Figure 8.40, p. 164)**. The most common type of packing material is made of rope fibers impregnated with graphite or lead. The material is pushed into a stuffing box by a packing gland driven by a packing adjustment mechanism.

As packing rings wear with the operation of the shaft, the packing gland can be tightened to control leaks. Where the packing rings contact the shaft, the friction develops heat. To overcome this, a lantern ring (spacer) supplied with the packing provides cooling and lubrication. The lantern ring allows a small amount of water to leak around the packing and prevent excessive heat buildup. If the packing is too tight, water cannot flow between the packing and the shaft,

Operating a centrifugal pump dry for any length of time will damage the shaft.

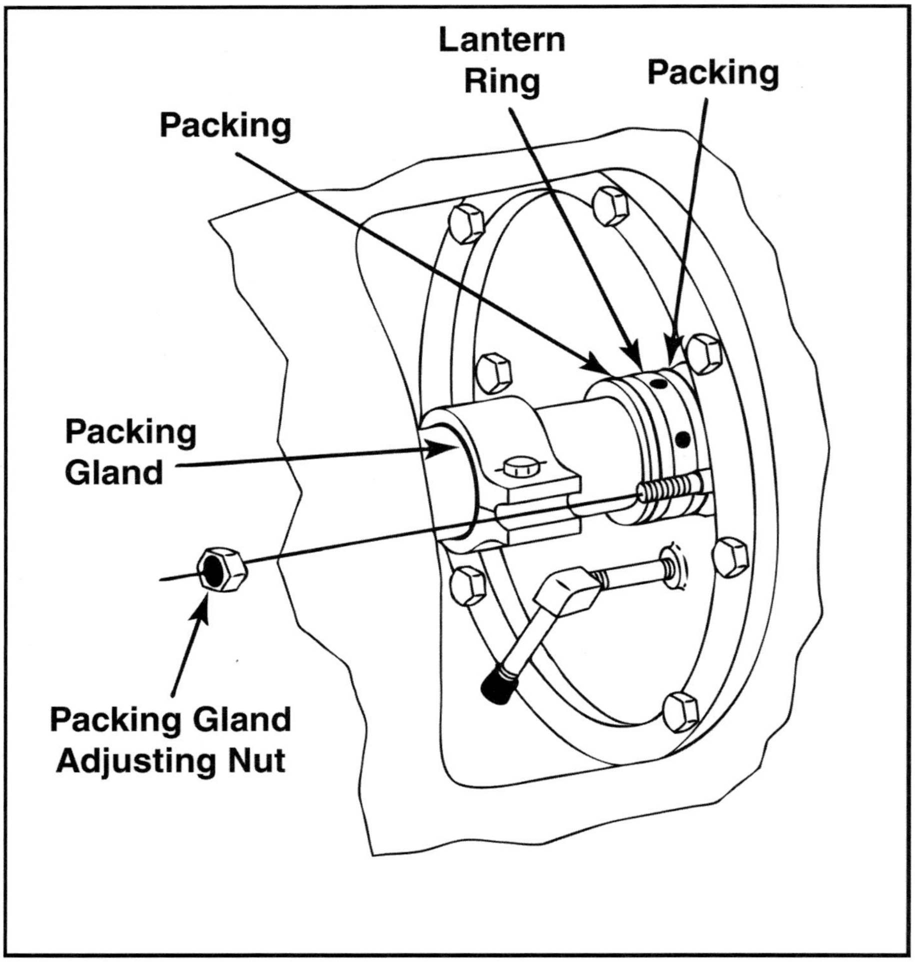

Figure 8.40 This diagram shows the position of the pump packing.

and excessive heat buildup results. A scored shaft prevents even new packing material from making a good seal. If the packing is too loose, as indicated by an excessive amount of water leaking from the pump during operation, air leaks will hinder the pump's ability to draft. The packing must be adjusted carefully according to the manufacturer's instructions. In general, however, when the pump is operating under pressure, water should drip from the packing glands but not run in a steady stream.

The packing receives the needed water for lubrication only if the pump is full and operating under pressure. ***Operating the pump dry for any length of time damages the shaft.*** This weakens the shaft and may cause it to fail in a future use.

Some fire departments keep the pump drained between fire calls. This is particularly common in cold climates. If the pump is not used for extended periods, however, the packing may dry out, and excessive leakage will result. After a period of dryness, the packing should not be adjusted until the pump is operating under pressure and the packing has had a chance to seal properly.

If the pump has mechanical seals instead of packing, they will not drip and will not require adjustment. Again, it is important to run the pump periodically to lubricate the seals. Mechanical seals are especially problematic if they freeze, which may cause damage that necessitates immediate repair. This repair process is not simple, and may require disassembling the entire pump and drive assembly.

Pump Piping and Valves

Of course, having a fire pump mounted on an apparatus in and of itself would be useless without a means to take in, distribute, and control the flow of water. Water flow through the pump and apparatus is controlled by a series of components, including:

- Intake piping
- Discharge piping
- Pump drains
- Valves

NFPA® 1901 requires all components of the piping system to be corrosion-resistant. Most piping systems are constructed of cast iron, brass, stainless steel, or galvanized steel. Some may include rubber hoses in certain locations. Apparatus commonly have a combination of these materials in their overall piping system. The piping system (as well as the pump itself) must be able to withstand a hydrostatic test of 500 psi before being placed into service. To minimize pressure loss within the apparatus, all piping and hoses should be designed to run as straight as possible, with a minimum of bends or turns.

Intake Piping

Water may enter the fire pump via two primary routes. The first is through piping that connects the pump and the apparatus water tank. The second is through piping that connects the pump to an external water supply.

Most fires are fought initially, if not completely, with the water carried in the apparatus water tank. NFPA® 1901 states that piping should be sized so that pumpers with a capacity of 500 gpm or less should be capable of flowing 250 gpm from their booster tank. Pumpers with capacities greater than 500 gpm should be able to flow at least 500 gpm from the tank.

Fire departments that want the capability of initiating a large-flow attack, often referred to as a blitz attack, from their apparatus water tank must specify larger than minimum piping between the water tank and the pump. Many apparatus today are equipped with 4-inch or larger tank-to-pump lines or even multiple tank-to-pump lines. The tank-to-pump lines must be equipped with check valves. The check valves prevent damage to the tank if the tank-to-pump valve inadvertently opens when water is being supplied to the pump under pressure, such as during a relay. If this line has a check valve, it is impossible to fill the tank through the pump by opening the tank-to-pump valve. As with all valves on the apparatus pumping system, the tank-to-pump valve must be maintained in good condition. If it leaks, priming the pump when the tank is

Figure 8.41 The main intake on a front-mount pump extends from the lower portion of the pump.

Figure 8.42 The main intakes on midship pumps usually are found on either side of the pump. *Courtesy of Ron Jeffers*

Figure 8.43 Large-volume pumpers, such as this 3,000 gpm industrial pumper, will have multiple large intakes on each side of the apparatus. *Courtesy of Celanese Corp., Clear Lake, Texas, Plant.*

empty is impossible because the pump draws air from the tank and cannot establish a partial vacuum. Conversely, a leak also will allow water to be drawn out of the tank during a drafting operation. The pump prime may then be lost when the tank is emptied of water and the pump starts drawing air.

If water is supplied to the fire pump through an external supply, the source may be pressurized or static. Because any air trapped in the pump during priming can prevent successful drafting from a static water supply source, all intake pipes to a centrifugal pump normally are located below the eye of the impeller, with no part of the piping above the impeller. The single exception to this may be the tank-to-pump line, where the water moves to the pump under the natural pull of gravity.

The primary intake into the fire pump is through large-diameter piping and connections. On a front-mount pump, this pipe and connection extend from the lower portion of the pump **(Figure 8.41)**. Midship pumps usually have a large intake connection on either side of the apparatus **(Figure 8.42)**. Intake piping is round at the intake hose connection. As the piping nears the pump, it typically tapers to a square cross section. This eliminates the vortex that may occur in water that flows through circular piping. If the vortex were not eliminated, it could introduce air into the pump. Pumps with a capacity of 1,500 gpm or greater may require more than one large connection at each intake location **(Figure 8.43)**. **Table 8.1** shows the intake sizes for various size pumps as required by NFPA® 1901.

Additional large-diameter intakes may be piped from the midship pump to the front or the rear of the apparatus **(Figure 8.44)**. To avoid a variety of obstacles between the pump and the front or rear of the apparatus (chassis components, the engine, the transmission, axles, etc.), this piping commonly contains a number of bends. These bends, in conjunction with the length of the pipe, frequently may prevent the pump from operating at its rated capacity, particularly from a draft. In some cases, the intakes may limit maximum flows to several hundred gpm less than the pump's rated capacity. These front and rear intakes should be considered auxiliary intakes in the same manner as the pump-panel-mounted gated intakes. In critical high-flow pumping situations or during pump testing, only the midship large-diameter intakes should be utilized.

Additional intake lines, often referred to as auxiliary intakes, are provided for use anytime the pump is receiving water through medium-diameter supply lines from a hydrant or another pumper **(Figure 8.45)**. Most auxiliary intake

Table 8.1 Required Number of Intake Hoses and Size For Fire Pumps			
Rated Capacity (gpm)	Maximum Suction Hose (in.)	Maximum Number of Suction Lines	Maximum Lift (ft)
250	3	1	10
300	3	1	10
350	4	1	10
450	4	1	10
500	4	1	10
600	4	1	10
700	4	1	10
750	4½	1	10
1000	5	1	10
1250	6	1	10
1500	6	2	10
1750	6	2	8
2000	6	2	6
2250	8	3	6
2500	8	3	6
2750	8	4	6
3000	8	4	6

Figure 8.44 Front or rear intakes on midship pumps are really auxiliary intakes. *Courtesy of Ron Jeffers*

Figure 8.45 Most pumpers have gated auxiliary intakes for medium-diameter hose. This one has two gated intakes.

openings are gated and have female 2½-inch hose couplings. The amount of flow that can be obtained is determined by the diameter of the pipe from the hose connection to the pump inlet and the straightness of its routing. NFPA® 1901 requires the piping to be at least 2½-inches in diameter. If 2½-inch pipe is used and if it contains 90-degree bends or T-fittings, the pressure loss in the piping may limit the flow through these intakes to 250 gpm or less. If 3-inch pipe is used and care is given to the fittings and design, flowing as much as 450 gpm through one of these intake openings may be possible.

Discharge Piping

Discharge piping supplies the hose lines and master stream appliances on the apparatus. It is constructed of the same materials as intake piping. Discharges usually have a locking ball valve, and they always should be locked when they are open to prevent movement. This is especially true when a line has been gated down to reduce its discharge pressure. NFPA® 1901 requires all valves to be designed so that they are easily operable at pressures of up to 250 psi.

When conditions require using multiple attack lines at different pressures, the only way to supply them is to set the engine pressure to the highest pressure needed. Then, the driver/operator can partially close each of the other lines' valves until the reduced flow provides the desired pressure at the hoseline. Individual line pressure gauges or flowmeters are essential to do this properly. Without individual line gauges, providing good fire streams becomes a matter of guesswork or constant feedback from the nozzle operator.

Figure 8.46 A typical 2 1/2-inch pump discharge connection.

According to NFPA® 1901, the apparatus must have enough 2 ½ -inch or larger discharge outlets to flow the rated capacity of the fire pump **(Figure 8.46). Table 8.2** lists the allowable discharge rates by outlet size specified in NFPA® 1901. (**NOTE:** Keep in mind that in fireground applications, each of these discharges can flow considerably more water than listed in this table.)

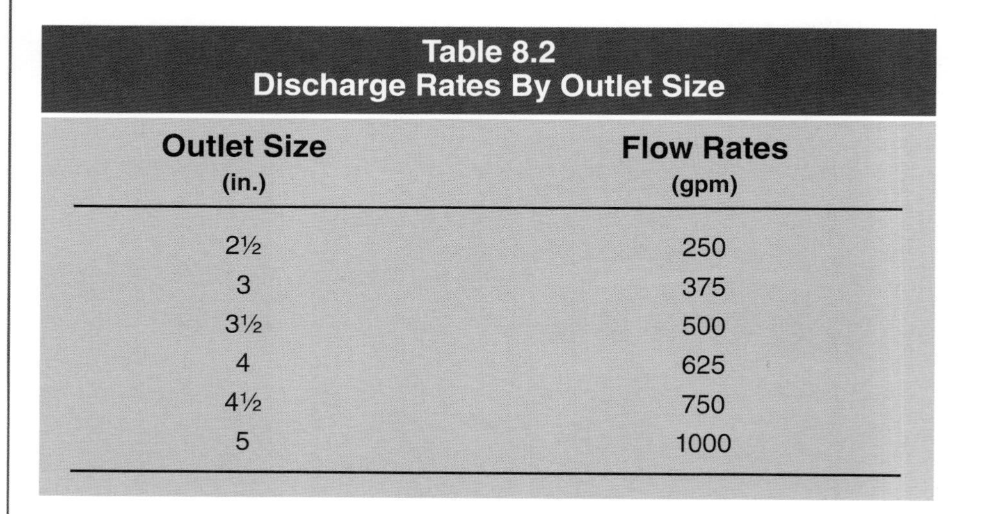

Table 8.2
Discharge Rates By Outlet Size

Outlet Size (in.)	Flow Rates (gpm)
2½	250
3	375
3½	500
4	625
4½	750
5	1000

To meet the NFPA® 1901 requirements, all fire apparatus with a rated pump capacity of 750 gpm or greater must be equipped with at least two 2½-inch discharges. Pumps rated at less than 750 gpm are required only to have one 2 ½-inch discharge. Most fire apparatus have one 2½-inch or larger discharge for every 250 gpm of rated pump capacity. Thus, a 1,000 gpm pumper will have four 2½-inch discharges and a 1,500 gpm pumper will have six. Discharges larger than 2½ inches must not be located directly on the pump operator's panel. This is due to the hazard that would be posed if such a large hose line were to fail near the pump panel.

In addition to the required 2½-inch or larger discharges, apparatus commonly have a number of smaller discharges at various locations for use with preconnected attack lines **(Figure 8.47)**. NFPA® 1901 requires that discharges to which 1½-, 1¾-, and 2-inch hand lines are attached must be supplied by at least 2-inch piping.

Figure 8.47 Preconnected hose lines may be supplied by discharges smaller than 2 1/2-inches in diameter.

In recent years fire departments that utilize large diameter (4- or 5-inch) supply hose commonly have specified one or more larger diameter discharges on their apparatus. These large diameter discharges often are comprised of 4- or 5-inch piping and valves. The larger piping and valves can add several thousand dollars per discharge to the cost of the apparatus. In general, piping and valves larger than 3-inches in diameter are not required for standard municipal pumpers. A 3-inch discharge equipped with a full-flow valve easily will supply all of the water that a single 5-inch supply hose can carry. Larger piping and valves typically are required only on industrial pumpers that supply multiple large-diameter hose lines or hose lines greater than 5-inch in diameter.

Though not technically considered a typical discharge line, a tank fill line (sometimes referred to as a pump-to-tank line) connects the discharge side of the pump to the apparatus water tank. This line allows the driver/operator to fill the tank without making any additional connections when supplying the pump from an external supply source. It may be used to replenish the water in the tank after the initial attack and a transition to an external supply source has been made. In the case of a two-stage pump, the tank fill line gets its flow from the first stage. This helps ensure that reduced pressures are supplied to the tank as a safety measure. Overpressurizing the tank could result in a dramatic tank failure. The diameter of the tank fill line will vary depending on the manufacturer and the fire department's specifications. NFPA® 1901 requires apparatus with a water tank of less than 1,000 gallons to have a tank fill line at least 1-inch in diameter. Apparatus with tanks of 1,000 gallons or more must have at least a 2-inch tank fill line.

The tank fill line can be used to circulate water through the pump to prevent overheating when no lines are flowing. However, in a two-stage pump with the fill line coming off the first stage, overheating can still result in the pump's second stage. In this case a specially-designed circulator valve more effectively prevents overheating. Like the tank fill line, the circulator valve is connected to the discharge side of the pump. The circulator line enables the driver/operator to dump water into the apparatus water tank or outside the tank on the ground.

Some pumpers have a booster-line cooling valve that serves the same function as the circulator valve by diverting a portion of the discharge water into the tank. While either of these arrangements is adequate for normal pumping operations, the piping consists of small copper tubing that limits flow to approximately 10–20 gpm. During prolonged operations with intermittent flows, or when operating at very high pressures, this may not be enough water to keep the pump cool. Doing this may necessitate discharging water through some type of waste or dump line.

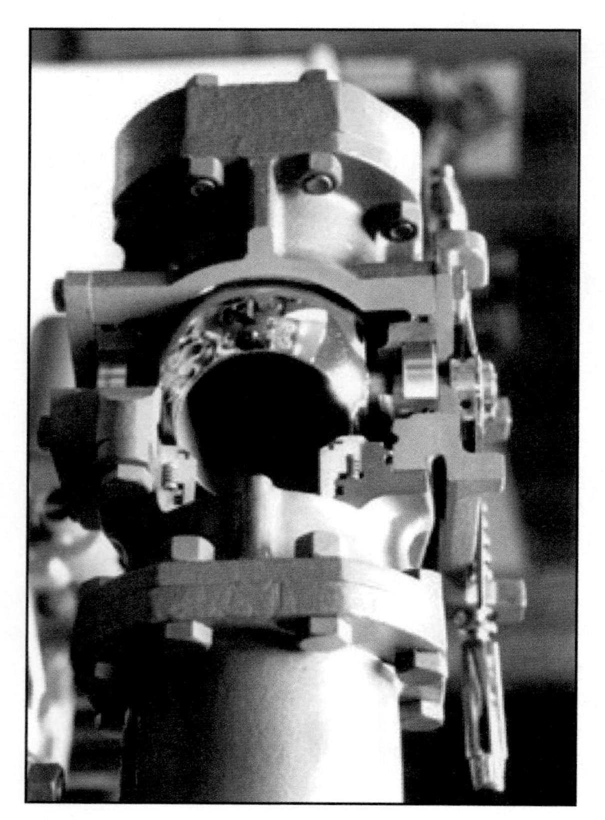

Figure 8.48 Ball valves control most fire pump pumping.

Valves

Valves allow the driver/operator to control the flow of water throughout the pump and piping system. Most of the intake and discharge lines from the pump are equipped with valves. When the pump is new, the valves must be airtight. Even though the valves are constructed to resist wear (in some cases, they are self-adjusting), they do require maintenance as they age and are subjected to frequent use.

NFPA® 1901 requires all valves on intakes or discharges that are 3 inches or greater to be equipped with slow-acting valve controls. This prevents movement from the fully opened to fully closed positions (or vice versa) in less than three seconds. This helps to minimize the risk of damage from water hammer when large volumes of water are moving.

The most common valve design on fire apparatus is the ball valve. The ball valve permits full flow through the piping with a minimum of friction loss when it is fully opened **(Figure 8.48)**. Historically, most ball valves are operated mechanically using either push-pull handles (commonly called T-handles) or quarter-turn handles **(Figure 8.49)**. The push-pull valve handle uses a sliding gear-tooth rack that engages a sector gear connected to the valve stem. This gear arrangement provides a mechanical advantage that makes the valve easier to operate under pressure. It also allows the driver/operator to set pressures precisely to

Figure 8.49 Many fire apparatus have push-pull handles **(a)** for controlling ball valves; others have quarter-turn handles **(b)**.

individual lines. The operating linkage design can allow the valve to be mounted remotely from the pump panel. The push-pull lever usually has a flat handle that can be turned 90 degrees to lock the valve in any position. When operating the push-pull lever, the driver/operator must pull it straight and level. Otherwise, the shaft will bind, and the valve will be inoperable from that point on.

The quarter-turn handle's mechanical linkage is simpler. The handle is mounted directly on the valve stem. A 90-degree turn of the handle opens or closes the valve. Some of the older quarter-turn valves are locked in position by raising and lowering the handle, but the more modern versions lock by rotating the handle clockwise. Some valves lock automatically when the handle is released, but locking most valves requires a positive action anytime the line is in use.

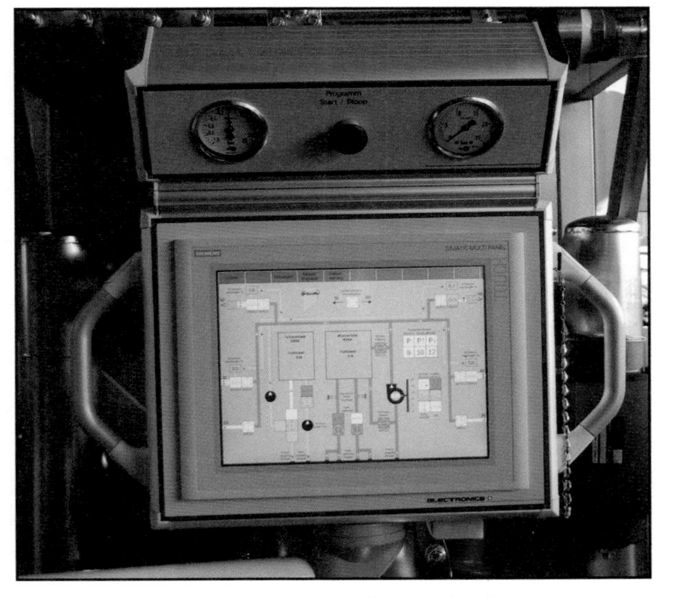

Figure 8.50 Some modern pumpers have hydraulic, pneumatic, or electrical valves that are controlled by a toggle switch.

Newer apparatus may have valves that are controlled hydraulically, pneumatically, or electrically. These are ball-type valves that a toggle switch or electronic control panel at the pump operator's control area **(Figure 8.50)**. A visual display indicates how far the valve is opened. Markings on the panel show which direction to operate the switch or electronic controls in order to open or close the valve.

Fire apparatus also use gate or butterfly valves, most commonly on large-diameter intakes and discharges. Gate valves usually are operated by a handwheel, and butterfly valves usually are operated by quarter-turn handles. Both may have hydraulic, pneumatic, or electric actuators. These typically are used as remote-operated dump controls on water tenders (tankers).

Pump Drains

Most connections to the pump have drain valves on the intake or outlet side of the control valve. On the discharge lines, the drain valve provides a way for the driver/operator to relieve the pressure from hose lines after the discharge valve and nozzle have both been closed. This is useful when the hose line was not bled off by opening the nozzle and the nozzle is a great distance from the apparatus. The drain valve allows the driver/operator to drain the unused line and disconnect the hose even while the pump is still in service.

The drain valve on a pump's gated intake also serves another purpose. These valves are also known as bleeder lines or bleeder valves. If a supply line is connected to the gated intake of a pump while the attack lines are being supplied from the tank, it is desirable to change over to the supply line without interrupting the fire streams. Because the uncharged supply hose is full of air when it is dry, the water coming through the hose would force air into the pump. At the very least, this would cause fluctuations in the nozzle pressure. Worse, it could cause the pump to lose its prime or to go into cavitation. Opening the bleeder valve on the line side of the intake valve permits the water to force the air from the hose through the bleeder as the line fills. When all the air has been evacuated from the line and the bleeder is discharging a steady stream of water, the driver/operator can close the drain valve, open the intake valve, and close the tank-to-pump valve. These coordinated actions can make the transition with no interruption to the flow.

In critical high-flow pumping situations or during pump testing, only the midship large-diameter intakes should be utilized.

Another purpose for pump and piping drains is to remove all water from the system in climates where freezing might occur. Water expands when it freezes and can damage pump and piping components, in addition to the fact that the ice creates barriers to water flow if the pump is needed. Drains must be supplied at the lowest point on the pump and at the lowest point on each line connected to it. A number of these drains are connected to a common master drain valve. Operating the control handle opens all of the drains simultaneously and drains the pump with one operation. This drain valve should not be opened when the pump is operating either with pressure or with vacuum on the intake. An O-ring gasket maintains an airtight connection when the valve is closed. Operating the valve under pressure or vacuum can damage the O ring. Some additional drain valves may be supplied if connecting all the drains to the master valve is not convenient. They all must be opened for the pump to drain completely. The driver/operator should close all drains immediately after use. Failure to do so could prevent the pump from priming and beginning drafting operations when necessary.

Failure to close all drains immediately after use could prevent the pump form priming and beginning drafting operations when necessary.

Whenever using the pump drains, the driver/operator also should open a discharge valve to allow air to replace the draining water. If not, the resulting vacuum will hold some of the water in the pump.

Whenever using the pump drains, the driver/operator should open a discharge valve to allow air to replace the draining water.

Manually unwinding and draining booster hose lines may be necessary during freezing weather. Simply opening the drain valve on the booster line supply piping does not allow all water to drain from the hose. The remaining water may freeze inside the hose, causing it to burst. Newer apparatus may have piping from the vehicle's air brake system to the booster reel. Opening the valve in this line discharges compressed air through the booster hose to remove the water. This negates the need to remove the hose manually to drain it.

PUMP PANEL INSTRUMENTATION

A variety of instruments and controls on the pump panel allow the driver/operator to operate the fire pump efficiently. Some of this instrumentation is specific to the pump operator's panel, and some of it duplicates gauges and controls in the cab. The duplication minimizes the pump operator's having to run back and forth between the pump panel and the apparatus cab. As a minimum, NFPA® 1901 requires the pump operator's panel to have the following controls and instruments:

- Master pump intake pressure indicating device
- Master pump discharge pressure indicating device
- Weatherproof tachometer
- Pumping engine coolant temperature indicator
- Pumping engine oil pressure indicator
- Pump overheat indicator
- Voltmeter
- Discharge gauges (pump pressure indicators)
- Pumping engine throttle
- Primer control
- Water tank to pump valve (discussed earlier in this chapter in the "Intake Piping" section)
- Tank fill valve (discussed earlier in this chapter in the "Discharge Piping" section)
- Water tank level indicator

Most of the listed "indicators" are gauges. Electronic indicators that provide digital readouts are becoming more commonplace on modern apparatus.

Master Intake and Discharge Gauges

The master intake and discharge gauges are the two primary gauges used to determine the water pressure entering and leaving the pump **(Figure 8.51)**. The *master intake gauge* (sometimes referred to as the *vacuum gauge* or *compound gauge*) must be connected to the intake side of the pump. This gauge must be capable of measuring either positive or negative pressure. It usually is calibrated from 0 to 600 psi for positive pressure and from 0 to 30 inches of mercury for vacuum. The master intake gauge indicates residual pressure when the pump is operating from a pressurized supply source.

Figure 8.51 The master intake and discharge pressure gauges.

When pumping from a draft, the master intake gauge indicates the amount of vacuum at the pump intake during priming or when operating from draft. This provides an approximation of how much pump capacity is not being used. As the flow from the pump increases, the vacuum reading increases because overcoming the friction loss in the suction hose requires more negative pressure. As the vacuum reading approaches 20 inches, the pump nears its maximum capacity and cannot supply any additional lines.

The pump also must have a master pump discharge pressure gauge. This gauge registers the pressure as it leaves the pump but before it reaches the gauges for individual discharge lines. It must be calibrated to measure up to 600 psi, and if the pumper is equipped to supply high-pressure streams, it may be calibrated up to 1,000 psi.

These gauges must have external connections to allow the installation of calibrated gauges for service tests **(Figure 8.52)**. These connections also are used for standard gauges during the initial acceptance tests of the pump.

Tachometer

The tachometer records the engine speed in revolutions per minute (rpm) **(Figure 8.53)**. It can give an experienced driver valuable information about the pump's condition and is a useful means of analyzing difficulties with the pump. The initial acceptance tests of the pump determine the rpm required for the pump to reach its rated capacity, and that information is recorded on an identification plate on the pump panel. A gradual increase in the rpm required to pump the rated capacity indicates wear in the pump and a need for repairs.

Pumping Engine Coolant Temperature Indicator

The pumping engine coolant temperature indicator shows the temperature of the coolant in the engine that powers the fire pump **(Figure 8.54)**. This may be the main vehicle engine or, in the case of an auxiliary engine-driven pump, the pump engine. The operating temperature of the engine is important: an engine that operates too cool is not efficient, and an operating temperature that is too high may damage the engine's mechanical parts.

Figure 8.52 The pump panel should provide test connections for the master intake and pump discharge pressure gauges.

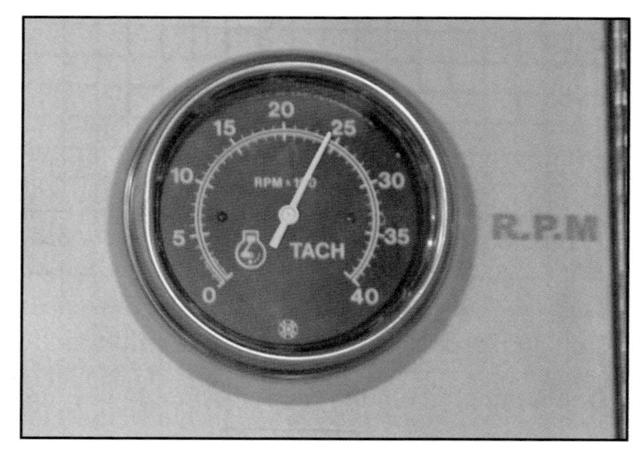

Figure 8.53 The tachometer allows the pump operator to monitor engine speed from the pump panel.

Pumping Engine Oil Pressure Indicator

The pumping engine oil pressure indicator shows whether an adequate supply of oil is reaching the critical areas of the engine that powers the fire pump. It is not a measure of the oil level in the crankcase, but if the oil level in the crankcase drops too low, the oil pump cannot maintain the required pressure. While the maintenance manual specifies normal operating oil pressures, the pressure for each particular engine will vary, so the operator needs to be familiar with its usual reading. Any significant deviation from the normal oil pressure reading indicates pending problems.

Figure 8.54 The engine coolant temperature gauge alerts the pump operator to overheating problems.

Pump Overheat Indicator

The pump panel may include an audible or visual indicator that warns the driver/operator when the pump overheats. Overheating may occur when the pump runs for prolonged periods during which no water is discharged.

Voltmeter

The voltmeter provides a relative indication of battery condition and alternator output by measuring the drop in voltage as some of the more demanding electrical accessories, such as the primer, are used **(Figure 8.55)**. It indicates the maximum voltage available when the battery is fully charged.

Figure 8.55 The voltmeter shows the status of the vehicle's electrical supply system.

Discharge Gauges (Pump Pressure Indicators)

Discharge gauges are connected to the individual discharge fittings of the pump **(Figure 8.56)**. These gauges must be connected to the outlet side of the discharge valve so that they report the pressure actually being applied to the hoselines after the valve. This allows the driver/operator to reduce pressure in individual discharges from the overall pump discharge pressure if necessary. If the nozzle is shut down on the hose line being supplied, the individual pressure gauge will read the same as the master pressure gauge because there is no flow through the valve to reduce the pressure.

Individual pressure gauges may also be supplied for master stream devices or the lines that supply them. These gauges are very important as maintaining effective master streams is not possible unless the pump is supplying proper pressure to the appliance. With the large flows required, friction loss in the supply lines is high. An individual line gauge is the only way to be certain that the pump has been adjusted properly. NFPA® 1901 allows flowmeter readouts to substitute for individual pressure discharge gauges **(Figure 8.57, p. 176)**. However, even if it has a flowmeter system, the apparatus still must have master intake and pressure gauges.

An individual line gauge is the only way to be certain that the pump has been adjusted properly.

Figure 8.56 Each pump discharge should have its own pressure gauge.

Pumping Engine Throttle

The pumper operator's panel must contain a pumping engine throttle. This device increases or decreases the speed of the engine that powers the fire pump. By increasing or decreasing engine speed, the driver/operator controls the amount of pressure that the fire pump supplies to the discharge. The most common type of throttle on the pump operator's panel is a rotating knob **(Figure 8.58, p. 176)**. The driver/operator turns the throttle knob one way or the other until the desired rpm/pressure is achieved. While earlier versions of the rotating knob

Figure 8.57 Flowmeters may be used in place of individual discharge gauges.

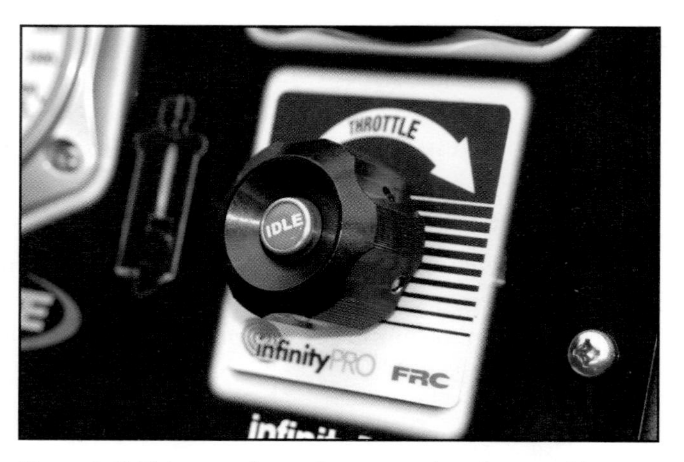

Figure 8.58 This type of pumping engine throttle control has been commonly used for many years.

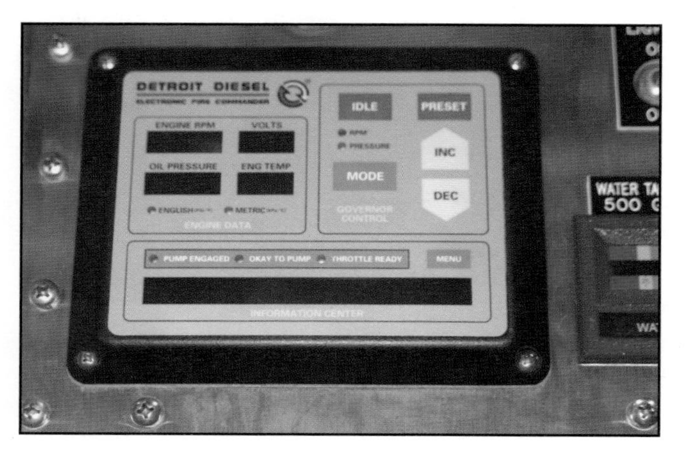

Figure 8.59 Newer apparatus may be equipped with toggle-type throttle controls.

type throttle manually operated the throttle, newer ones are actually electronic. Automatic throttle controls, such as those operated by a toggle switch, are also available and in use on newer apparatus **(Figure 8.59)**.

Primer Control

The primer control operates the priming device for drafting from a static water supply. This control is generally a push button, toggle switch, or pull lever.

Water Tank Level Indicator

The water tank level indicator tells the driver/operator how much water is in the onboard water tank. This allows the driver/operator to anticipate how much longer the tank may supply attack hoselines before an external water supply source is needed. This is particularly crucial when the water tank is the pump's only supply for interior fire attack hoselines. Interior crews need to be evacuated if the onboard supply will run out before an external water supply has been established.

The most common type of water tank level indicator uses a series of lights on the pump operator's panel **(Figure 8.60)**. Sensors within the tank send signals that indicate the amount of water in the tank by quarters (empty, 1/4, 1/2, 3/4, full). These lights may be small LCD-type lights on a pump panel plate, or they may be larger lights mounted where the driver/operator and other personnel on the scene can view them **(Figure 8.61)**.

Some apparatus have sight gauges that allow the driver/operator to view the actual level of water in the tank through a clear tube. As the level of water in the tank decreases, so does the level in the tube. Newer apparatus may be equipped with digital readouts that indicate the level of water in the tank.

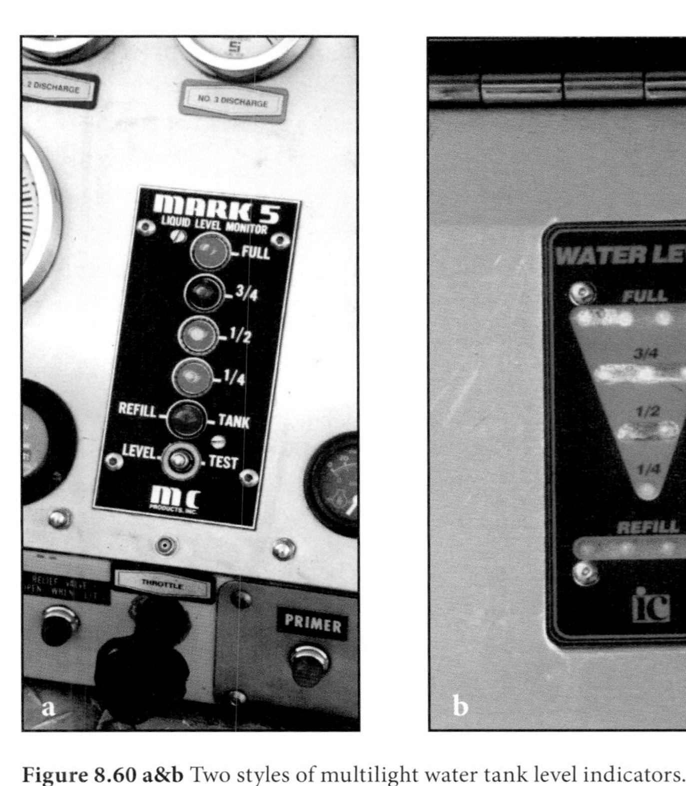

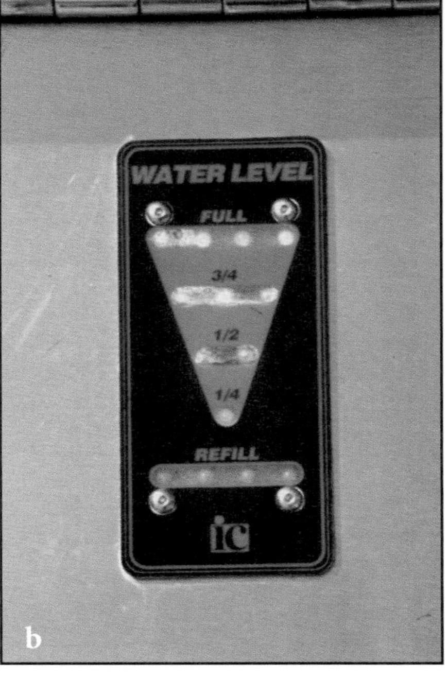

Figure 8.60 a&b Two styles of multilight water tank level indicators.

Figure 8.61 Some apparatus are equipped with large water tank level displays that are visible well away from the truck.

Other Recommended Gauges

Although not required by NFPA® 1901, the pump operator's panel should include a pumping engine fuel gauge. The pump operator's panel fuel gauge is important during extended pumping operations when fuel may run low. It allows the driver/operator to check the fuel level without having to climb into the apparatus cab. As well, the panel may have an automatic transmission temperature gauge. The driver/operator can monitor this gauge to make sure transmission damage does not occur during pumping operations.

Flowmeters

Until the late 1970s, the only way a pump operator could determine the volume of water flowing through a fire pump was to read the pressure gauges and use that information to calculate the flow rate. This mathematical conversion is not easy, and performing it accurately on the fireground is nearly impossible. However, modern technology provides an alternative. Flowmeters reduce the number of pressure calculations the driver/operator must perform.

Rather than a readout of the pressure going through a discharge, flowmeters indicate the water flow in gallons per minute. The number displayed on the flowmeter requires no further calculation because it shows how much water is moving through the discharge valve and consequently the nozzle. The quantity of water diminishes before it reaches the nozzle only if the hoseline has a leak or a break. While supplying the proper pressure to a hoseline is important, ultimately the most important thing is to supply the proper volume of water.

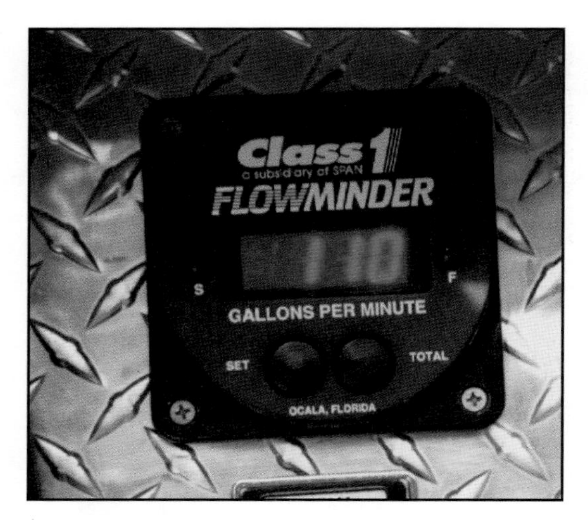

Figure 8.62 Flowmeters should provide a readout in increments no larger than 10 gpm.

Flowmeters are particularly advantageous when supplying hoselines or master stream devices equipped with automatic nozzles. Automatic nozzles maintain pressure at a predetermined level and thus may lead driver/operators to supply insufficient discharge pressures. The automatically maintained pressures can create a fire stream that appears to be acceptable but in reality has a very low flow in gallons per minute. A flowmeter makes it easier for the driver/operator to ensure that the flow rate is adequate. Flowmeters enable driver/operators to supply (within the limits of the pump) the correct volume of water to nozzles without having to know the length of the hoseline, the amount of friction loss, or the elevations of the nozzles relative to the pump. This relieves the driver/operator from having to make calculations that provide, at best, only close approximations of the amount of water reaching the nozzle.

NFPA® 1901 allows flowmeters to be used instead of pressure gauges on all discharges 1½ to 3 inches in diameter. Discharges that are 3½ inches or larger may be equipped with flowmeters, but they also must have an accompanying pressure gauge. The flowmeter must provide readouts in increments no larger than 10 gpm **(Figure 8.62)**.

Types of Flowmeters

The fire service commonly uses two basic types of flowmeter sensors, the paddlewheel and the spring probe. Paddlewheel flowmeters are mounted on top of a straight section of pipe so that they extend very slightly into the waterway **(Figure 8.63;** *Courtesy of Fire Research Corporation***)**. This reduces the problems of impeded flow and damage by debris. Because the paddlewheel is located at the top of the pipe, sediment does not deposit on the paddlewheel. As passing water spins the paddlewheel, a sensor measures the paddlewheel's rpm and translates that information into a flow measurement.

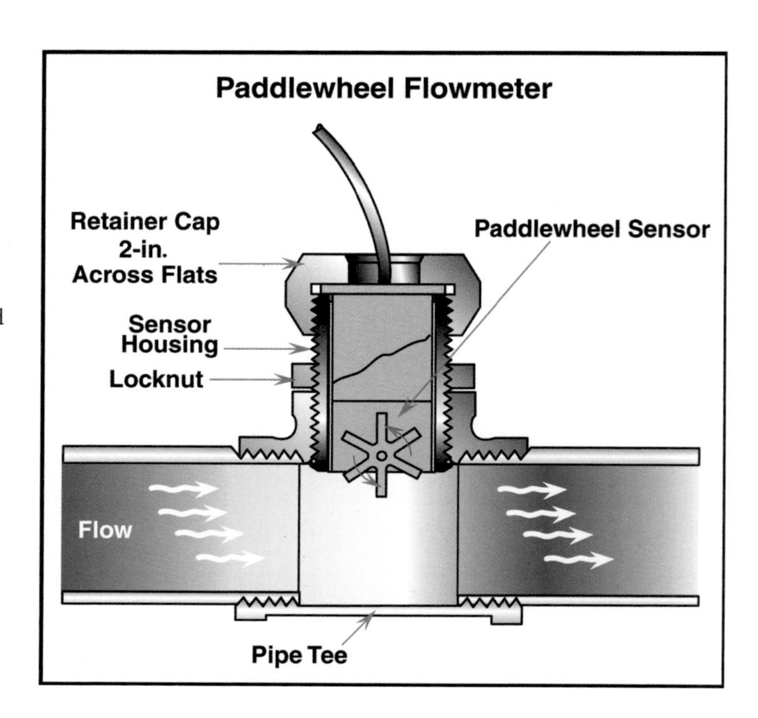

Figure 8.63 The earliest fire service flowmeters utilized a paddlewheel design. *Courtesy of Fire Research Corporation.*

The spring probe flowmeter uses a stainless steel spring probe to sense water movement in the discharge piping **(Figure 8.64;** *Courtesy of Class 1/Span Instruments***)**. The greater the flow of water through the piping, the more it forces the spring to bend. The bending sends a corresponding electrical charge to the digital display unit. Because the spring probe is their only moving part, these devices are relatively maintenance free.

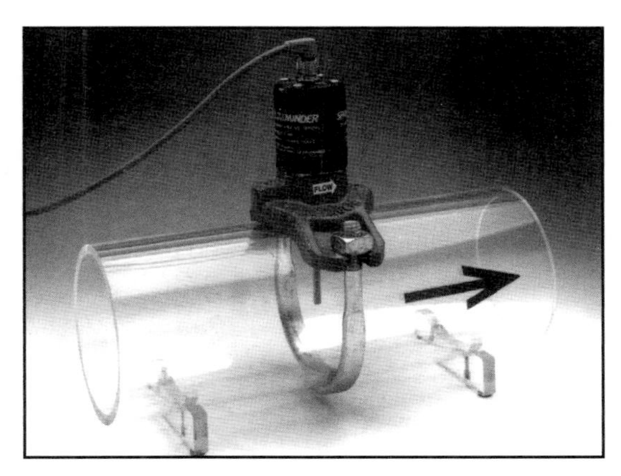

Figure 8.64 The spring-probe flowmeter has become increasingly popular in recent years. *Courtesy of Class 1/Span Instruments.*

When properly calibrated and in good working condition, flowmeters should be accurate to a tolerance of ±3 percent. This means that the readout should not be more than 3 gpm high or low for every 100 gpm flowing.

Each discharge equipped with a flowmeter has a digital readout display mounted within 6 inches of its valve control. In addition, some apparatus have a central flowmeter monitoring device that allows the driver/operator a number of options. Some of the information that the central monitor may provide includes the following:

- The flow through any particular discharge at that time
- The total amount of water being flowed through the pump at that time
- The total amount of water that has been flowed through the pump for the duration of that incident

SUMMARY

Virtually any firefighter can be taught to drive a fire apparatus and operate the fire pump on that vehicle. The mechanical skills needed to perform those actions are not particularly difficult. However, many fire service personnel perform those very skills for their entire careers without ever having a thorough understanding of the design and operating principles of the equipment they are running. A true fire service *professional* knows not only which levers to pull and what buttons to push but also the consequences of those actions and how the equipment actually works.

In this chapter we have reviewed the operating principles and design of fire service pumps past and present. The earliest fire service pumps were manually operated, positive displacement pumps. They were replaced by mechanically operated positive displacement pumps of the piston or rotary gear/vane design. Positive displacement pumps were simple in design and operation, but their numerous limitations ultimately led to their demise as the primary pump for fire fighting operations. Most notable was their inability to substantially take advantage of incoming pressure. As municipal water supply systems grew and improved, this deficiency became crucial.

Modern fire service pumps are of the centrifugal design. Centrifugal pumps impart pressure on the water by taking it in through the eye of a spinning disk (the impeller) and casting it outward toward the edges of the disk and the casing of the pump. Centrifugal pumps are flexible in design. They also can take advantage of water being supplied under pressure and are can increase the intake pressure. The only major disadvantage of centrifugal pumps is their inability to create the partial vacuum necessary to draft water from a static supply source. For this reason, centrifugal pumps are supplemented with rotary vane or rotary gear pumps that draw air from the pump and piping to create the necessary vacuum. Once the centrifugal pump is primed, it is a reliable mechanism for drafting from a static water supply.

To operate effectively, the modern pump requires a wide variety of other equipment including primers, pressure control devices, gauges, valves, and other devices. As with the main pump, the fire service professional also must understand the purpose and operation of all of these auxiliary devices. To use a pump or apparatus to its fullest potential, the driver/operator must understand all of its components and use them at the appropriate time and in the correct manner. Of course, operating the equipment is, at best, only half the battle for a pump operator. As we will learn in the remainder of this book, determining the operating pressures and setting the pump to meet those demands requires as many or more skills than physically operating the pump.

REVIEW QUESTIONS

1. State the principle of pressure upon which the operation of positive displacement pumps is based.

   ____

2. Name the earliest operating design for positive displacement pumps.

   ____

3. List at least three disadvantages of piston-type positive displacement pumps.

   1. ______________________________
   2. ______________________________
   3. ______________________________

4. What are the two designs for rotary-type positive displacement pumps?

   1. ______________________________
   2. ______________________________

5. What two drive mechanisms for rotary pumps are used for primers on modern centrifugal fire pumps?

   1. ______________________________
   2. ______________________________

6. Explain why the centrifugal fire pump is not classified as a positive displacement-type pump.

   ____

   ____

7. What are the two primary parts of a centrifugal fire pump?

   1. ______________________________
   2. ______________________________

8. What is the term for the process of switching between the pressure and volume modes of operation on a two-stage centrifugal fire pump?

   ____

9. List the three main factors that determine a centrifugal fire pump's discharge pressure.

   1. ______________________________
   2. ______________________________
   3. ______________________________

10. Describe the basic difference between single-stage and two-stage centrifugal fire pumps.

    ____

    ____

11. In most cases, what is the maximum operating pressure at which a changeover valve on a two-stage fire pump should be operated?

    ____

12. What is the primary inherent disadvantage of an auxiliary-driven fire pump?

____

____

13. From where does a power take-off pump receive its power?

____

____

14. What devices other than a fire pump may be powered by the PTO on a modern fire apparatus?

____

____

15. What are the primary differences between a PTO and front-mount fire pump?

____

____

____

16. How is power supplied to a midship transfer-driven fire pump?

____

____

17. What is a major disadvantage of midship transfer-driven fire pumps?

____

____

18. What are the two basic types of pressure control devices used on modern fire apparatus?

    1. ____

    2. ____

19. What is the most common design for a pressure relief valve?

____

20. Describe the basic operating principle of a pressure governor.

____

____

21. List the three basic categories of priming devices.

    1. ____

    2. ____

    3. ____

22. What are the two types of auxiliary cooling devices commonly found on fire apparatus?

    1. ____

    2. ____

23. What are the two primary ways that water may enter the fire pump?

    1. ______________________________

    2. ______________________________

24. Name the most common design of valve used on fire apparatus.

    ____

25. When pumping from a draft, what information does the master intake gauge provide?

    ____

    ____

26. What are the two basic designs for flowmeters used in the fire service?

    1. ______________________________

    2. ______________________________

Answers found on page 428.

FESHE Course Objectives

The information in this chapter is intended to meet some of the objectives outlined for the Fire Protection Hydraulics and Water Supply course by the United States Fire Administration's National Model Core Curriculum as developed by the Fire and Emergency Services Higher Education (FESHE) initiative. The information in this chapter also allows the student to meet the requirements of NFPA® 1002, *Standard for Fire Apparatus Driver/Operator Professional Qualifications* (2009 edition) listed below.

FESHE Course Objectives

1. List the various types of preservice tests that are used to determine the viability of fire apparatus equipped with fire pumps.
2. Describe the procedures for performing road, hydrostatic, pumping, pumping overload, pressure control system, priming, water tank-to-pump flow, and vacuum preservice tests on fire apparatus equipped with fire pumps.
3. Explain the benefits of acceptance tests and describe the types of acceptance tests that are commonly required.
4. Explain the importance of regular service testing of fire apparatus equipped with a fire pump.
5. Perform the following service tests on a fire apparatus equipped with a fire pump:
   - Engine speed check
   - Vacuum test
   - Pumping test
   - Pressure control test
   - Gauge and flow meter test
   - Tank-to-pump flow rate test
6. List some of the site considerations for performing service testing of fire apparatus equipped with a fire pump.

NFPA® 1002 Requirements

5.1.1 Perform the routine tests, inspections, and servicing functions specified in the following list in addition to those in 4.2.1, given a fire department pumper and its manufacturer's specifications, so that the operational status of the pumper is verified:

(2) Pumping systems

Chapter 9

Fire Department Pumper Testing

INTRODUCTION

Apparatus testing is a vital part of any fire department apparatus purchasing and maintenance program. Testing a new apparatus before accepting its delivery from a manufacturer ensures that the apparatus was *built* to the fire department's specifications (specs). It also verifies that the apparatus will *perform* in the manner that the fire department intended when it authored the specs. Testing the apparatus periodically after it has been placed into service ensures that it is being maintained in optimum working condition with little or no loss of its original operational capability. As well, communities that rely on good insurance industry ratings to minimize their residents' insurance costs must test apparatus regularly in order to receive full credit in the rating process.

Tests conducted before the purchasing fire department accepts the apparatus are referred to as *preservice* tests. Preservice tests include the manufacturer's test, pump certification test, and acceptance tests. Fire department personnel usually do not actually participate in these tests, but they may be required to witness them. Knowledge of how the tests are performed also allows the fire protection professional to appreciate and understand the service tests that will be conducted later in the apparatus's life.

Once a fire apparatus is placed into service within the department, s*ervice* tests should be conducted at least yearly and after all major repairs to ensure that it is performing properly. An independent testing company may conduct these tests, or they may be performed entirely by fire department personnel and mechanics. All fire protection professionals should have a working knowledge of the service testing procedure so that the tests will be performed properly and the readiness and capabilities of the apparatus will be evaluated accurately.

This chapter reviews the principles of both preservice and service tests for fire department pumpers. It provides important insight into crucial testing issues such as correcting net pump discharge pressure for tests, the sequence of tests, equipment needed, safety precautions, and troubleshooting during the tests. Because the scope of this text is limited to fire service hydraulics, the chapter does not cover preservice or service testing of fire department aerial devices; information on testing those devices can be found in the IFSTA *Fire Department Aerial Apparatus Driver/Operator Handbook*. As well, information on the operation and testing of foam proportioning systems is covered in both the IFSTA *Fire Department Pumping Apparatus Driver/Operator Handbook* and *Foam Fire Fighting Practices* manuals.

By following the information in this chapter, the fire protection professional will ensure that an organized system of apparatus testing is in place. This is the best assurance a fire department has that an apparatus will perform to its design capabilities.

PRESERVICE TESTING

Preservice tests assure that the apparatus has been constructed according to the purchaser's specs and that it will perform as intended. To ensure that preservice tests are performed, they must be required in the apparatus bid specifications used to purchase the vehicle. Otherwise the manufacturer may elect to skip them, and the purchasing department will have no way of knowing if the apparatus meets the requirements of its intended standard and specifications. Preservice tests can be grouped into three categories: manufacturer's tests, pump certification tests, and acceptance tests. These tests may include evaluations of the vehicle and chassis, the pumping system, and any other equipment carried on the apparatus if desired.

To ensure that preservice tests are performed they must be required in the apparatus bid specifications used to purchase the vehicle.

Most pumping fire apparatus are designed to meet the requirements of either NFPA® 1901, *Standard for Automotive Fire Apparatus*, or NFPA® 1906, *Standard for Wildland Fire Apparatus.* To ensure this, the bid specifications should contain a clause that requires the apparatus to meet the pertinent chapters of NFPA® 1901 or 1906. The specs should state that failure to meet these requirements will be one cause for rejecting acceptance of the apparatus. As with all NFPA® standards, the requirements for preservice and service tests are *minimum* requirements. The fire department has the option of placing more stringent requirements in its specs. However, the department must realize that the more stringent the requirements are, the fewer manufacturers will be likely to bid on the apparatus, and the price of the apparatus may increase greatly.

The manufacturer and/or representatives of Underwriters Laboratories (UL) typically perform preservice tests. Fire department personnel usually only observe these tests, though in some cases they may be physically involved with acceptance testing. More detailed information on who performs the testing is highlighted in each of the following three sections on preservice testing.

Manufacturer's Tests

The purchaser of an apparatus may require any type and number of manufacturer's tests. If the apparatus bid specifications include the requirements of NFPA® 1901, as a minimum the manufacturer is required to perform two specific tests, the road test and the hydrostatic test. NFPA® 1901 further requires the manufacturer to perform the pump certification tests explained in the next section.

Road Test

NFPA® 1901 requires the manufacturer to perform a road test on the vehicle to ensure that it will operate safely once it is fully loaded with equipment and personnel and placed into service. Before starting this test, the apparatus is fully loaded in the same manner as it would be in service. This includes making sure that all agent tanks are full and the weight of personnel, hose, and equipment that will be carried on the apparatus is accounted for. Test weights, such as bags of sand, may be used in lieu of personnel, hose, and equipment. However, the sum of the test weight must equal the sum of the real people and equipment. The test weights also should be located where the real weight would be. For example, to pile all of the sand bags on the hose bed would be unacceptable; they should be distributed appropriately among the hose bed, compartments, seats, and other areas of the apparatus.

Choosing a good location for the road test is important. Most manufacturers have a closed-course driving area where they conduct these tests. If this is not the case, care should be used to choose a location that will not cause the driver to violate any applicable traffic laws or motor vehicle codes (**Figure 9.1**). The test surface should be a flat, dry, paved road surface that is in good condition.

Figure 9.1 The manufacturer's road test should cover a variety of travel conditions.

NFPA® 1901 requires apparatus to meet the following minimum criteria during road tests:

- The apparatus must accelerate to 35 mph from a standing start within 25 seconds. This test must consist of two runs, in opposite directions, over the same surface.
- The apparatus must achieve a minimum top speed of 50 mph. This requirement may be dropped for specialized wildland apparatus not designed to operate on public roadways.
- The apparatus must come to a full stop from 20 mph within 35 feet.
- The apparatus parking brake must conform to the specifications listed by the braking system manufacturer.

The fire department may choose to include in the specs special road test requirements that exceed the NFPA® minimums. For example, departments that protect hilly jurisdictions may have special requirements for apparatus acceleration, deceleration, and braking abilities on severe grades.

To address special situation, performance-based specifications are better than engineering-based specifications.

When addressing these types of special situations, performance-based specifications are better than engineering-based specifications. Performance-based specs simply say what you expect the apparatus to do. You leave it up to the manufacturer to engineer the apparatus to meet those requirements. Engineering-based specifications are simply lists of vehicle components with which the purchaser is requiring the manufacturer to construct the apparatus. The pitfall of engineering-based specs is that the manufacturer may follow them to the letter, but the finished apparatus may not be able to meet the purchaser's operational requirements. As long as the manufacturer can demonstrate that it followed the engineering-based specs, the purchaser has no recourse. With performance-based specs the situation is clearly defined. Either the apparatus performs as desired or it does not. If not, the purchaser can refuse to accept the vehicle until corrections are made and the apparatus performs as desired.

Figure 9.2 All discharge valves should be open, all intakes should be closed, and the caps should remain on all connections during a hydrostatic test.

Hydrostatic Test

The second manufacturer test required by NFPA® 1901 is the hydrostatic test. The hydrostatic test determines whether the pump and pump piping can withstand normal operating pressures. Hydrostatic testing involves filling the pump and its piping with water and then placing an external pressure source on it. NFPA® 1901 requires the pump to be hydrostatically tested at 250 psi for 3 minutes. During the test, the tank fill line, tank-to-pump line, and bypass line valves all should be closed. Discharge valves should be opened and capped. Intake valves should be closed and/or capped (**Figure 9.2**). If any component of the pump fails during the three minutes of the test, the apparatus will not be certified.

Pump Certification Tests

The hydrostatic test determines only if the apparatus fire pump and piping were constructed and assembled properly. In order to determine if that assembly *operates* properly, a series of *pump certification tests* must be performed. These tests ensure that the fire pump system operates as designed after it is installed on the apparatus chassis. Certification tests must be required in the apparatus bid specifications. Though pump certification tests frequently are conducted on the manufacturer's site, they are not considered manufacturer's tests because most purchasers require an independent testing organization, such as Underwriters Laboratories (UL), to conduct them. Some fire departments have these tests performed after the apparatus has been delivered; however, if the truck fails and requires substantial repair or reengineering, it will have to be returned to the manufacturing facility.

Pumping certification tests assure both the fire department and insurance rating companies that the apparatus will perform as expected after being placed into service. The results of these tests must be stamped into a plate that is affixed to the apparatus's pump panel.

NFPA® 1901 requires the following pump certification tests:

- Pumping test
- Pumping engine overload test
- Pressure control system test
- Priming device test
- Vacuum test
- Water-tank-to-pump flow test

The pumping engine overload test may be omitted for apparatus equipped with a fire pump having a rated capacity less than 750 gpm.

NFPA® 1901 specifies the conditions under which the pump certification tests must be conducted. The tests must be carried out from a static water supply source. The static water supply source must be at least 4 feet deep, and the strainer must be submerged at least 2 feet below the surface of the water. The surface of the water may be no more than 10 feet below the centerline of the pump intake, and 20 feet of hard intake hose should be used for drafting during testing (**Figure 9.3**).

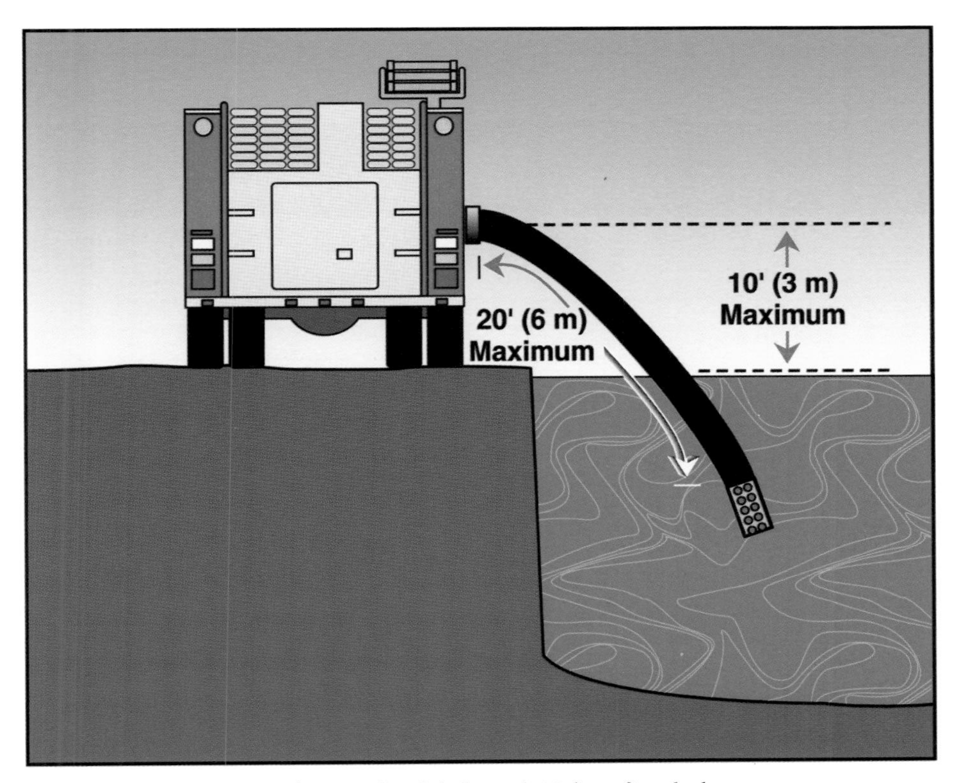

Figure 9.3 Pumps are tested at a 10-foot lift through 20 feet of intake hose.

Requirements for the environmental conditions under which the tests must be conducted include:

- Air temperature: between 0°F and 110°F
- Water temperature: between 35°F and 90°F
- Barometric pressure: at least 29 inches of mercury, corrected to sea level

Of these three variables, the one that most people consider least on a daily basis is the barometric pressure. However, barometric pressure also is the one that is likely to have the greatest effect on the tests' outcome. This is because each 1-inch drop in barometric pressure reduces a pumper's maximum possible lift by about 1 foot. Elevation affects the pump's performance in much the same manner. A pump's ability to lift water from a static water source decreases by about 1 foot for every 1,000-foot increase in elevation. Elevation also affects gasoline engines adversely, though they are rare in modern fire apparatus. On average, gasoline engines operate about 3.5 percent less efficiently for every 1,000-foot increase in elevation.

NFPA® 1901 also specifies minimum hard intake hose arrangements for tests on pumps of varying capacities (**Table 9.1**). Conducting the tests at an elevation more than 2,000 feet above sea level may necessitate increasing intake hose diameter or the number of intake hoses used in order to pump the rated capacity. Laying out a variety of discharge hoses and nozzles also will be necessary in order to complete the pumping certification tests. An overview of these requirements appears later in this chapter in the section on performing service tests.

Table 9.1
Required Number and Size of Intake Hoses

Rated Capacity (gpm)	Suction Hose Size (in.)	Number of Suction Lines	Maximum Lift (ft)
250	3	1	10
300	3	1	10
350	4	1	10
450	4	1	10
500	4	1	10
600	4	1	10
700	4	1	10
750	4½	1	10
1000	5	1	10
1250	6	1	10
1500	6	2	10
1750	6	2	8
2000	6	2	6
2250	8	3	6
2500	8	3	6
2750	8	4	6
3000	8	4	6

NFPA® 1901 requires that during testing the entire apparatus be maintained in an operating condition similar to that which would be expected during normal fireground operations. All normal electrical loads on the chassis and equipment should be imposed during pumping tests. Any engine-driven accessories must also be operational. Structural enclosures, such as floorboards, heat shields, grills, and gratings that would not be opened or removed during normal operations must remain in place.

During testing, an apparatus must be maintained in an operating condition similar to that which would be expected during normal fireground operations.

The following sections briefly overview the pump certification tests required by NFPA® 1901. More detailed information can be found by consulting that standard directly.

Pumping Test

The pumping test verifies that the pumping system will perform at its rated capacity over an extended time. For apparatus equipped with a fire pump rated at 750 gpm or greater, the entire pumping test takes three hours, as follows:

- The apparatus shall continuously pump its rated volume capacity at 150 psi net pump discharge pressure for two hours.
- The apparatus shall continuously pump 70 percent of its rated volume capacity at 200 psi net pump discharge pressure for one-half hour.
- The apparatus shall continuously pump 50 percent of its rated volume capacity at 250 psi net pump discharge pressure for one-half hour.

The pump must not be stopped at any time during the two-hour portion of the test except to clear the suction strainer. The pump may be stopped between the three various tests in order to clear the strainer, change hose and nozzle arrangements, or add fuel to the apparatus. The volume, discharge pressure, intake pressure, and engine speed (rpms) should be recorded every 15 minutes for the duration of the testing. At the end of the test these figures should be averaged to determine the rated capacity of the apparatus. These are the figures that are then stamped onto the pump-panel test plate.

The test process for apparatus equipped with pumps rated at less than 750 gpm is essentially the same except for the time intervals. The time parameters for testing these smaller pumps are:

- The apparatus shall continuously pump 100 percent of its rated volume capacity at 150 psi net pump discharge pressure for 30 minutes.
- The apparatus shall continuously pump 70 percent of its rated volume capacity at 200 psi net pump discharge pressure for 10 minutes.
- The apparatus shall continuously pump 50 percent of its rated volume capacity at 250 psi net pump discharge pressure for 10 minutes.
- The volume, discharge pressure, intake pressure, and engine speed (rpms) should be recorded every 10 minutes for the duration of the testing.

Pumping Engine Overload Test

The pumping engine overload test is performed only on apparatus equipped with fire pumps whose rated capacity is 750 gpm or greater. NFPA® 1901 requires this test to be performed immediately following the two-hour portion of the pumping test. During the overload test the apparatus is required to pump 100 percent of its rated volume capacity at a pressure of 165 psi net pump discharge pressure for at least 10 minutes. The volume, discharge pressure, intake pressure, and engine speed (rpms) should be recorded at least three times during this test period.

Pressure Control System Test

Fire pumps are equipped with pressure control systems to ensure that dangerously high pressures do not build in the pump when discharge lines are closed and the pump remains engaged. The pressure control system test ensures that pressure relief systems activate appropriately and reduce the dangers of overpressurization. The procedure for conducting the pressure control system test during pump certification testing is exactly the same as for conducting it during annual service testing. This procedure is outlined later in this chapter in the Service Testing section.

Priming Device Test

The priming device test checks the pump's ability to evacuate water from the pump, piping, and intake hose so that it may draw water from a static water supply source. Because all of the pumping certification tests are conducted from a static water supply source, most testing personnel perform this test before the pumping test, which requires that the pump be primed.

Before this test begins, all intake valves must be open, all intakes must be capped or plugged, and all discharge caps must be removed. The primer may then be operated. For pumps with a capacity of 1,250 gpm or less, the time required to achieve a prime should not exceed 30 seconds. For pumps with a capacity of 1,500 gpm or more, the time required to achieve a prime should not exceed 45 seconds. An additional 15 seconds is allowed if the pump is equipped with a 4-inch or larger auxiliary intake that has a piping volume in excess of 1 cubic foot.

During this test the pump must obtain a maximum vacuum of at least 22 inches of mercury. The only exception to that rule is if the test is being conducted at an elevation greater than 2,000 feet. In that case the maximum vacuum can be reduced by 1 inch of mercury for every 1,000 feet above 2,000 feet *(not above sea level).*

Vacuum Test

The vacuum test is performed to ensure that there are no air leaks within the pump, piping, and intake hoses when preparing to draft from a static water supply source. Air leaks are the bane of drafting operations. In nearly every case, the failure to achieve prime and draft water from a static source is due to an air leak somewhere in the system. The air leak prevents the primer from creating the partial vacuum required to draw water into the pump. The procedure for conducting the vacuum test during pump certification testing is exactly the same as for conducting it during annual service testing. This procedure is outlined later in this chapter in the Service Testing section.

Water-Tank-to-Pump Flow Test

The majority of fire incidents are handled successfully by simply utilizing the water that is carried onboard the apparatus water tank. Thus, it is crucial that the apparatus be tested to ensure that water will flow from the water tank into the pump at a rate sufficient to supply adequate fire streams. NFPA® 1901 states that the tank-to-pump piping should be sized so that pumpers with a capacity of 500 gpm or less can flow 250 gpm from their booster tank. It further states that pumpers with capacities greater than 500 gpm should flow at least 500 gpm. The procedure for conducting the tank-to-pump flow test during pump certification testing is exactly the same as for conducting it during annual service testing. This procedure is outlined later in this chapter in the Service Testing section.

Acceptance Testing

Acceptance tests are not required or outlined in NFPA® 1901 or any other standard. Acceptance tests are conducted at the discretion of the purchaser to ensure that the apparatus meets the bid specifications once it is delivered. Though these tests could be conducted at the manufacturing facility, it is advisable to conduct them within the jurisdiction where the apparatus will be placed into service. This ensures that the apparatus will perform as specified in that jurisdiction. This is particularly important when the fire department's jurisdiction varies widely from the manufacturer's location in terms of elevation, terrain, temperature, or similar factors, as these may impact the apparatus's pumping ability.

Conducting acceptance testing within the jurisdiction where the apparatus will be placed into service ensures that the apparatus will perform as specified in that jurisdiction.

The types of tests, test criteria, and personnel actually performing the tests vary widely with local jurisdiction preference and conditions. All of these details must be covered in the bid specifications and contract for constructing the apparatus so that the manufacturer knows what will be expected. Some departments' specifications simply replicate the manufacturer's and pump certification tests. Many other departments specifications require special tests based on unique requirements for the apparatus. Regardless of the tests selected, representatives of the manufacturer, purchaser, and testing agency (if any) must all be present during testing.

At a minimum, acceptance tests should include another pumping test, even if a certification test was performed at the factory. In many instances a pump certified at the factory has failed to perform as desired or rated once it was delivered. The pump test does not have to mirror the certification test but rather may follow the service test procedure outlined later in this chapter. This should be proof enough that the certification test was accurate.

If the apparatus fails to perform according to the acceptance testing requirements detailed in the bid specifications, it should be rejected. In rejecting the apparatus, the purchaser is letting the manufacturer know that the apparatus will not be accepted (and paid for) until the original specifications are met. In some cases the corrections can be made by the manufacturer's representative on site. In serious cases the apparatus may have to be returned to the factory for additional work. Regardless, the fire department should not accept the apparatus until it clearly meets the requirements of the bid specifications.

SERVICE TESTING

The need to verify operational capabilities certainly does not end once the apparatus is placed into service. That the apparatus continues to function in the required manner must be verified periodically throughout its service life. The process of performing this verification is referred to as *service testing.* Service testing of the apparatus periodically and after any major repair work is crucial for a number of reasons. First and foremost, the lives of the firefighters who use the apparatus are highly dependent on its ability to perform its job, as is the property that the fire department protects. As well, insurance rating systems used to evaluate protection levels in communities require proof of testing in order to give the fire department and community the maximum allowable insurance grade or rating.

The need to verify the operational capabilities does not end once the apparatus is placed into service.

The requirements for service testing fire department pumping apparatus are contained in NFPA® 1911, *Standard for the Inspection, Maintenance, Testing, and Retirement of In-Service Automotive Fire Apparatus.* NFPA® 1911 requires a pumper to be service tested at least once a year and anytime it has undergone extensive pump or power train repairs. These service tests are necessary to ensure that the pumper will perform as it should and to check for defects that otherwise might go unnoticed until too late.

A pumper must be service tested at least once a year and anytime it has undergone extensive pump or power train repairs.

The following section on service testing fire department pumps is adapted from IFSTA's *Fire Department Pumping Apparatus Driver/Operator Handbook.* It organizes all of the required elements and outcomes contained in NFPA® 1911 into a logical sequence in which they may be carried out. These tests include:

- Engine speed check
- Vacuum test
- Pumping test
- Pressure control test
- Gauge and flow meter test
- Tank-to-pump flow rate test

Site Considerations for Pumper Service Tests

The site considerations for conducting service testing are generally the same as those for pump certification testing. Because many fire departments do not have the fixed facilities that manufacturers typically have for conducting these tests, some elements, such as safely laying out the discharge hose line, will require extra thought and planning.

Performing service tests will necessitate laying out enough discharge hoses and nozzles to pump the rated capacity of the fire pump; the minimum size hose that may be used for this application is 2½ -inches. Larger hoses may be used if available. Make sure that whatever hose is used has been tested to ensure that it can withstand the discharge pressures during the pump test. As you would when service testing the hose, scribe a mark where the hose and couplings meet (**Figure 9.4**). While the pump testing is proceeding, regularly check the couplings to make sure that the hose is not starting to pull loose of the coupling. If your mark moves more than 3/8-inch from the coupling, stop the test and replace the hose (**Figure 9.5**). **Table 9.2** shows the minimum hose and nozzle arrangements needed to discharge sufficient water for various size pumps being tested.

Figure 9.4 Mark the hose before testing

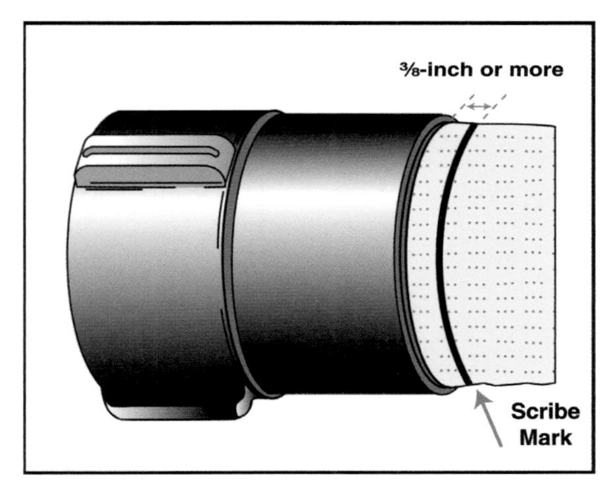

Figure 9.5 If the mark moves more than 3/8-inch from the coupling, stop the test and remove the hose from service for repair.

Table 9.2
Hose and Nozzle Layouts for Pump Tests (U.S.)

Pump Capacity in gpm	Hose and Nozzle Layout (all hose 2½-inch in diameter)
250-350	One 50-foot line with a 1⅛-inch or 1¼-inch nozzle
400-500	One 50-foot line with a 1⅜-inch or 1½-inch nozzle
600-750	Two 100-foot lines with a 1½-inch or 1¾-inch nozzle
1,000	Two or three 100-foot lines with a 2-inch nozzle
1.250	Two 100-foot lines with a 1¾-inch nozzle and one 50-foot line with a 1½-inch nozzle
1,500	Three 100-foot lines with a 2-inch nozzle and one 50-foot line with a 1½-inch nozzle
1,750	Two sets of twin 100-foot lines, each set supplying a 2-inch nozzle
2,000	Two sets of twin 100-foot lines, each set supplying a 2-inch nozzle
2,250	Two sets of three 100-foot lines, each set supplying a 2¼-inch nozzle
2,500	Two sets of three 100-foot lines, each set supplying a 2¼-inch nozzle

Correcting Net Pump Discharge Pressure for the Tests

NFPA® 1911 requires elements of the service test to be conducted at pressures of 150, 165 (overload), 200, and 250 psi net pump discharge pressure. Net pump discharge pressure is the total work done by the pump to get the water into, through, and out of the pump. When pumping from a pressurized water source, the net pump discharge pressure is *less* than the pressure shown on the discharge pressure gauge. For example, if the intake pressure coming into the pump is 30 psi and the discharge pressure is 150 psi, the *net* pump discharge pressure (the amount of pressure actually being developed or added by the pump) is 120 psi.

When at draft, the net pump discharge pressure is *more* than the pressure shown on the discharge gauge. This is due to the pressure losses from lift and friction inside the hard intake hose being used to draw water into the pump. Because most pump testing is done from a draft, when the tests are conducted the allowances for friction loss in the hard intake hose and the height of lift must be determined (**Table 9.3**). These allowances are then used to ascertain the correct pump discharge pressure during each test. The following formula is used to calculate the pressure correction:

Equation 9.1

$$\text{Pressure correction} = \frac{\text{lift (ft.)} + \text{intake hose friction loss}}{2.3}$$

Table 9.3
Friction Loss in 20 ft of Intake Hose, Including Strainer

	Intake Hose Size (inside diameter)				
	3 in.	3½ in.	4 in.	4½ in.	5 in.
Flow Rate gpm	ft water*	ft water*	ft water*	ft water*	ft water*
250	5.2 (1.2)				
175	2.6 (0.6)				
125	1.4 (0.3)				
300	7.5 (1.7)	3.5 (0.8)			
210	3.8 (0.8)	1.8 (0.4)			
150	1.9 (0.4)	0.9 (0.2)			
350		4.8 (1.1)	2.5 (0.7)		
245		2.4 (0.5)	1.2 (0.3)		
175		1.2 (0.3)	0.7 (0.1)		
450			4.1 (1.0)	2.7 (0.4)	
315			2.0 (0.5)	1.2 (0.2)	
225			1.0 (0.2)	0.6 (0.1)	
500			5.0 (1.3)	3.6 (0.8)	
350			2.5 (0.7)	1.8 (0.4)	
250			1.3 (0.4)	0.9 (0.3)	
600			7.2 (1.8)	5.3 (1.0)	3.1 (0.6)
420			3.5 (1.0)	2.5 (0.5)	1.6 (0.3)
300			1.8 (0.4)	1.3 (0.2)	0.6 (0.1)
700			9.7 (2.7)	7.3 (1.3)	4.3 (0.8)
490			4.9 (1.1)	3.5 (0.7)	2.0 (0.4)
350			2.5 (0.7)	1.6 (0.3)	0.9 (0.2)

The corrected pressure is subtracted from the desired net pump discharge pressure to determine what the pump's actual discharge pressure can be. For example, 150 psi – 8 = 142 psi.

Example 9.1

A 1,000-gpm pumper is being serviced tested. The lift is 9 feet through 20 feet of 5-inch intake hose. Find the necessary pressure correction for this test.

From **Table 9.3**, the friction loss allowance for 20 feet of 5-inch intake hose is 8.4 feet; therefore:

$$\text{Pressure correction} = \frac{\text{lift (9 ft.)} + \text{intake hose friction loss}}{2.3}$$

$$\text{Pressure correction} = \frac{9 + 8.4}{2.3} = \frac{17.4}{2.3}$$

Pressure correction = 7.56 or 8 psi

Table 9.3 (continued)
Friction Loss in 20 ft of Intake Hose, Including Strainer

	Intake Hose Size (inside diameter)					
	4 in.	4½ in.	5 in.	6 in.	Two 4½ in.	Two 5 in.
Flow Rate gpm	ft water*	ft water*	ft water*	ft water*	ft water*	ft water*
750	11.4 (2.9)	8.0 (1.6)	4.7 (0.9)	1.9 (0.4)		
525	5.5 (1.5)	8.9 (0.8)	2.3 (0.5)	0.9 (0.2)		
375	2.8 (0.7)	2.0 (0.4)	1.2 (0.2)	0.5 (0.1)		
1000		14.5 (2.8)	8.4 (1.6)	3.4 (0.6)		
700		7.0 (1.4)	4.1 (0.8)	1.7 (0.3)		
500		3.6 (0.8)	2.1(0.4)	0.9 (0.2)		
1250			13.0 (2.4)	5.2 (0.9)	5.5 (1.2)	
875			6.5 (1.2)	2.6 (0.5)	2.8 (0.7)	
625			3.3 (0.7)	1.3 (0.3)	1.4 (0.3)	
1500				7.6 (1.4)	8.0 (1.6)	4.7 (0.9)
1050				3.7 (0.7)	3.9 (0.8)	2.3 (0.5)
750				1.9 (0.4)	2.0 (0.4)	1.2 (0.2)
1750				10.4 (1.8)	11.0 (2.2)	6.5 (1.2)
1225				5.0 (0.9)	5.3 (1.1)	3.1 (0.7)
875				2.6 (0.5)	2.8 (0.6)	1.6 (0.3)
2000					14.5 (2.8)	8.4 (1.6)
1400					7.0 (1.4)	4.1 (0.8)
1000					3.6 (0.8)	2.1 (0.4)
2250						10.8 (2.2)
1575						5.3 (1.1)
1125						2.8 (0.5)
2500						13.0 (2.4)
1750						6.5 (1.2)
1250						3.3 (0.7)

Table 9.3 (continued)
Friction Loss in 20 ft of Intake Hose, Including Strainer

	Intake Hose Size (inside diameter)		
	Two 6 in.	Three 6 in.	8 in.
Flow Rate gpm	ft water	ft water*	ft water
1500	1.9 (0.4)		
1050	0.9 (0.3)		
750	0.5 (0.1)		
1750	2.6 (0.5)		
1225	1.2 (0.3)		
875	0.7 (0.2)		
2000	3.4 (0.6)		
1400	1.7 (0.3)		
1000	0.9 (0.2)		
2250	4.3 (0.8)		
1575	2.2 (0.4)		
1125	1.1 (0.2)		
2500	5.2 (0.9)		
1750	2.6 (0.5)		
1250	1.3 (0.3)		
3000	7.6 (1.4)	3.4 (0.6)	8.5 (1.6)
2100	3.7 (0.7)	1.7 (0.3)	4 (0.8)
1500	1.9 (0.4)	0.9 (0.2)	1.9 (0.4)

All equipment used to perform service tests must be in good shape, properly calibrated, and regularly tested.

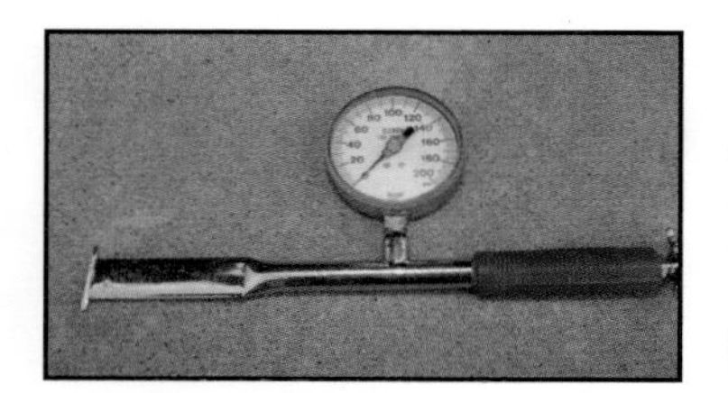

Figure 9.6 A pitot tube and gauge are used to measure the velocity pressure of the stream of water. This figure may then be used to calculate the volume of flow.

Equipment Needed for Service Tests

A variety of both special and standard fire fighting equipment will be required in order to perform all the elements of the service test. Any equipment used to perform service tests must be in good shape, properly calibrated, and tested regularly, or else accurate test results cannot be expected. The following equipment is needed to perform the service test for pumpers:

- A gauge to check the pump intake pressure. This gauge should have a range of 30 inches of mercury to zero for a vacuum gauge or 30 inches of mercury vacuum to 150 psi for compound gauges.
- A gauge to check the pump discharge. This gauge should be capable of a range of at least from 0 to 400 psi.
- A pitot tube with knife edge and air chamber rated at least from 0 psi to 160 psi (**Figure 9.6**). (**NOTE:** This is not needed if a flowmeter is used.)
- Solid stream nozzles of correct sizes to match the volumes pumped for the different tests. (**NOTE:** If a flowmeter is used, fog nozzles may be used provided they are rated for the necessary flows.)
- Rope, chain, or a test stand for securing test nozzle(s) (**Figure 9.7**).
- Revolution counter or hand tachometer.
- Fire department or insurance agency forms.

Several other pieces of equipment, although not required, may be useful during service testing and are recommended by IFSTA. These include:

Figure 9.7 This nozzle test stand is held in place. by parking a vehicle with one of its tires between the two boards at the base of the stand.

- Two 6-foot lengths of 1/4-inch, 300 psi hose with screw fittings. (**NOTE**: These are used to connect the test gauges to the test gauge fittings at the pump operator's panel.)
- A clamp to hold the pitot tube to the test nozzle.
- A test stand for gauges.
- A thermometer to verify water temperatures.
- A stopwatch or wristwatch that shows time in seconds.

A *flowmeter*, which reads the flow directly in gallons per minute, may be used instead of a pitot gauge arrangement to determine the flow from the nozzles (see Chapter 8). Flowmeters allow much more flexibility and help to complete the tests more quickly. When a flowmeter is used, all the pump tests can be run without shutting down the pump, without changing nozzles, and without having to convert pitot pressure readings to gpm. If flowmeters are used, they must be calibrated to the manufacturer's specifications.

Safety Precautions During Service Tests

The following safety precautions should always be followed when performing fire pump service testing:

- All personnel should wear protective headgear and hearing protection (if exposed to noise in excess of 90 dB).
- Prevent water hammer by opening and closing all valves and nozzles slowly.
- Never stand over or straddle hose.
- Manipulate the engine throttle slowly. Prevent sudden pressure changes that can damage equipment and injure personnel.
- Tie down test nozzles and devices securely.
- Cover all open manholes at the test pit.
- Be aware of the location of all personnel in the test area in relation to hose lines.

Engine Speed Check

The first test that should be performed is the engine speed check. The engine speed may be checked by the tachometer on the engine and/or a properly calibrated handheld tachometer or revolution counter. The engine speed should be checked under no-load conditions to ensure that the engine is still running at the governed speed for which it was rated when the apparatus was new. If it is not running at the correct speed, no further testing should be started until a qualified mechanic corrects the situation.

Vacuum Test

The *vacuum test* is performed to check the priming device, pump, and hard intake hose for air leaks. This test is performed first because it will be difficult to proceed if the apparatus cannot hold an appropriate vacuum.

Step 1: Make sure that the pump is completely drained of all water.

Step 2: Inspect all gaskets of intake hose and caps.

Step 3: Look for foreign matter in the intake hose. Clean the hose if necessary.

Step 4: Connect 20 feet of the correct intake hose to the pump intake connection (check original test records for correct hose diameter).

Step 5: Cap the free end of the intake hose (**Figure 9.8**).

Figure 9.8 Place a cap on the free end of the intake hose.

Step 6: Make sure that all intake valves are open and the intake connections are tightly capped. Make sure as well that all discharge valves are closed and their caps are removed.

Step 7: Connect an accurate vacuum gauge (or mercury manometer) to the threaded test-gauge connection on the intake side of the pump (**Figure 9.9**).

Figure 9.9 The test gauges are attached to the test connections on the pump panel.

Caution:
If the gauge is mistakenly connected to the discharge side, it will be irreparably damaged.

Step 8: Check the oil level in priming pump reservoir and replenish if necessary.

Step 9: Make the pump packing glands accessible for checking (raise floorboards or open compartment doors).

Step 10: Run the priming device until the test gauge shows 22 inches of mercury developed. (**NOTE:** Reduce the amount of mercury developed 1 inch for each 1,000 feet of elevation.)

Step 11: Compare readings of the apparatus intake gauge and test gauge. Record any difference.

Step 12: Shut off the engine. Listen for air leaks. No more than 10 inches of vacuum should be lost in 5 minutes. Excessive leaks will affect the results of subsequent tests and should be located and corrected before performing the rest of the tests.

If the apparatus cannot reach 22 inches (560 mm) of mercury, it should be removed from service and repaired as soon as possible.

Following the completion of the vacuum test, the following preparations should be made for conducting the remainder of the tests:

Step 1: Open a discharge valve or drain to allow the pressure in the pump to equalize.

Step 2: Replace the cap at the end of the intake hose with the intake strainer.

Step 3: Use standard departmental procedure to tie off the intake hose in preparation for drafting; then lower the hose into the water. The strainer should be at least 2 feet below the surface. At least 2 feet of water also should surround the sides and bottom of the strainer.

Step 4: Connect the discharge pressure test gauge to the pressure side of the pump at the test fitting on the operator's panel.

Step 5: Connect enough hose lines to carry the capacity of the pump to the test nozzle. The test nozzle must be the correct size to handle the capacity of the pump (**Table 9.4**).

Step 6: Make sure that the nozzle is secured so that it cannot come loose and injure personnel. Never hold the test nozzle by hand during a test.

Step 7: Connect the pitot gauge and test stand gauges. It is recommended that the pitot gauge be clamped in position at the nozzle.

Table 9.4
Flow in gpm from Various Sized Solid Stream Nozzles

Nozzle Pressure in psi	Nozzle Diameter in Inches									
	1	1⅛	1¼	1⅜	1½	1⅝	1¾	1⅞	2	2¼
50	209	265	326	396	472	554	643	740	841	1065
55	219	277	342	415	495	581	674	765	881	1118
60	229	290	357	434	517	607	704	810	920	1168
65	239	301	372	451	537	631	732	843	958	1215
70	246	313	386	469	558	655	761	875	994	1260
75	256	324	399	485	578	678	787	905	1030	1305
80	264	335	413	500	596	700	813	935	1063	1347

Pumping Test

The next test in the service test sequence is the pumping test. The pumping test checks the overall condition of the engine and the pump. This test is similar to the pump certification test. The primary difference is the duration of each of the various steps in the testing process.

Obtaining the correct engine and nozzle pressures for the capacity test requires making adjustments and readjustments. All changes must be made slowly to prevent damage to the pump and hose, to avoid possible injury to personnel, and to allow time for the resulting pressure changes to register on the test gauges. The procedure for the pumping test is as follows:

Step 1: Gradually speed up the pump until the net pump discharge pressure is 150 psi, adjusted for intake hose friction loss and elevation. If the pump is a two-stage pump, the transfer valve must be in the *volume* (*parallel*) position. This portion of the test, known as the capacity test, measures the apparatus's ability to pump its rated volume capacity at 150 psi.

Step 2: Check the flow at the nozzle, using either a pitot gauge or a flowmeter. If the flow is too great, close a valve further. Readjust (lessen) engine speed to correct discharge pressure. If the flow is too low, open a valve further. Readjust (increase) engine speed to correct discharge pressure. (**NOTE:** All these adjustments must be made with the engine speed not exceeding 80 percent of its peak.)

Step 3: When both the pump discharge pressure and the volume flowing are satisfactory, the test officially begins. Record the following readings at the beginning of the test and at 5-minute intervals thereafter until the 20 minutes for the test are over. (**NOTE:** Fluctuations in pressure necessitate more frequent readings.)

- Pump discharge pressure
- Nozzle pressure (or flow)
- Engine tachometer
- Rpm using portable rpm counter
- Engine coolant temperature (optional)
- Oil pressure (optional)
- Automatic transmission fluid temperature (optional)

Step 4: Once the 20-minute capacity test has been completed, the net pump discharge pressure should be increased to 200 psi. At this point, the pump should be delivering at least 70 percent of its rated volume capacity. According to NFPA® 1911, two-stage pump transfer valves may be in either the *volume* (*parallel*) or *pressure* (*series*) position for this portion of the test. It is usually best to use the same position that was used during the certification test (to determine which position the transfer valve was in during the certification test, consult the pump certification information either in the paperwork or on the pump-panel data plate). The pump should be allowed to run at this setting for 10 minutes, with the 200 psi/70 percent readings listed above verified at 5 minutes and again at the conclusion of the 10-minute period.

Step 5: Once the 200 psi test has been completed, the net pump discharge pressure should be increased to 250 psi. At this point, the pump should be delivering at least 50 percent of its rated volume capacity. Two-stage pump transfer valves must be in the *pressure* (*series*) position for this portion of the test. The pump should be allowed to run at this setting for 10 minutes, with the 250 psi/50 percent readings listed above verified at 5 minutes and again at the conclusion of the 10-minute period.

Remember the following additional guidelines while performing the pumping test:

- Hold the pitot gauge with the blade opening in the center of the stream. The distance from the blade's tip to the end of the nozzle should be about one-half the diameter of the nozzle (**Figure 9.10**). If the pitot is too close to the nozzle, the reading will be erroneously increased.
- Keep the engine temperature within the proper range.
- Check the oil pressure to be sure that proper engine lubrication is maintained.
- Record any unusual vibration of pump or engine.
- Record any other defect in the performance of pump or engine. Correct minor defects immediately if possible.

Figure 9.10 The pitot tube should be in the center of the water stream.

Pressure Control Test

The pressure control device(s) should be tested to make sure that they maintain a safe level of pressure on the pump when valves are closed at a variety of discharge pressures. All pressure control devices should be operated according to manufacturer's instructions during this testing. The pressure control test follows a three-part sequence:

Part I

Step 1: Set the fire pump so that it is discharging its rated volume capacity at a net pump discharge pressure of 150 psi.

Step 2: Set the pressure control device to maintain the discharge pressure at 150 psi.

Step 3: Once the device is set, close each of the flowing valves, one at a time. Close each valve in no less than 3 seconds and no more than 10 seconds. (**NOTE:** Closing a valve in less than 3 seconds can damage the pump, piping, or pressure control device. Closing it in longer than 10 seconds will not test the pressure control device realistically.)

Step 4: Observe the pump discharge pressure gauge. It should rise no more than 30 psi when all gauges are closed.

Part II

Step 1: Set the fire pump so that it is discharging its rated capacity at a net pump discharge pressure of 150 psi.

Step 2: Reduce the pumping engine throttle until the net pump discharge pressure drops to 90 psi with no change to the discharge valve or nozzle setting(s).

Step 3: Set the pressure control device to maintain the discharge pressure at 90 psi.

Step 4: Once the device is set, close each of the flowing valves, one at a time. Close each valve in no less than 3 seconds and no more than 10 seconds. (**NOTE:** Closing a valve in less than 3 seconds can damage the pump, piping, or pressure control device. Closing it in longer than 10 seconds will not test the pressure control device realistically.)

Step 5: Observe the pump discharge pressure gauge. It should rise no more than 30 psi when all gauges are closed.

Part III

Step 1: Set the fire pump so that it is discharging 50 percent of its rated capacity at a net pump discharge pressure of 250 psi.

Step 2: Set the pressure control device to maintain the discharge pressure at 250 psi.

Step 3: Once the device is set, close each of the flowing valves, one at a time. Close each valve in no less than 3 seconds and no more than 10 seconds. (**NOTE:** Closing a valve in less than 3 seconds can damage the pump, piping, or pressure control device. Closing it in longer than 10 seconds will not test the pressure control device realistically.)

Step 4: Observe the pump discharge pressure gauge. It should rise no more than 30 psi when all gauges are closed.

Discharge Pressure Gauge and Flowmeter Operational Tests

The next portion of the service test procedure involves checking the discharge pressure gauges and flowmeter (if the pump is equipped with one) to make sure that the driver/operator is getting accurate discharge information when the pump is operating. If these devices are not working properly, the driver/operator could supply dangerously insufficient or excessive amounts of water to firefighters operating hose streams.

Testing the apparatus pressure discharge gauges is a relatively quick and simple process. Each of the discharges on the apparatus must be capped to perform this test properly. This means that preconnected hose lines must be disconnected and caps or closed nozzles screwed onto their discharges (**Figure 9.11**). Once all the discharges are capped, each discharge valve should be opened slightly. The throttle should then be increased until the test discharge pressure gauge reads 150 psi. A quick visual inspection of the master discharge pressure gauge and each individual line discharge gauge should reveal all to be at 150 psi as well. The gauges should then be checked at 200 psi and 250 psi in the same manner. Any gauges that are off by more than 10 psi should be recalibrated, repaired, or replaced.

Testing discharges equipped with flowmeters is not as simple a process as testing pressure gauges. To test the flowmeter, a hose line equipped with a solid stream nozzle must be connected to each discharge being tested (the discharges do not have to be tested all at once). Refer to Table 9.2, p. 195, to determine appropriate hose and nozzle arrangements for this test. **Table 9.5** shows the minimum flow rates that must be achieved for each of the listed pump discharge piping sizes in order for the test to be valid. The actual flow will be calculated using pitot tube readings taken from the discharge of the solid stream nozzle. (To compute flows from a solid stream nozzle, see Chapter 11.) The flow measured from the nozzle and the reading on the flowmeter should not differ by more than 10 percent. If the difference is more than 10 percent, the flowmeter must be recalibrated, replaced, or repaired.

Figure 9.11 The discharge should be capped in some manner.

Tank-to-Pump Flow Test

The tank-to-pump flow test must be conducted on all apparatus that are equipped with a water tank, regardless of size. The purpose of this test is to ensure that the piping between the water tank and the pump is sufficient to supply the minimum amount of water specified by NFPA® 1901 and the design of the manufacturer. NFPA® 1901 states that the tank-to-pump piping should be sized so that pumpers with a capacity of 500 gpm or less can flow 250 gpm from their booster tank and pumpers with capacities greater than 500 gpm can flow at least 500 gpm. Some departments may specify greater flow rates when they order apparatus from a manufacturer. If that was the case with the apparatus being tested, this test should be performed for the higher figure. Use the following test procedure to check the operation of the tank-to-pump line:

Table 9.5
Minimum Flow Measuring Points for Flowmeters

Pipe Size in Inches	Test Flow in gpm
1½	128
2	180
2½	300
3	700
4	1,000

Step 1: Make sure that the water tank is filled until it is overflowing.

Step 2: Close the tank fill line, bypass cooling line, and all pump intakes.

Step 3: Attach sufficient hose lines and nozzles to flow the desired discharge rate.

Step 4: With the pump in gear, open the discharge(s) to which the hose(s) is (are) attached, and begin flowing water.

Step 5: Increase the engine throttle until the maximum consistent pressure is obtained on the discharge gauge.

Step 6: Close the discharge valve without changing the throttle setting and refill the tank (usually through the top fill opening or a direct tank fill line). The bypass valve may be temporarily opened during this operation to prevent pump overheating.

All calculations and figures determined during apparatus tests should be recorded so that they can be filed according to departmental record-keeping procedures.

Step 7: Reopen the discharge valve and check the flow through the nozzle using a pitot tube or flowmeter. Adjust the engine throttle if the pressure needs to be brought back to the amount determined in Step 5.

Step 8: Compare the flow rate being measured to the NFPA® minimum or the manufacturer's designed rate. If the flow rate is less than this, a problem exists in the tank-to-pump line. Remember that the minimum flow rate should be discharged continuously until the tank has been at least 80 percent emptied.

Reviewing the Test Results

At no time during the service testing procedures should the pumping system or pumping engine show signs of overheating, power loss, or any other mechanical problems. All calculations and figures determined during the tests should be recorded so that they can be filed according to departmental record-keeping procedures.

If the fire pump underwent certification testing and now tests to less than 90 percent of its capabilities when it was new, two options are available:

- Take the pump out of service and restore it to its designed capabilities (obviously, it will be necessary to test it again after the repairs). Most jurisdictions prefer this option.
- Give the pump a lower rating based on the test results.

Troubleshooting During Service Testing

An almost endless array of pitfalls may occur during service testing. Personnel performing the tests should be familiar with these potential problems and their causes and/or appropriate solutions. **Table 9.6** lists potential problems and solutions that may be used when troubleshooting during service testing operations. Every effort should be made to correct any problem that is found. The portion of the test that was unsuccessful should be redone to ensure that the problem has been corrected.

Table 9.6
Service Test Troubleshooting

Problem	Cause/Solution
Turbulence or whirlpooling in the static water supply source	1. Insufficient depth of water (must be at least 4 feet deep) or strainer too close to the surface (must be submerged at least 2 feet). 2. Insufficient baffles in drafting pit between the discharge input and the suction output.
Pump cavitation or erratic gauge readings	1. Insufficient volume of water in the drafting pit. There should be at least 10 gallons of water for each gpm of pumping rating. For example, if testing a 1,500 gpm pump, the pit should contain at least 15,000 gallons of water. 2. Debris sucked against the strainer; clear debris.
Low maximum capacity readings	Check for high spots or humps in the suction hose that are higher than the level of the intake on the pump.
Inability to hold prime	Check for air leaks (lose couplings, open drains, open valves, leaky intake relief valve discharges, etc.
Pump panel tachometer shows significant lower rpms than dashboard tachometer	Check for automatic transmission lock-up problems, clutch slippage, and make sure the transmission is in the proper gear for pumping.
Inability to perform the 250 psi at 50% volume test	Check for worn pump wear rings.
Lower than normal pressure and volume output	Check the relief valve to make sure it is not open.
Any reduction from proper pressure and volume capacity to less than that during the course of the test	Not likely to be a pump problem. Check for air leaks debris in the strainer, intake hose liner collapse, or engine problems.

SUMMARY

The importance of apparatus testing, both before and after entering service, cannot be overstated. With the lives of citizens and firefighters alike resting on the proper performance of the fire apparatus, every check must be made to ensure that the apparatus is mechanically reliable and capable of performing as it was designed. Preservice tests ensure that the apparatus was built according to the specifications developed by the fire department. They also verify that the apparatus is ready to be placed into service. After the apparatus is placed into service, periodic service testing is conducted to make sure that the apparatus continues to function as it was designed. Fire protection professionals have a working knowledge of all of these test procedures as they may be required to perform them or observe them to ensure they are carried out correctly.

REVIEW QUESTIONS

1. List the three general parts of the preservice test.

   1. ______
   2. ______
   3. ______

2. Requirements for preservice tests are contained in which NFPA® standard(s)?

   ______

   ______

3. Who usually performs preservice tests?

   ______

4. What are the two subparts of the manufacturer's test?

   1. ______
   2. ______

5. List the four minimum criteria that apparatus must meet during the road test according to NFPA® 1901.

   1. ______
   2. ______
   3. ______
   4. ______

6. State the required pressure and time length for the preservice hydrostatic pump test.

   ______

7. What is the minimum depth of water for a static source used for pump testing and what is the maximum distance the surface of the water may be below the level of the pump?

   ______

   ______

8. What is the total duration of pumping time for the preservice pump certification test?

   ______

   ______

9. List the percentages of volume capacity that a fire pump must pump at 150 psi, 200 psi, and 250 psi.

   150 psi: ______

   200 psi: ______

   250 psi: ______

10. During the pumping overload test, the pump must supply 100 percent of its rated volume capacity at ______ psi for 10 minutes.

11. List the NFPA® 1901 requirements for tank-to-pump flow rates.

____

____

12. Which NFPA® standard contains the requirements for service testing fire department pumpers?

____

____

13. Assume that a 1,500 gpm pumper is being service tested. The lift is 8 feet through 20 feet of 6-inch intake hose. Find the necessary pressure correction for this test.

____

____

14. What is the total duration of pumping time for the service pumping test?

____

____

15. What is the maximum allowable amount that any discharge pressure gauge may vary from the master pressure gauge?

____

____

Answers found on page 430.

Part 4

Photo courtesty Ron Jeffers

Fire Streams

Putting the "wet stuff on the red stuff" remains the basic mission of the fire service. Applying water (the wet stuff) in an appropriate manner and at an adequate volume is crucial from both a tactical standpoint and a safety standpoint. Sending firefighters into a working fire (the red stuff) with an improper nozzle or insufficient amount of water reduces the chances of successful fire extinguishment and places firefighters in danger as they may not have enough water to protect themselves if something goes wrong.

The previous section of this text explored fire apparatus and their pumps. In this section we move to the other end of the hoseline. While the pump operator is unlikely to spend much, if any, time at the working (nozzle) end of the hoseline, he or she must understand the various fire streams that may be needed at an emergency scene and the types of nozzles used to produce them. The types of streams needed will dictate the types of nozzles used. In turn, the type of nozzle used will impact the hydraulic calculations required of the pump operator as each nozzle has a specific optimum operating pressure.

Chapter 10: This chapter explores the three basic types of fire streams available to firefighters: solid, fog (or spray) and broken. It covers the advantages and disadvantages of each.

Chapter 11: This chapter gives detailed information on the various types of nozzles that are needed to produce these three fire streams including their impact on hydraulic requirements for the pump operator.

By understanding the basic principles associated with fire streams and the nozzles that produce them, the pump operator will be able to meet the hydraulic requirements of the incident more effectively when operating the pump.

FESHE Course Objectives

The information in this chapter is intended to meet some of the objectives outlined for the Fire Protection Hydraulics and Water Supply course by the United States Fire Administration's National Model Core Curriculum as developed by the Fire and Emergency Services Higher Education (FESHE) initiative. The information in this chapter also allows the student to meet the requirements of NFPA® 1001, *Standard for Fire Fighter Professional Qualifications* (2008 edition) listed below.

FESHE Course Objectives

1. Explain the characteristics of solid streams, including their flow and reach characteristics.
2. Explain the characteristics of fog streams, including their volume, stream velocity, reach, and water particle size characteristics.
3. Explain the characteristics of broken streams, including their flow and water particle size characteristics and their various special uses.

NFPA® 1001 Requirements

5.3.10 (A) Requisite Knowledge. Principles of fire streams.

6.3.2 (A) Requisite Knowledge. Selection of the nozzle and hose for fire attack given different fire situations.

Chapter 10

Types of Fire Streams

A fire stream can be defined as a jet of water or other extinguishing agent from the time it leaves the nozzle until it reaches the point of its desired application. Technically, a fire stream may be comprised of water, a water-foam solution, a dry powder or chemical agent, a wet chemical agent, or a gaseous extinguishing agent. For the purposes of this text, the term *fire stream* refers to a stream of plain water used for fire fighting purposes.

Selecting the appropriate fire stream for the situation is extremely critical from both a tactical standpoint and a safety standpoint. Tactically, if the wrong stream is selected, the incident may not be contained or controlled in an optimum time or manner. From a safety standpoint, selection of the wrong type of stream can place firefighters in a hazardous location without providing the ability to adequately protect them, much less to handle the incident.

A number of factors affect the overall condition or performance of a fire stream. At the point of discharge, the nozzle pressure, design, adjustment, and the condition of the nozzle influence the condition of a fire stream. As the stream passes through space, it is affected further by velocity, gravity, wind, and friction with the air. This chapter focuses on several aspects of water-based fire streams, including the elements required for their production and the different types of streams. The nozzles used to produce fire streams are covered in Chapter 11.

FIRE STREAM PRODUCTION

To produce an appropriate and effective fire stream requires four basic elements:

1. A water supply source. This may be from the apparatus tank, a pressurized source (such as a hydrant), or a static source (such as a pond).
2. A means to impart sufficient pressure on the water. In nearly all municipal fire fighting operations, this is provided by a fire apparatus equipped with a fire pump.
3. A means to transport the water from the source to the desired point of application. The fire service uses a variety of equipment in order to get water from the source to the fire, including fire hose, appliances, adapters, nozzles, and so on.
4. Personnel trained in using the first three elements effectively. The equipment cannot deploy and operate itself. Trained, competent firefighters and driver/operators are required to assemble and operate the equipment appropriately.

The following section expands on each of these elements.

Water Supply

Water for producing or developing fire streams may be taken from an apparatus water tank, municipal water supply distribution system, or a static water supply source. As discussed in Chapter 7, NFPA® 1901 and 1906 both require fire apparatus equipped with fire pumps to carry an onboard supply of water. The amount of water carried on the apparatus is a matter of preference for each individual fire department. It often is based on the availability of a municipal water supply system. Jurisdictions that lack total coverage by the water supply system typically carry greater amounts of water on their apparatus.

Static supply sources include ponds, lakes, streams, swimming pools, and portable tanks. To draw water from a static supply source, the pumper must be able to draft the water through hard suction hose into the pump, as discussed in Chapter 8.

The most commonly used exterior water supply is the municipal water supply distribution system. A properly designed, constructed, and maintained municipal water supply distribution system should be able to supply all the water needed for both normal domestic and fire requirements plus a reserve for heavy demands. The water supply connection, usually a fire hydrant, is an important component in producing fire streams. For more detailed information on municipal water supply distribution systems, refer to Chapter 4.

Fire Apparatus

The fire department pumper or other apparatus equipped with a fire pump is another essential component in the production of fire streams. The fire pump takes water from the supply source and increases its flow pressure to the level required to produce an effective fire stream. Many factors limit a pumper's ability to perform to its maximum potential, but under normal conditions a pumper should be capable of delivering its rated capacity. A good supply source may allow the pumper to achieve flows that exceed its rated capacity. For more information on the operation of the types of fire apparatus equipped with fire pumps, refer to Chapter 6. Information on fire pumps is detailed in Chapter 8.

Fire Equipment

A variety of fire equipment is required to deliver the water from the supply source to the pump and then from the pump to the point of application. This equipment falls into four basic categories: fire hose, nozzles, fire hose appliances, and related hardware. Nozzles are discussed in detail in the next chapter.

Fire Hose

One of the most important articles of equipment in fire stream production is the fire hose. The fire hose serves as the conduit between the water supply source and the pump as well as between the pump and the nozzle. Fire streams are influenced by fire hose and its resistance to the flow of water. This resistance to flow is commonly called friction loss. A variety of factors influence the amount of resistance that a hose will offer water moving through it.

The first factor is the diameter of the hose. The smaller the diameter of the hose, the greater the ratio of the hose surface to the water volume. That means that a higher percentage of the water will be in contact with the hose liner as it moves down the hose. This translates into higher friction loss. The higher the friction loss is, the more pressure the pump will have to supply in order to produce an adequate discharge pressure at the end of the hose equipped with the nozzle. In simple terms, for any given flow, the smaller the diameter of the hose, the more friction loss will occur. Increasing the diameter of the hose will reduce the friction loss and allow the pump operate at lower pressures while still producing adequate fire streams.

The smaller the hose diameter, the greater the friction loss.

The length of the hose also is important. Friction loss within the hose is additive. Thus, the longer the hose, the greater the friction that the pump will need to overcome to produce an effective fire stream. In many cases nothing can be done about this on the fire scene. The hose needs to be long enough to put the nozzle and fire stream in position for effective application. However, as we will see in Chapters 12 and 13, hose length is important in determining the pressures necessary to produce an effective fire stream.

The longer the hose, the greater the friction loss.

The age and condition of the hose play a significant factor in the amount of friction loss within a fire hose. In general, older fire hose tends to create more friction loss than new hose. There are two basic reasons for this. First, the manufacturing technology and materials used to make fire hose have improved significantly over the years. Older hose, particularly hose manufactured before the 1980s, tends to contain rubber liners that have a rougher surface (or higher coefficient of friction) than modern hose. This rougher surface translates into greater friction loss. Newer hose tends to be smoother and impart less friction on the water moving across it. The second basic reason is that the more a hose is used, the more likely it is to develop creases, cracks, and other imperfections. All of these impede the water traveling within the hose and tend to increase pressure loss. Proper maintenance and care of the hose can minimize these imperfections; however, they cannot be completely eliminated.

The improvements in modern hose manufacturing technology have resulted not only in smoother liner materials but also in hose that tends to expand in diameter when charged with water under pressure. For example, one current manufacturer markets a hose that when uncharged has an internal diameter of 1¾-inches. However, when the hose is deployed and charged, the internal diameter may expand to over 2 inches. This means that this hose will have greater flow volume ability and less friction loss than a 1¾-inch hose constructed using older technology.

The impact of hose diameter and length on pressure loss and fire fighting capabilities is discussed in more detail in Chapters 12 and 13. For more information on the construction, care, and use of all types of fire hose, see the IFSTA manuals *Essentials of Fire Fighting* and *Hose Practices*.

Fire Hose Appliances

Fire hose appliances are devices that combine or separate multiple hoselines. The four principle types of fire hose appliances are:

- Wyes
- Siameses
- Water thiefs
- Manifolds

Wyes divide one hoseline into two separate hoselines of equal size **(Figure 10.1)**. The two new hoselines may be of the same size or smaller than the original hoseline. The wye's two discharges may or may not be equipped with ball-type valves that individually control the two new hoselines. Those that are equipped with valves are referred to as *gated wyes*. Larger wyes are designed to break one large-diameter supply hose into two smaller, medium-diameter supply hoses that may send water to different destinations. Smaller wyes, sometimes referred to as *leader line wyes*, are used to break one medium-diameter supply hose into two 1 3/4-inch or smaller attack hoses.

Figure 10.1 A gated wye.

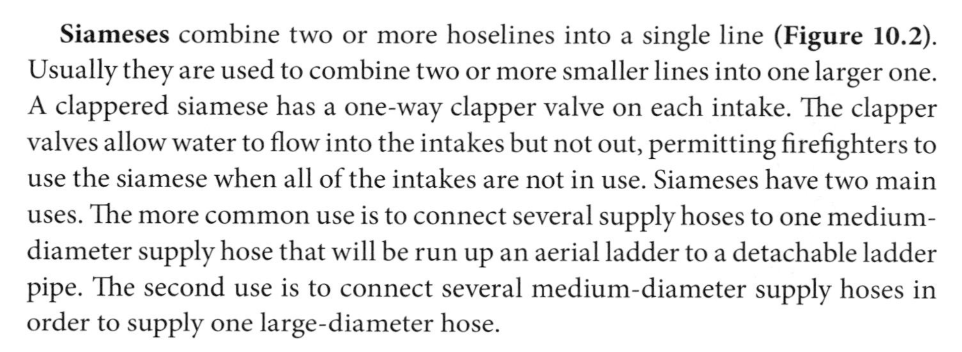

Siameses combine two or more hoselines into a single line **(Figure 10.2)**. Usually they are used to combine two or more smaller lines into one larger one. A clappered siamese has a one-way clapper valve on each intake. The clapper valves allow water to flow into the intakes but not out, permitting firefighters to use the siamese when all of the intakes are not in use. Siameses have two main uses. The more common use is to connect several supply hoses to one medium-diameter supply hose that will be run up an aerial ladder to a detachable ladder pipe. The second use is to connect several medium-diameter supply hoses in order to supply one large-diameter hose.

A **water thief** is similar to a wye. It enables firefighters to take three attack lines off the original attack or supply line **(Figure 10.3)**. The water thief has one intake and three gated discharges. One discharge is the same size as the intake, and the other two are smaller. The most common water thief has one 2½-inch intake with one 2½-inch and two 1½-inch discharges.

Manifolds are used primarily with large-diameter (4-inch or larger) hose **(Figure 10.4)**. They generally are placed at the end of a large-diameter supply line and enable firefighters to divide the line into smaller lines that can be sent to several different locations. Manifolds also can be used to join several small lines into a one large line.

Figure 10.2 A siamese.

Figure 10.3 A water thief.

Figure 10.4 An LDH manifold.

The important thing to remember about all of these appliances, from a hydraulics standpoint, is that they all will create additional friction loss within the hose lay. This additional friction loss will affect the fire stream and must be compensated for in calculating the required pump discharge pressure. Information on how to accommodate hydraulically for these appliances is covered in Chapters 12 and 13.

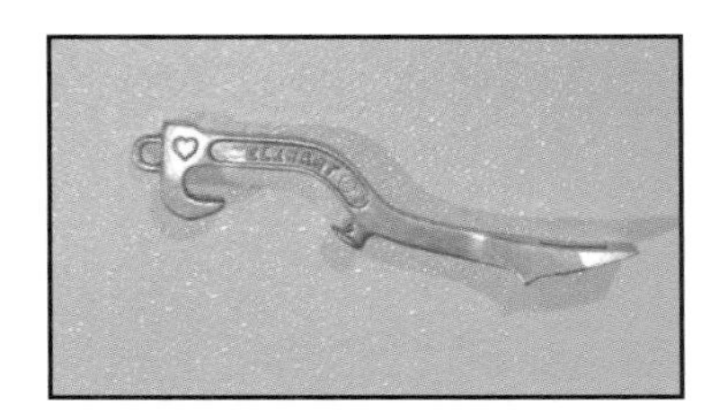

Figure 10.5 Spanner wrenches commonly are used to make and break coupling connections. Several kinds are available for specific sizes and types of couplings.

Related Hardware

A variety of other hardware is related to but does not directly affect the quality of a fire stream. These include tools used to make hose and nozzle connections, to handle hoselines, and to protect hoselines from damage. Spanner wrenches commonly are used to make and break coupling connections. Several kinds are available for specific sizes and types of couplings (**Figure 10.5**). Hose straps and rope hose tools are designed to help firefighters move hoselines from one place to another (**Figure 10.6**). They are particularly useful for charged 2½-inch and 3-inch handlines. When firefighters must lay hoselines across roadways, hose bridges will allow vehicles to pass over the hoselines without damaging them (**Figure 10.7**).

Figure 10.6 Hose straps and rope hose tools are designed to help firefighters move hoselines from one place to another.

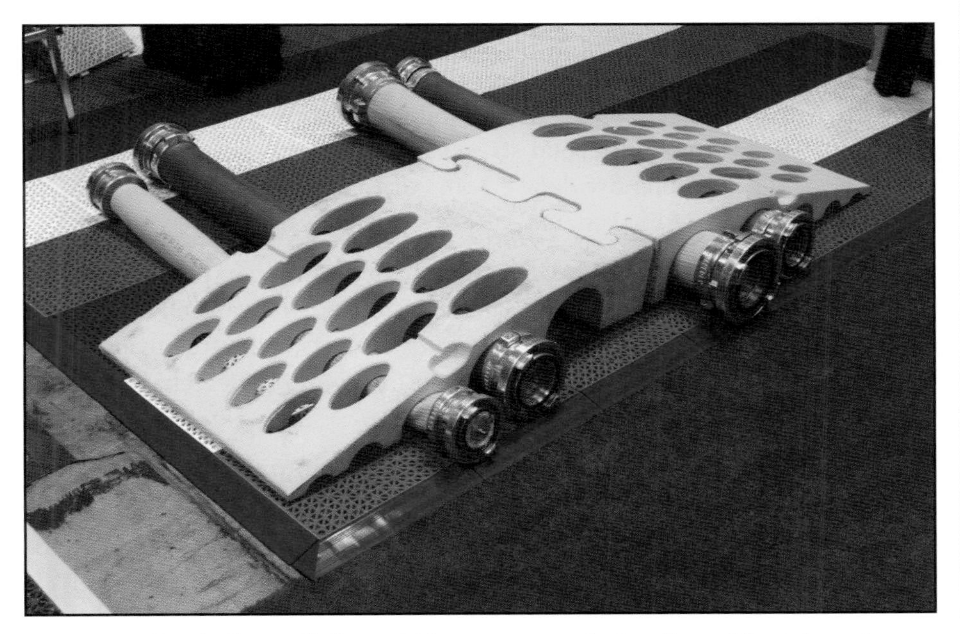

Figure 10.7 Hose bridges will allow vehicles to pass over the hoselines without damaging them.

Human Ability

The best water supply, fire apparatus, and fire equipment will not produce effective fire streams if personnel are poorly trained or make errors in selecting and using equipment. Some of the most common mistakes are:

- Using supply hose or intake hose that is too small.
- Overtaxing a pumper with too many hoselines for the supply available.
- Using large capacity nozzles that demand a total volume greater than the rated capacity of the pump.

Personnel must be properly trained and educated to produce effective fire streams.

- Developing excessive nozzle pressure, which may damage the hoseline or equipment. Excessive nozzle pressure also can endanger firefighters if they lose control of the hoseline.

Many other errors are possible, of course. The important point is that proper training and education of personnel should prevent most of these errors.

SOLID STREAMS

The oldest type of fire stream, dating back to the very origins of the fire service, is the solid stream. Of the three basic types of fire streams, the solid stream is the easiest to produce and requires the simplest type of nozzle. This section examines the characteristics of solid streams, their advantages and disadvantages, and other pertinent information regarding their use.

Figure 10.8 A good solid stream appears to shoot nine-tenths of the whole volume of water inside a circle 15 inches in diameter and three-quarters of it inside a 10-inch circle, as nearly as can be judged by the eye.

Characteristics of Solid Streams

A solid stream is a fire stream that is produced from a fixed-orifice, smoothbore nozzle. The solid stream is generally preferred in situations where a powerful, long-range, high-volume stream is desired. Many jurisdictions use the solid stream primarily for master streams in defensive fire fighting operations. However, some jurisdictions do use solid streams on handlines for interior, direct fire attacks. In recent years the use of solid stream nozzles has increased significantly among fire departments that employ compressed air foam (CAFS) systems.

Over the years, numerous tests and trial studies have been conducted to determine the best qualities and effective range of solid streams. Based on these tests, solid streams are classified as good if they have the following physical characteristics at the point of breakover:

- The stream has not lost continuity by breaking into showers of spray.
- The stream appears to shoot nine-tenths of the whole volume of water inside a circle 15 inches in diameter and three-quarters of it inside a 10-inch circle, as nearly as can be judged by the eye **(Figure 10.8)**.
- The stream is stiff enough to attain the required height and/or reach, even though a moderate breeze is blowing.

If a solid stream appears to be having the desired effect on the fire, then it should be considered a good stream at that point.

A solid stream that has these characteristics certainly will remain effective beyond the breakover point. However, beyond the breakover point the solid stream tends to take the form of a heavy rain, which can be carried away easily in wind currents. It is not possible to state the extreme limit at which a solid stream of water can be classified as a good stream. It is difficult to say within 5 feet, or in some instances even within 10 feet, exactly where the stream ceases to be good. In a fire fighting setting, that determination must be based on the stream's observed effectiveness. If the stream appears to be having the desired effect on the fire, then it should be considered a good stream at that point.

Flow Capacity of Solid Streams

The fire service measures the rate of discharge of any fire stream in gallons per minute (gpm). The flow rate from a nozzle depends on the velocity of the stream and the area of the discharge opening. Any increase in the area of the discharge orifice or the velocity of the stream of water will result in a corresponding increase in the flow rate.

Given equal discharge pressures, the flow capacity of a smoothbore, solid-stream nozzle quadruples when the diameter of the discharge opening is doubled. For example, given equal discharge pressures, four 1-inch nozzles are required to flow the same capacity as a single 2-inch nozzle. Replacing two 1-inch nozzles with a single nozzle that flows the same amount of water as both nozzles combined would require only a 1⅜-inch nozzle.

A given size nozzle will discharge a maximum quantity of water at a definite stream velocity. To increase the discharge, a firefighter must either use a larger nozzle or increase the stream velocity. Both of these options have their limits, particularly when using handlines. In a real world sense, increasing the velocity means increasing the discharge pressure from the fire pump. The pressure can be increased only so much before the hose becomes unmanageable for firefighters and the pressure nears or exceeds the hose's rated burst pressure. Most fire departments and fire training authorities including IFSTA advocate that the nozzle pressure on a handline solid-stream nozzle should not exceed 50 psi. The recommended nozzle pressure for master-stream solid-stream nozzles is 80 psi.

The maximum recommended pressure for handline nozzles is 50 psi, for master stream nozzles 80 psi.

When working with plain water fire streams, increasing the diameter of the nozzle discharge opening has its limits as well. For years, many authorities have followed the rule of thumb that the tip diameter of solid-stream nozzles used on handlines should not exceed one-half the diameter of the hose. For example, the maximum diameter of the nozzle tip used on 2½-inch hoseline would be 1¼-inches. This rule of thumb has no scientific basis, and some departments' standard operating procedures call for solid stream nozzles whose diameter exceeds the "one-half the hose diameter" rule. Each department needs to determine what works best for it. If the nozzle whose diameter exceeds the one-half the hose diameter rule produces an effective fire stream and firefighters can safely manage the hoselines, the department has no reason not to use this combination as a regular practice. After all, in most cases more water is better.

The Reach of a Solid Stream

The distance a solid fire stream can project effectively from a nozzle is its reach. Of all the types of fire streams available to firefighters, solid streams tend to have the longest reach. In many cases, such as defensive "surround and drown" operations, it is specifically reach that leads to their selection.

A solid stream, or any other type of stream for that matter, can be controlled mechanically only until it leaves the nozzle. From that point forward the stream is influenced by a number of factors, including gravity, friction of the air, and wind. The stream's effectiveness will be guaranteed only so long as it can overcome these forces.

A fire stream can be controlled mechanically only until it leaves the nozzle.

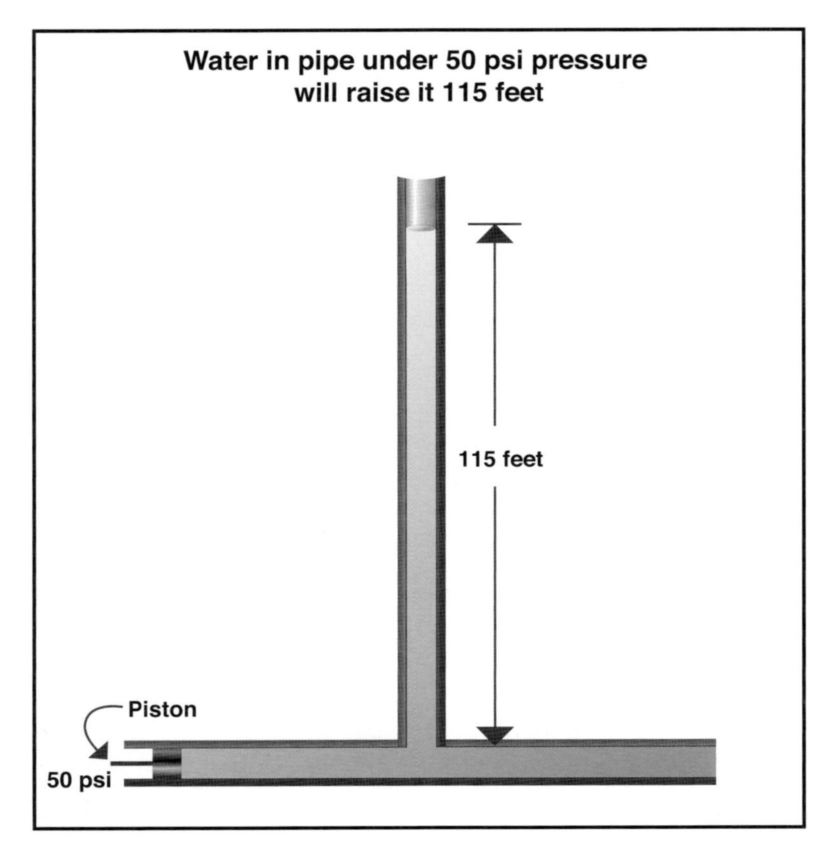

Figure 10.9 Water encased within a pipe and under 50 psi of pressure at the base of the pipe will fill the pipe to a height of 115 feet.

Recall from Chapters 2 and 3 that water encased within a pipe and under 50 psi pressure at the base of the pipe will fill the pipe to a height of 115 feet **(Figure 10.9)**. Previously we learned that each 1 psi will raise water in the pipe 2.3 feet. Thus, 50 psi will raise the water 115 feet.

Suppose we remove the pipe and direct the stream of water straight up into the air. After the water leaves the end of the pipe or nozzle, the stream holds its shape for some distance and then begins to flare and break into drops. These drops soon lose their momentum and fall to the ground. This process continues until only the water column's inner part reaches the maximum height. The friction of the air on the water's outer surface breaks the stream, and gravity pulls the drops to the ground.

Returning to our initial example, if the pipe encasing the water is inclined and its length increased to the same vertical height, the water column will again reach 115 feet, even though the length of the pipe may be 150 or 200 feet depending on its angle of inclination. The water level still rises to a height of 115 feet due to the fifth principle of pressure on fluids: *The pressure of a liquid on the bottom of a vessel is independent of the shape of the vessel* (see Chapter 2). Removing the pipe and inclining the nozzle lowers the water level because it introduces air friction as an opposing force.

Through the years, tests and practical observations have revealed the following characteristics of solid streams:

- The greater the discharge pressure at the nozzle, the greater the reach of the stream.
- Given equal discharge pressures, the reach of the stream will increase as the diameter of the discharge orifice is increased.
- The solid stream's maximum horizontal reach is at an angle of 32 degrees to the earth's surface.
- The closer the directed stream's angle is to 70–75 degrees, the greater will be its vertical reach into structures. The stream's greatest vertical reach is at 90 degrees from, or perpendicular to, the earth's surface.

While understanding these characteristics is valuable to the firefighter, each of them has its practical limits.

- While it is true that more pressure equates to a greater reach, the maximum discharge pressure of 50 psi for a solid stream handline should never be exceeded. Discharge pressures above this point create a dangerous amount of nozzle and hose reaction for the firefighters handling the line. If the reach is not sufficient at 50 psi discharge pressure, firefighters must either move the hose closer to the target or consider using a master stream device.

- The nozzle tip's diameter can be increased only so far before the quality of the stream is reduced. For handlines, typically if the nozzle tip's diameter exceeds about 60 percent of the diameter of the hose to which it is attached, the stream's quality will be lowered. The exception to this rule is when using compressed air foam streams, which will be relatively effective even if flowed from the end of a hose without a nozzle.
- The stream's theoretical maximum horizontal reach will be achieved at an angle of 32 degrees, but the conditions at the incident scene will dictate the actual angle of discharge. Frequently that angle will be less than ideal. The same holds true for the stream's maximum vertical reach into structures at 70–75 degrees.

Tables that show ranges of fire streams usually are computed for streams flowing while the air is comparatively still. If a breeze is blowing, it reduces the stream's reach markedly. For this reason, some margin of safety is desirable. **Table 10.1** gives effective ranges for solid streams with a moderate wind of 18 mph.

Table 10.1
Effective Range of Solid Fire Streams (in feet) Nozzle Diameter (inches)

	1		1⅛		1¼		1½		1½	
Nozzle Pressure PSI	Vertical[a] Distance	Horizontal[b] Distance	Vertical Distance	Horizontal Distance	Vertical Distance	Horizontal Distance	Vertical Distance	Horizontal Distance	Vertical Distance	Horizontal Distance
20	35	37	36	38	36	39	36	40	37	42
25	43	42	44	44	45	46	45	47	46	49
30	51	47	52	50	52	52	53	54	54	56
35	58	51	59	54	59	58	60	59	62	62
40	64	55	65	59	65	62	66	64	69	66
45	69	58	70	63	70	66	72	68	74	71
50	73	61	75	66	75	69	77	72	79	75
55	76	64	79	69	80	72	81	75	83	78
60	79	67	83	72	84	75	85	77	87	80
65	82	70	86	75	87	78	88	79	90	82
70	85	72	88	77	90	80	91	82	92	84
75	87	74	90	79	92	82	93	84	94	86
80	89	76	92	81	94	84	95	86	96	88
85	91	78	94	83	96	87	97	88	98	90
90	92	80	96	85	98	89	99	90	100	91

Advantages and Disadvantages of Solid Streams

No fire stream is perfect for every situation. Each stream has its advantages and disadvantages. For solid fire streams those advantages and disadvantages are:

- Advantages
  - Because they are compact and produce less spray than other types of fire streams, solid streams allow firefighters better visibility than the other types of streams.
  - Historically, solid streams have always been considered to have greater reach than other types of streams. However, advances in fog-stream nozzle technology in recent years allow some of those nozzles to produce straight streams that have a reach comparable to standard solid stream nozzles.
  - Solid streams operate at lower nozzle pressures (50 psi for handlines) per gallon than other types of streams (100 psi for handlines), thus reducing nozzle reaction. This means that manually controlling a solid stream handline flowing 150 gpm is generally easier than controlling a fog stream flowing the same amount.
  - Solid streams have greater penetration power than other types of streams. This allows solid streams to reach further into structures, penetrate and/or break apart tightly compacted fuels, and penetrate through large areas of flame that otherwise would vaporize streams of lesser power or made of droplets.
  - Because solid streams are very compact and contain little entrained air, they are less likely to disturb normal thermal layering of heat and gases during interior structural attacks than other types of streams. Thermal layering is disturbed when significant air movement is directed into the layer. Solid streams will not do this.
- Disadvantages
  - Solid stream nozzles do not allow for different stream pattern selections. Thus, once firefighters commit to using a solid stream, they cannot change to a different stream without shutting down the line and replacing the nozzle. This could become crucial to firefighter safety should a situation occur that required a wide fog pattern to protect firefighters from high levels of radiant heat.
  - Solid streams cannot be used for Class B foam application. The solid stream will not entrain enough air into the fire stream to aerate the foam solution to form an effective finished foam blanket. Of course, solid stream nozzles are perfectly acceptable for use with both low- and high-energy Class A foam streams.
  - As explained earlier in this text, water's heat absorption ability is most efficient when the surface area of the water applied to the heated area is maximized. In other words, many little droplets of water can absorb more heat than fewer, larger droplets. Thus, solid streams provide less heat absorption per gallon delivered than other types of streams because their stream is a relatively solid column of water versus a column made up of many smaller droplets.

— Solid streams will conduct electricity back to the nozzle if they contact an energized electrical source. This may pose an electrocution hazard for the firefighters handling the hose.

FOG STREAMS

In Chapter 6, we explored the development of the fog nozzle and the use of fog streams dating back to Griswold, Layman, Royer, and Nelson. Since the pioneering efforts of those fire service legends, the fog stream has become firmly entrenched in the modern fire service. It is the most flexible of the three types of fire streams, though it is not perfect for every situation. This section examines the important characteristics of fog streams and highlights it's advantages and disadvantages.

Characteristics of Fog Streams

The term *fog stream* is used to describe a patterned fire stream composed of many individual droplets of water. The water droplets, in either a shower or spray, are formed to expose the maximum water surface for heat absorption. Some jurisdictions and publications, such as NFPA® standards, use the term *spray* stream instead of *fog stream*. Adjustable fog stream nozzles create virtually all fog streams. These nozzles can develop a wide range of stream patterns, from a straight stream to a narrow fog pattern of 15–45 degrees to a wide fog pattern of 45–80 degrees.

Many fire service professionals incorrectly use the terms *solid stream* and *straight stream* interchangeably. Though from a quick glance the two may appear to be the same, they are distinctly different. A *solid stream* is a solid stream of water discharged from a smoothbore nozzle. A *straight stream* is meant to emulate a solid stream, but it is discharged from a fog-stream nozzle. A straight stream actually is composed of water droplets and contains some entrained air.

The fire service uses a wide variety of types and designs of nozzles to produce fog streams. They are available for both handlines and master streams. These nozzles may be of fixed gallonage, adjustable gallonage, or automatic designs. The following chapter describes them in more detail.

Velocity of the Fog Stream

Before discussing the velocity of a fog fire stream, we must review two scientific terms:

- **Velocity** is the rate of motion of a particle in a specific direction.
- **Speed** is the rate of travel regardless of direction.

The exact measurement of water speed or velocity toward its objective is of little practical value to firefighters. As long as the stream reaches the objective in the desired pattern, it does not matter whether it takes one or five seconds for a particle of water to travel from the nozzle to the fire. The exception to this rule is when using the fog stream to perform hydraulic ventilation of an enclosed area. In this situation, the greater the stream's forward velocity, the more air it will move and the quicker it will vent the area.

Water for the fire stream is given velocity by the pressure that the pump exerts upon it. This is a proportional relationship: as the pressure in a hoseline increases, so does the velocity of the water traveling through it, and vice versa. Momentum carries the water forward after it is discharged from the nozzle. As with solid streams, friction with the air retards the fog stream's velocity and the force of gravity pulls it downward. Obviously, the greater the velocity of the water when it leaves the nozzle, the farther it will travel before gravity pulls it to earth. Forward velocity, therefore, is one factor that governs the fog stream's reach.

When a stream strikes an obstacle, its forward velocity is reduced in proportion to the shape of the obstacle and the angle of impact. If the obstacle is at a right angle to the stream, the entire forward velocity will be lost (the stream is stopped). If the obstacle is at less than a right angle, the loss will be in proportion to the angle and the stream will simply be deflected.

The effect of nozzle design or adjustment upon stream velocity can be seen by operating two identical adjustable pattern nozzles at identical pressures. Each nozzle, however, is adjusted to produce different stream patterns. Since the pressures are equal, we can assume that the water is discharging from both nozzles at the same velocity; nevertheless, the difference in nozzle pattern produces markedly different streams **(Figure 10.10)**. This pattern adjustment changes both the angle of the discharge from the orifice and the forward velocity of the stream. Regardless of the nozzle setting, fog streams should divide into an even spray with a uniform discharge pattern around their "cone."

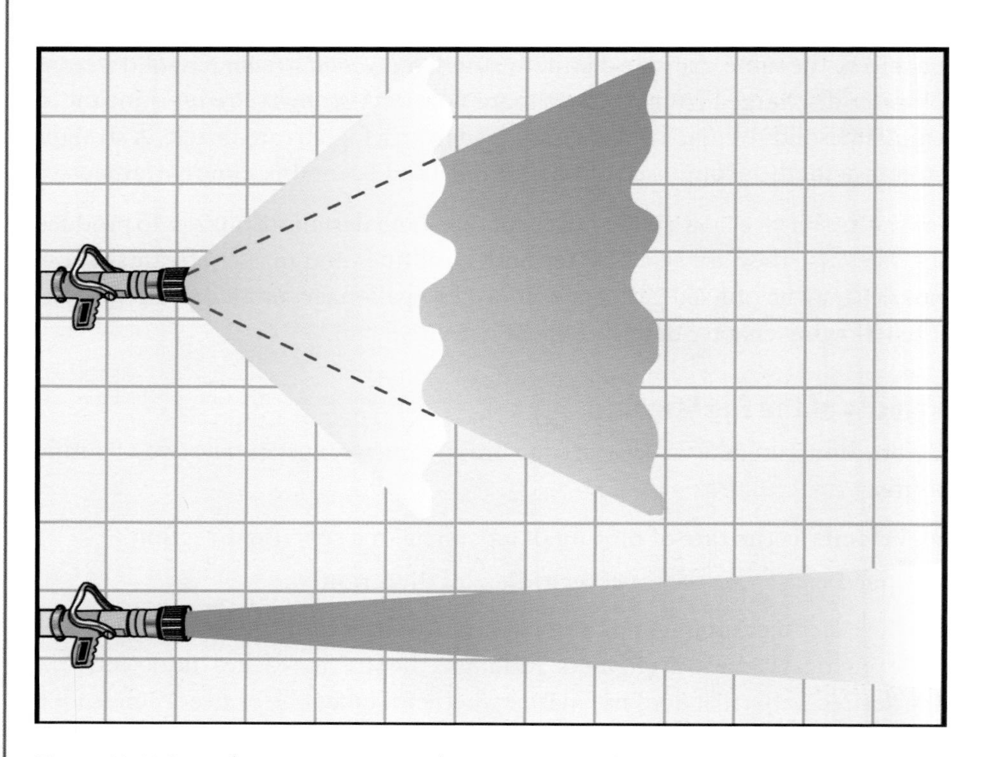

Figure 10.10 Since the pressures are equal, we can assume that the water is discharging from both nozzles at the same velocity; nevertheless, the difference in nozzle pattern produces markedly different streams.

Reach of Fog Streams

A fire stream's effective reach is one of the basic tactical considerations when planning a fire attack. It goes without saying that if the stream cannot reach the seat of the fire effectively, its usefulness in making the fire attack will be nil. Determining the effective reach of a fog stream is not as simple and straightforward as determining the reach of a solid stream. More factors affect the fog stream's reach than affect the solid streams' reach. Fog-stream selection must consider five factors:

- Gravity
- Water velocity
- Fire stream pattern
- Water droplet friction with air
- Wind

Noticeably, more factors may hinder a fog stream than a solid stream. Common sense suggests that the more hindrances there are, the shorter the stream reach is likely to be. Thus, these factors' interactions on a fog stream result in a fire stream whose reach is typically less than that of a solid stream. This reduced reach is one reason that fog streams are seldom useful for outside, defensive fire fighting operations. The other primary reason for this is that in many cases the fog stream will be converted to steam in the thermal column above the fire and the water will never reach the seat of the fire.

Fog streams are most useful for fighting enclosed fires. What the fog stream lacks in reach is made up for by the larger space it occupies and the greater water surface it exposes to the heat. The virtues of using fog streams for interior fire attacks date back to the early research conducted by Royer, Nelson, Layman, and others.

Fog streams are most useful for fighting enclosed fires and are seldom used for outside, defensive fire fighting operations.

In addition to the factors affecting a solid stream's reach, fire stream pattern and water droplet friction with air affect the reach of the fog stream. Obviously, these two factors do not exist with solid streams because their "pattern" is limited to a very narrow jet of water exiting the nozzle. This jet of water is essentially solid, so there are no droplets subject to friction with the air.

Every fog stream is composed of droplets of water surrounded by air. As the stream travels from the nozzle, it draws in air. This process is known as entraining air. Any stream of water is retarded in its travel through the air by friction between the stream's outer surface and the air. A fog stream also is retarded by the friction between the outer surface of the fog cone and the surrounding air, turbulence within the fog stream, and the process of entraining air.

To a great extent, the fog stream pattern and water droplet friction with air are interrelated. Because the very nature of fog streams means that they will have a larger diameter than solid streams, they have more area subject to friction with air. The wider the stream, the more the friction created as it moves through the air. And the greater the friction, the slower the forward velocity of the stream. The slower the forward velocity is, the less the reach of the stream will be.

A wide-angle fog stream has little forward velocity and a short reach, while a narrow-angle fog pattern has greater forward velocity and reach.

The fire professional thus understands that a wide-angle fog stream has little forward velocity and a short reach. A narrower-angle fog pattern will have a greater forward velocity and reach. The forward velocity and reach will vary depending on the width of the stream and the nozzle pressure. Of course, any fog pattern has a maximum reach, as with any other stream. Once the nozzle pressure has produced a stream with maximum reach, further increases in nozzle pressure have little effect upon the stream except to increase the flow volume.

If a fog stream has a broken or otherwise unsatisfactory pattern, the likely culprit is insufficient nozzle pressure. Reducing the nozzle pressure decreases both the initial forward velocity of the water droplets and the amount of water being discharged. This reduces the fog stream's reach and overall effectiveness. Most standard fog nozzles, regardless of whether they are fixed gallonage, adjustable gallonage, or automatic nozzles (as described in Chapter 11), are designed for optimum flow and stream operation at a discharge nozzle pressure of 100 psi. The exceptions to this rule are special low-pressure nozzles whose most common use is in high-rise fire fighting. These nozzles may be designed for optimum use at discharge pressures as low as 50 psi. Firefighters and driver/operators must know the appropriate discharge pressures for the nozzles they are using and calculate the pump discharge pressures accordingly. (More information on calculating these pressures is contained in Chapters 12 and 13.)

Space Occupied by the Fog Stream

A key factor in fog streams' effectiveness in controlling fires within confined spaces is maximizing the ratio of the stream's water surface to the air in the space. In other words, the more of the confined fire space that the stream occupies, the greater the rate of conversion of water to steam will be and the faster extinguishment will occur.

The size of the area occupied by fire within a building and the firefighters' ability to safely access that area both affect the width and reach of the fog stream that will be used. If the situation requires a far-reaching, narrow fog stream to cover the entire heated area, the water-surface-to-air ratio will be low, and the steam conversion rate will be reduced. If firefighters can get closer to the fire area and use a wider fog pattern, the water-surface-to-air ratio will be greater and the steam conversion will be increased.

A very large fire area may require one or more far-reaching fog streams in order to project enough water into the heated area to achieve maximum heat absorption. This is because a wide angle stream will reduce the temperature in the immediate area but will absorb little heat beyond its reach. The displacement of the heated gases by the wide-angle fog stream may assist in extinguishment.

Size of Water Particles

Much research has been done over the years regarding the optimum size of water droplets in a fog stream. Both fire service personnel and nozzle manufacturers have studied the issue to determine the perfect stream and droplet size. Some of this research has resulted in scientific recommendations on the size of each drop of water in microns and other fascinating facts. However, much of this information has no practical value in everyday operations. Firefighters are not likely to show up on the scene of a fire and base their decision on the type of stream pattern to use on the optimum micron size of water droplets for the scenario.

Much of what firefighters and other fire service professionals need to know about water droplet size can be related simply to common sense based on the information covered up to this point. Several times we have stressed the influence of the water-surface-to-air ratio on effective steam conversion when fighting fires in confined spaces. The greater this ratio is, the quicker and more completely water will covert to steam. Achieving a high water-surface-to-air ratio requires creating water droplets that are as small as practically possible.

Steam conversion will be optimal when the water droplets are so fine that they appear to be mist suspended in air. This is the principle upon which low-volume water-mist fire extinguishing systems work. It is also the principle upon which high-pressure fog fire pumps were sold to the fire service in the 1960s and 1970s. These pumps produced powerful cones of water fog at pressures of up to 1000 psi. They had enormous physical power in that they could strip sheet rock off walls, peel shingles off roofs, and perform other destructive feats. However, their downside was that they were flowed through booster hose and had flow rates of only 8 to 12 gallons per minute. While the high-pressure fog would convert quickly to steam when exposed to a heated atmosphere, this low volume of water often was insufficient to absorb all the heat of a fire and left firefighters in the potentially dangerous situation of not having enough water to extinguish the fire or even to protect themselves in forward positions. These streams also had limited use in exterior fire fighting because they would often convert to steam in the thermal column of a fire and never deliver water to the seat of the fire. Firefighters soon learned that these high-pressure fog streams had a limited number of practical applications. These included small exterior flammable- and combustible-liquid fires and wildland fires in short to medium grasses.

While scientific studies may be cited that state droplets of this or that size are the best, there really is no single optimum size of water droplet for fire fighting operations. The firefighter's goal should always be to apply a fog stream that breaks the water into the finest droplets possible. However, the ability to do so will vary depending on the required reach of the stream, the wind conditions being faced, the available nozzle pressure, and the size and volume of the fire. The droplets need to be large enough to ensure that they reach the seat of the fire or the heated space in a condition that allows them then to convert to steam and cool the area or fire.

The firefighter's goal should always be to apply a fog stream that breaks the water into the finest droplets possible.

Advantages and Disadvantages of Fog Streams

Fog fire streams have the following advantages and disadvantages:

- Advantages
  - Fog streams dissipate heat by exposing the maximum water surface for heat absorption.

  - Fog streams have adjustable stream patterns that may be changed to suit the situation.
  - The stream may vary from a straight stream (similar to a solid stream) up to a wide fog pattern that approaches a 90° discharge angle to the nozzle.
  - Some fog stream nozzles have adjustable settings to control the amount of water being used.
  - Wide-angle fog streams can be used to provide a protective barrier for firefighters exposed to high levels of convective or radiated heat.
  - Fog streams can be used to aid ventilation by creating air movement out of a building opening (a process known as hydraulic ventilation).
  - Fog streams can be applied to exposures in a gentle enough form to prevent physical damage to the structure.
  - Because the fog stream is not a contiguous stream of water, it typically will not conduct electricity back to the nozzle and therefore is relatively safe to use around energized electrical equipment.
  - Both Class A and Class B foam concentrates can be used with fog streams.
- Disadvantages
  - Fog streams do not have the reach or penetrating power of solid streams.
  - Fog streams are more susceptible than solid streams to wind currents.
  - Fog streams may contribute to fire spread, create heat inversion, and cause steam burns to firefighters when improperly used during interior attacks.

BROKEN STREAMS

A *broken stream* is a stream of water that has been divided into coarse drops. Broken streams actually have characteristics of both solid and fog streams. Most broken stream nozzles form the stream by directing the water from the hose through a series of small holes in the end or sides of the nozzle. The result is a series of small solid streams that quickly break up into large droplets of water. The coarse drops of a broken stream absorb more heat per gallon (liter) than a solid stream, and a broken stream has greater reach and penetration than a fog stream.

The main differences between a fog stream and a broken stream are that the fog stream has a definite pattern, generally is composed of small droplets, and usually is adjustable, whereas the broken stream does not always have a definite pattern, usually is composed of larger drops, and generally is not adjustable. Another difference between fog streams and broken streams is that a broken stream may have sufficient continuity to conduct electricity, so its use around Class C fires is not recommended.

The broken stream's use typically is limited to fighting fires in confined or otherwise inaccessible spaces.

When using broken streams, the volume of water applied should be comparable to that from a solid stream for the same situation. In most cases, when broken stream nozzles are used, water damage is secondary to extinguishing the fire since a considerable amount of water is projected over the general area, sometimes not in the most efficient manner.

Of the three types of fire streams discussed in this text, the broken stream is by far the least used by most firefighters. Its use typically is limited to special situations such as fighting fires in confined or otherwise inaccessible spaces. As we will see in Chapter 11, broken streams are produced by special nozzles such as attic or cellar nozzles, piercing nozzles, and chimney nozzles.

Advantages and Disadvantages of Broken Streams

Broken streams typically have very specialized uses. Within these uses they have a number of general advantages and disadvantages:

- Advantages
  - Broken streams are often the only means of applying water to fires in otherwise inaccessible spaces, such as attics, basements, shipping containers, trailers, and chimneys.
  - Broken streams have more heat absorbing capabilities than solid streams.
- Disadvantages
  - Broken streams may conduct electricity and therefore should not be used around energized electrical equipment.
  - Broken streams' method of water application may not be optimal and can produce more water damage than other types of streams.
  - Each broken stream nozzle typically has one specific application and may not have any other uses.

SUMMARY

Fire service personnel have at their disposal three distinctly different types of fire streams to use when facing any fire situation: solid streams, fog or spray streams, and broken streams. Each has its advantages and disadvantages. Fire protection professionals must understand the basic principles of each stream so that they will choose the most effective stream for the given fire situation. However, selecting the correct type of stream is only part of the overall decision. Once the fire protection professional knows the desired type of stream, he or she must select an appropriate type of nozzle to deliver that stream and must ensure that an adequate volume of water at sufficient pressure is delivered to the nozzle. The remaining chapters of this text will explain how to answer these additional questions.

REVIEW QUESTIONS

1. What are the three principle types of fire streams used by the fire service?

   1. ______________________________

   2. ______________________________

   3. ______________________________

2. What factors affect the overall condition or performance of a fire stream at the point of discharge from the nozzle?

   ____

   ____

3. List the four basic elements that must come together in order to produce a fire stream.

   1. ______________________________

   2. ______________________________

   3. ______________________________

   4. ______________________________

4. Name the hose appliance that is used to divide one hoseline into two separate hoselines of equal or smaller size.

   ____

5. Name the hose appliance that is used to combine two or more hoselines into a single line.

   ____

6. List the three physical characteristics a solid stream must have at the point of breakover to be classified as good.

   1. ______________________________

   2. ______________________________

   3. ______________________________

7. Given equal discharge pressures, the flow capacity of a smoothbore, solid stream nozzle increases by ______ times when the diameter of the discharge opening is doubled.

8. The maximum horizontal reach of a stream is achieved when the stream is directed at an angle of ______ degrees to the earth's surface.

9. What is the maximum recommended nozzle pressure for solid stream handlines?

   ____

10. What is the difference, if any, between a solid stream and a straight stream?

    ____

11. What five factors affecting the reach of a fog stream must firefighters consider during stream selection?

   1. ______________________________

   2. ______________________________

   3. ______________________________

   4. ______________________________

   5. ______________________________

12. At what nozzle pressure are most fog stream handline nozzles designed to be operated?

   ____

13. What two factors in the quality of fog streams are not applicable to solid streams?

   1. ______________________________

   2. ______________________________

14. How are broken streams formed?

   ____

   ____

15. Which of the three types of fire streams are firefighters least likely to use?

   ____

Answers found on page 430.

FESHE Course Objectives

The information in this chapter is intended to meet some of the objectives outlined for the Fire Protection Hydraulics and Water Supply course by the United States Fire Administration's National Model Core Curriculum as developed by the Fire and Emergency Services Higher Education (FESHE) initiative. The information in this chapter also allows the student to meet the requirements of NFPA® 1001, *Standard for Fire Fighter Professional Qualifications (*2008 edition) listed below.

FESHE Course Objectives

1. Describe the construction and explain the operation of handline and master stream solid stream nozzles.
2. Calculate the flow from a solid stream nozzle.
3. Describe the construction and explain the operation of handline and master stream fog stream nozzles, including constant flow, variable flow, and automatic fog nozzles.
4. Describe the construction and explain the operation of broken stream nozzles, including cellar, water curtain, chimney, and piercing nozzles.
5. List the appropriate nozzle discharge pressures for solid, fog, and broken stream handline and master stream nozzles.
6. Calculate nozzle reaction forces on solid and broken stream nozzles.
7. Describe various appliances that may be used with hoselines and nozzles.

NFPA® 1001 Requirements

5.3.10 (A) Requisite Knowledge. Types, design, operation, nozzle pressure effects, and flow capabilities of nozzles.

6.3.2 (A) Requisite Knowledge. Selection of the nozzle and hose for fire attack given different fire situations.

Chapter 11

Fire Hose Nozzles

INTRODUCTION

In Chapter 10 we examined the three basic types of fire streams used in manual fire fighting operations: solid stream, fog or spray stream, and broken stream. Each has its advantages and disadvantages, and producing each requires a specific type of nozzle. As the fire service has discovered needs and uses for these different types of fire streams over the years, correspondingly the specific types of nozzles necessary to develop them have been produced.

Importantly, in our world of advanced technology and multiuse tools and equipment, fire hose nozzles remain relatively locked into specific uses. Nozzle use and design still fall into one of the three main fire stream categories; no nozzle can develop both a true solid stream and a true fog stream. Most fog stream nozzles can produce a straight stream that has similar properties to a solid stream, but it is not a true solid stream.

Thus, to produce the variety of different fire fighting streams that today's fire service may require, most fire apparatus are still equipped with a variety of different nozzles to be used depending on the needs of the incident. The professionals who operate off of that apparatus need to understand which size and type of nozzle is appropriate for the incident. Once they have made that determination, they must know the nozzle's operational principles including how much pressure the nozzle needs to operate efficiently. The varying correct nozzle pressures for each type of nozzle will affect the pump operator's hydraulic calculations.

As with most types of equipment and apparatus in the fire service, a variety of manufacturers produce nozzles for one or all of the fire stream types. This text's purpose is not to evaluate the various design merits of different manufacturer's nozzles. In fact, it would be difficult to conclude a definitive discussion on this issue, because as with all aspects of society, the technology changes almost daily. By the time this text was published it would be out of date. Rather, we will discuss some of the most basic design and operating principles of each of the three types of nozzles. We also will examine some of the basic operating principles that apply to all nozzles of a certain type, such as the correct operating pressures and the amount of nozzle reaction created by water flowing through them.

This chapter will examine nozzle designs for both handline and master stream uses. First we must define both of these terms for purposes of clarity. Throughout the fire service there exist no definitive, universally agreed upon definitions for these two terms. The most commonly accepted definitions are those in IFSTA's various texts that address nozzle usage and design. IFSTA defines a *handline* nozzle as any nozzle designed primarily for attachment to a single, moveable, hoseline and manual control and operation by one or more firefighters **(Figure 11.1)**. IFSTA limits the maximum amount of water that may be safely flowed through a handline nozzle to 350 gpm or less. In contrast, *master streams* are designed to be located and operated from a specific location and to flow large volumes of water beyond that which firefighters could control manually **(Figure 11.2)**. IFSTA limits any flow above 350 gpm to master stream applications. The most common flow capacities of master streams used by the municipal fire service range from 500 to 1,000 gpm, although some municipalities have higher capacities. Specialized applications, such as fire boats or industrial fire protection use master stream nozzles capable of flowing up to 20,000 gpm. Solid, fog, and broken stream nozzles all are available in both handline and master stream designs.

Figure 11.1 IFSTA defines a *handline* nozzle as any nozzle designed primarily for attachment to a single, moveable, hoseline and manual control and operation by one or more firefighters.

Figure 11.2 In contrast, *master streams* are designed to be located and operated from a specific location and to flow large volumes of water beyond that which firefighters could control manually.

This chapter is limited to nozzles that typically are used to fight standard Class A, ordinary combustibles, fires. This may be achieved with either plain water or a foaming additive, such as a Class A foam concentrate. While fog nozzles, in conjunction with Class B foam concentrates, may also be used with limited effectiveness on flammable and combustible liquids fires, that type of fire fighting is not the focus of this text. For more information on using standard nozzles to attack Class B fires or for information on specially designed foam or chemical agent nozzles consult the IFSTA *Principles of Foam Fire Fighting* manual.

SOLID STREAM NOZZLES

The oldest style of nozzle used by the fire service is the solid stream nozzle. Solid stream nozzles date to the very origins of mechanized fire fighting. They are designed to produce a stream as compact as possible with little shower or spray. Solid stream nozzles may be used on handlines, portable or apparatus-mounted master streams, or elevated master streams.

The design of a solid stream nozzle is extremely important in producing the desired compact fire stream. Research on the best design for solid stream nozzles has been conducted since their introduction into the fire service more than 100 years ago. Some of the early experiments involved comparing streams from different kinds, shapes, and sizes of nozzles. One famous series of these experiments was named the "Freeman Experiments" after the leader of the project. The initial results of these comparisons were disappointing to Freeman and his assistants. None of the tested nozzles produced the desired polished stream of water with a clear, glassy surface and little spray. These results spurred Freeman to conduct additional tests, tinkering with various designs, until he finally found a nozzle shape and design that produced an optimum stream. Freeman's design, sometimes referred to as the "standard tip," remains the basis for solid stream nozzle design even today **(Figure 11.3)**.

Figure 11.3 Most solid stream nozzles have a standard design for the shape of the waterway.

Standard solid stream nozzles are designed to gradually reduce the shape of the water in the nozzle until it reaches a point near the outlet. At this point, the nozzle becomes a cylindrical bore whose length is from one to one and one-half times its diameter. This short, truly cylindrical bore rounds the water before discharge. A smooth-finished waterway contributes to both the shape and reach of the stream. Alteration or damage to the nozzle can significantly alter stream shape and performance.

Historically, the majority of the fire service held fast to the rule that solid stream nozzles on handlines should have a discharge diameter no greater than one-half the diameter of the hose to which they are attached. For example, a 2½-inch hoseline would be equipped with a solid stream nozzle whose tip diameter is no larger than 1¼ inches. This rule of thumb has no real scientific or practical basis. In fact, many fire departments, including large metropolitan departments, such as the Fire Department of New York (FDNY), routinely equip 2½-inch handlines with solid stream nozzles whose tips exceed 1 1/4 inches. Though little scientific research on this issue has been conducted, many fire departments have learned that a 2½-inch handline can be equipped with a solid stream nozzle whose tip is as large as 1⅝-inches without reducing the quality of the stream beyond effectiveness.

This information on nozzle tip sizes applies only to low energy, traditional fire streams. Handlines attached to high energy, compressed air foam systems (CAFS) are routinely equipped with nozzle tips equal to the hoseline's diameter with little loss of effectiveness **(Figure 11.4, p. 236)**.

Figure 11.4 Handlines attached to high energy, compressed air foam systems (CAFS) are routinely equipped with nozzle tips equal to the hoseline's diameter with little loss of effectiveness.

Figure 11.5 Stacked tips are a series of threaded nozzle tips whose diameters decrease as they extend from the playpipe to the tip.

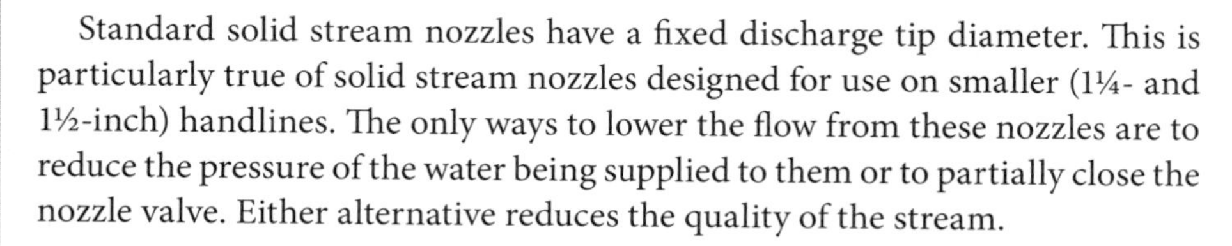

Standard solid stream nozzles have a fixed discharge tip diameter. This is particularly true of solid stream nozzles designed for use on smaller (1¼- and 1½-inch) handlines. The only ways to lower the flow from these nozzles are to reduce the pressure of the water being supplied to them or to partially close the nozzle valve. Either alternative reduces the quality of the stream.

Larger handlines and most master stream solid stream nozzles are equipped with "stacked tips." Stacked tips are a series of threaded nozzle tips whose diameters decrease as they extend from the playpipe to the tip (**Figure 11.5**). A common 2½-inch handline stacked tip nozzle has a playpipe with a 1½-inch tip to which 1⅜-, 1¼-, and 1-inch tips are attached. Before opening the nozzle, firefighters select the diameter of the tip based on the conditions they face and the available water supply.

The stream velocity and the size of the discharge opening determine the flow from a solid stream nozzle.

The stream velocity (nozzle pressure) and the size of the discharge opening determine the flow from a solid stream nozzle. As either or both increases, so does nozzle flow. From a practical standpoint, both velocity and discharge opening size have limitations. In addition to the limitations on the discharge opening's size described in the preceding paragraphs, there are practical limitations on the amount of nozzle pressure that can be supplied to solid stream nozzles as well.

When solid stream nozzles are used on handlines, they should be operated at no more than 50 psi nozzle pressure. Above that pressure the nozzle reaction on the hoseline may become too great for firefighters to control safely. Most solid stream master stream devices should be operated at nozzle pressures no greater than 80 psi. This is particularly true of portable and elevated master stream devices. Operating a portable master stream device above 80 psi may cause excessive nozzle reaction. The excessive nozzle reaction may in turn cause the portable master stream device to move, in some cases violently, and endanger personnel working on the scene. Operating an elevated master stream nozzle at an unsafe pressure may place excessive stress on the aerial device. This excessive stress may damage the water system, dislodge a detachable master stream nozzle from the ladder, or damage the ladder itself.

The maximum operating pressure for solid streams nozzles is 50 psi for handlines and 80 psi for master streams.

For special situations, some solid stream master stream nozzles may be designed to operate at discharge pressures in excess of 80 psi. These situations include nozzles connected to fixed piping systems in industrial facilities, large nozzles on fire boats, and specially designed deluge fire apparatus. Deluge fire apparatus with larger master stream nozzles may even be equipped with outriggers similar to those used on aerial apparatus to keep the apparatus from moving when the nozzle is being flowed at high pressures. Whatever the case, fire protection professionals using these special nozzles should be familiar with their operation and should not exceed their designed safe operational limits.

Determining the Flow Volume from a Solid Stream Nozzle

Determining the flow from a solid stream nozzle is not complicated. The procedure works for solid stream nozzles of all sizes and designs. Assuming that the equipment used to measure the nozzle discharge pressure is accurately calibrated the actual volume from the nozzle can be calculated fairly closely.

The procedure for determining the flow volume from a solid stream nozzle is essentially the same as that for determining the flow from a fire hydrant discharge opening (Chapter 5). By inserting a pitot tube and gauge into the fire stream at a distance from the tip equal to one-half the diameter of the discharge opening, the fire protection professional can get an accurate reading on the discharge pressure exiting the nozzle (**Figure 11.6**). Given this pressure, calculating the flow volume is relatively simple.

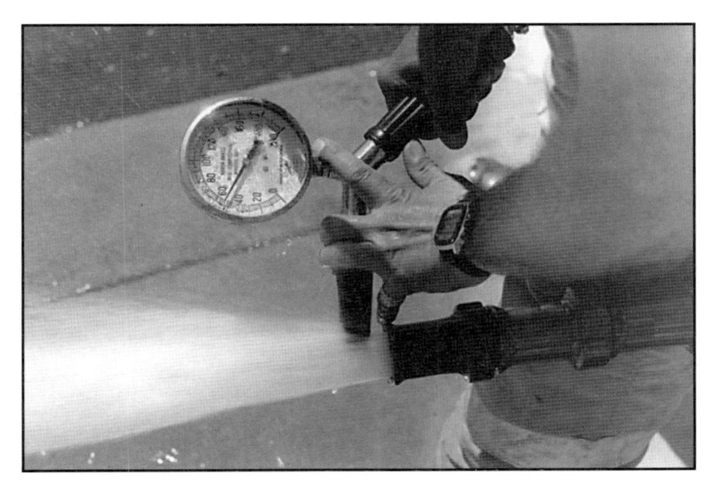

Figure 11.6 By inserting a pitot tube and gauge into the fire stream at a distance from the tip equal to one-half the diameter of the discharge opening, the fire protection professional can get an accurate reading on the discharge pressure exiting the nozzle.

The formula to determine the flow from a solid stream nozzle is a simplified version of the formula used to determine the flow from a hydrant discharge (Equation 5.1: Flow in gpm = $(29.83)(C_d)(d^2)(\sqrt{P})$. Because the design of most solid stream nozzles is very similar, a standard coefficient of discharge (C_d) is factored into the equation with the following results:

Equation 11.1

Flow from a solid stream nozzle in GPM = $29.7 \times d^2 \times \sqrt{NP}$

Given:

GPM = Discharge in gallons per minute

29.7 = A constant

d = Nozzle tip diameter in inches

NP = Nozzle pressure in psi

Example 11.1

Determine the flow volume from a 1 1/4-inch solid stream nozzle tip operating at 50 psi.

Solution:

$gpm = (29.7)(d)^2(\sqrt{NP})$

$gpm = (29.7)(1.25)^2(\sqrt{50})$

$gpm = (29.7)(1.5625)(7.07)$

gpm = 328

Example 11.2

A detachable ladder pipe is attached to an aerial device and is equipped with a solid stream nozzle having a 2-inch nozzle tip. The nozzle is being operated at a discharge pressure of 80 psi. What is the flow volume being discharged from that nozzle?

Solution:

$gpm = (29.7)(d)^2(\sqrt{NP})$

$gpm = (29.7)(2)^2(\sqrt{80})$

$gpm = (29.7)(4)(8.94)$

gpm = 1,063

Example 11.3

A specially designed deluge fire apparatus is being supplied by multiple large diameter hoselines and is equipped with a solid stream nozzle having a 4-inch nozzle tip. The nozzle is being operated at a discharge pressure of 110 psi. What is the flow volume being discharged from that nozzle?

Solution:

$gpm = (29.7)(d)^2(\sqrt{NP})$

$gpm = (29.7)(4)^2(\sqrt{110})$

$gpm = (29.7)(16)(10.49)$

gpm = 4,984

FOG STREAM NOZZLES

The most flexible nozzle available to the fire service is the fog stream nozzle. These nozzles are sometimes also referred to as spray nozzles. They allow firefighters to create a variety of streams including a solid-stream-like straight stream and a range of fog patterns from narrow to very wide. Each stream has its own uses and advantages. In general, straight streams and narrow fog streams are used for aggressive, up-close fire fighting operations. Wide fog patterns are most commonly used to protect exposures and to protect firefighters from high levels of radiant heat.

Straight streams and narrow fog streams generally are used for aggressive, up-close fire fighting, while wide fog patterns are most commonly used to protect exposures and to protect firefighters from radiant heat.

Mechanically, fog stream nozzles differ from solid stream nozzles in that the fog stream nozzle does not have an unobstructed waterway through the nozzle. All fog stream nozzles use some type of mechanism to break the column of water into a series of droplets as it travels through the nozzle. In most cases the column of water is driven against an obstruction with sufficient velocity to shatter the mass. The angle at which the obstruction deflects the stream of water determines the reduction in the stream's forward velocity and the pattern or shape the stream assumes. A wide-angle deflection produces a wide-angle fog pattern, and a narrow-angle deflection produces a narrow-angle fog pattern.

Most modern fog stream nozzles operate on the principle of periphery deflection. Periphery-deflected fog streams are produced by deflecting water from the periphery of an inside circular stem in the nozzle **(Figure 11.7)**. The exterior barrel of the nozzle again deflects this water before it leaves the nozzle. The relative positions of the deflecting stem and the exterior barrel determine the fog stream's shape.

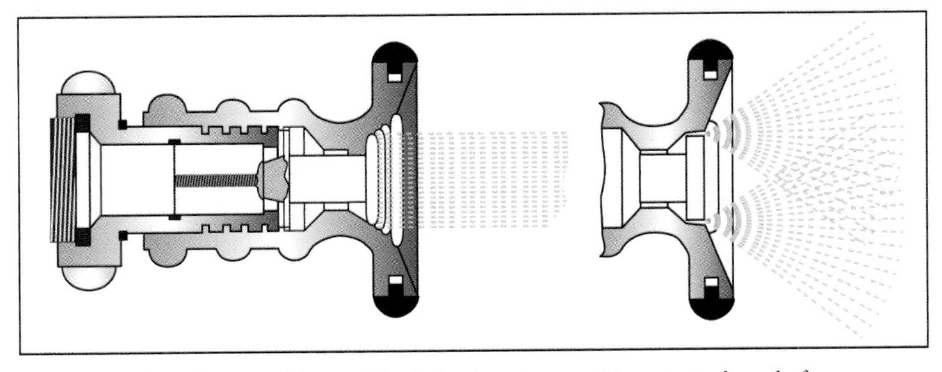

Figure 11.7 The relative positions of the deflection stem and the exterior barrel of a periphery-deflected fog nozzle determine the stream pattern.

Some of the earliest fog stream nozzles operated on the principle of impinging streams. Impinging stream nozzles drove several jets of water together at a specific angle, breaking the water into finely divided particles **(Figure 11.8, p. 240)**. Impinging stream nozzles produced very effective quality wide fog streams but less effective narrow fog patterns and straight streams. The modern fire service rarely uses them.

An advantage of fog nozzles is that when water is discharged at angles to the direct line of flow, the reaction forces tend to counterbalance each other, thus reducing nozzle reaction. This is why fog patterns are easier to handle than solid or straight stream patterns when the nozzle pressures are the same. For this reason most modern fog stream nozzles are designed to operate at a nozzle pressure of 100 psi, regardless of the manufacturer. Operating the fog nozzle at discharge pressures lower than the designed 100 psi pressure will impair the fire

Operating the fog nozzle at discharge pressures lower than the designed 100 psi pressure will impair the fire stream from every type of fog nozzle.

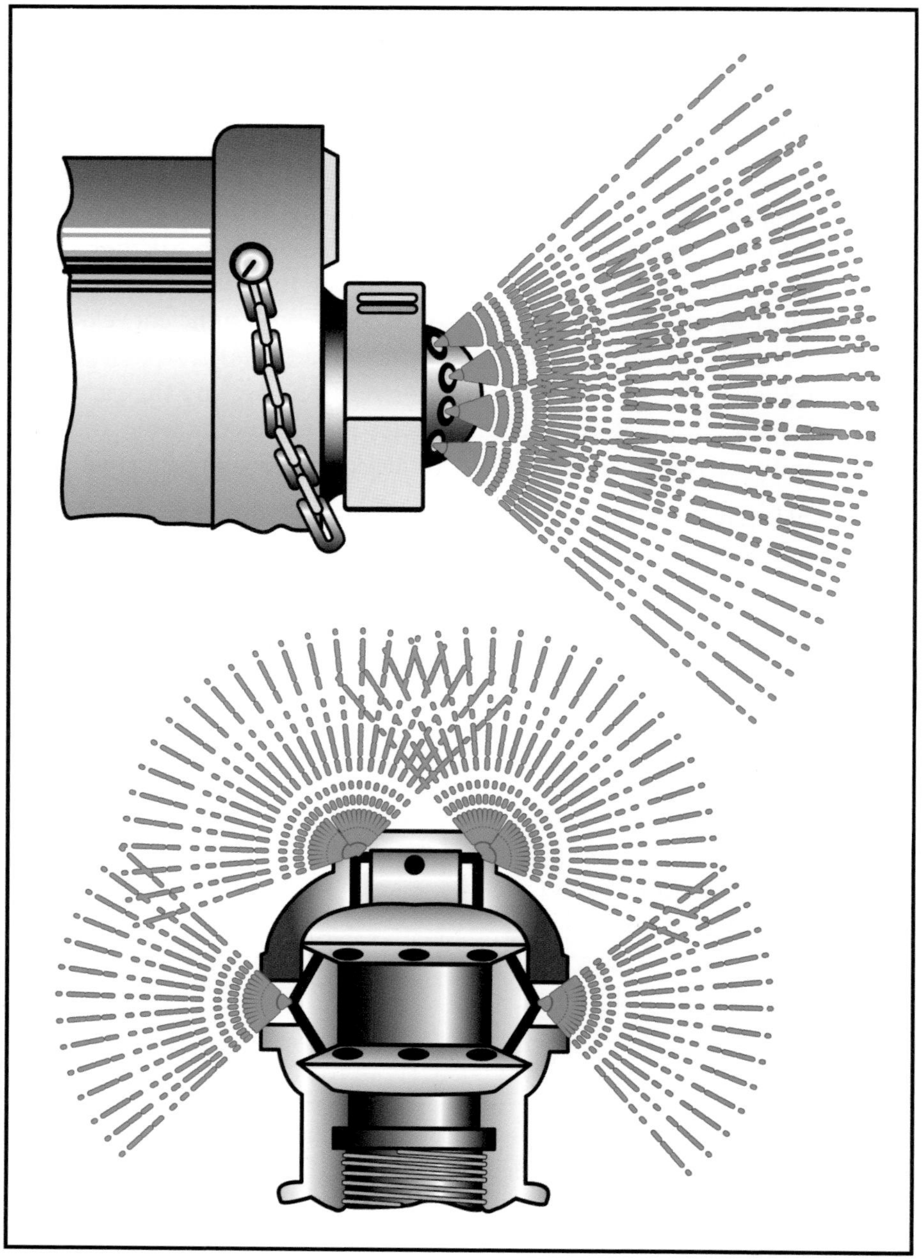

Figure 11.8 Impinging stream nozzles drove several jets of water together to form a fog pattern.

stream from every type of fog nozzle, although the exact consequence will vary depending on the type of nozzle. The exceptions to this rule are the specially designed low-pressure fog stream nozzles that have been introduced into the fire service in recent years. These nozzles were designed primarily for high-rise fire fighting. In many cases it is difficult to get sufficient pressure to the upper stories of very tall buildings in order to supply fog stream nozzles with 100 psi. Thus, low-pressure fog nozzles are designed to have effective stream patterns and safe flow volumes at nozzle pressures of 50 to 75 psi. Though these nozzles originally were designed for high-rise fighting, many fire departments are using them in everyday, standard fire fighting situations. They allow firefighters and apparatus to deploy effective fire streams with less nozzle reaction and lower pump discharge pressures.

The fire service uses several basic design types of fog or spray nozzles. Each major nozzle manufacturer generally produces several models of each basic design. Each has its respective place in the fire service. Fire protection professionals who understand the basic principles of each type of nozzle will be able to select the best one for any emergency situation.

Variable Flow Nozzles

The earliest fog nozzle designs were variable flow. Variable flow nozzles produced different flow volumes depending on the particular variables of their design. These variables included the stream pattern, the water pressure supplied to the nozzle, and the size of the supply line feeding the nozzle. While few of the variable flow fog nozzles are still in service today, firefighters should understand their operating principles in the event that they encounter one. As well, from a historical standpoint it is interesting to see exactly how far the fire service's equipment technology has advanced since the early days of the fog stream nozzles.

One type of variable flow fog nozzle used by the fire service was the rotary control nozzle. There were two basic designs of the rotary control nozzle; both used the periphery-deflected stream. One type was turned on and off by means of a ground seat within the nozzle barrel. When the exterior barrel was screwed against the ground seat, it shut off the water. When the nozzle was first opened, water flowed in a wide pattern fog. As the exterior barrel was screwed forward, the deflection stem was drawn into the exterior barrel and produced a straight stream.

The second type of rotary control nozzle contained a deflection stem that fit the ground seat at the orifice of the exterior barrel. When the exterior barrel was screwed forward against the deflection stem, the water shut off. This nozzle differed from the first type in that when it was opened water first flowed in a straight stream. As the barrel was screwed backward, the deflection stem was pushed out of the exterior barrel and produced a wide pattern fog stream.

Both styles of rotary control nozzles have variable flow volumes depending upon the stream pattern selected. The flow volume from rotary control nozzles is greatest when the stream pattern is widest. Some of these nozzles are capable of a near 90-degree fog pattern. The flow volume decreases as the fog pattern is narrowed. The flow volume is lowest when the nozzle is set at a straight stream. Depending on the particular model of rotary control nozzle, flow volume may decrease as much as 50 percent when a wide fog pattern is changed to a straight stream.

An improvement on the early rotary control fog stream nozzles was a nozzle that eventually became nicknamed the "mystery nozzle." The mystery nozzle was similar in design to the constant flow nozzles described in the next section. Mystery nozzles were variable pattern nozzles with a ball valve shutoff that also used the periphery-deflected stream. However, this nozzle got the "mystery nozzle" nickname because the discharge changed each time the fog stream pattern changed. Mystery nozzles received their flow volume rating at a 100 psi nozzle pressure using a narrow pattern fog stream (about 30 degrees). If the 100 psi nozzle pressure were maintained and the pattern changed to a straight stream, the flow volume decreased approximately 15 percent. If the pattern of the stream changed from a straight stream to the widest possible fog pattern, the flow increased approximately 15 percent.

Constant Flow Nozzles

While fire service pioneers quickly realized the value of the fog stream nozzle's flexibility, they also realized the hazards that those early variable flow nozzles presented. The fluctuations in flow volume potentially could place firefighters in the dangerous situation of having an insufficient flow to safely extinguish a fire or to protect themselves. The earliest solution to this problem was to develop a nozzle that would maintain the same flow volume regardless of stream pattern setting. Thus, the constant flow nozzle was born. Constant flow nozzles flow a specific amount of water at a specific nozzle discharge pressure on all stream patterns. Most constant flow nozzles utilize a periphery-deflected stream. The constant flow volume is maintained because as the exterior barrel is rotated to change the fog stream pattern, the space between the deflecting stem and the internal throat remains the same. As a result, the same quantity of water passes through the nozzle.

While the earliest constant flow nozzles provided the security of a constant flow volume at any stream pattern setting, they were somewhat inflexible in that they were rated for a single flow volume at the 100 psi nozzle pressure. For example, a 1½-inch nozzle may have been rated for 95 gpm at 100 psi nozzle pressure. Firefighters recognized that different situations called for different flow volumes, and their feedback to nozzle designers resulted in a refinement of the constant flow nozzle. The refined constant flow nozzle has a number of constant flow settings, enabling the firefighter to select the flow rate that best suits conditions (**Figure 11.9**). The nozzle supplies the selected flow at the rated nozzle discharge pressure. One common model of these nozzles was a 1½-inch nozzle with a collar that the operator turned to choose flow volume ratings of 35, 60, 95, or 125 gpm at 100 psi. Another was a 2½-inch nozzle that had 150, 175, 225, and 250 gpm settings. Keep in mind that if the pump operator cannot supply the proper pressure, the actual flow will differ from that indicated at the nozzle.

Both the constant flow and adjustable constant flow nozzles have proved highly reliable since their introduction into the fire service some 40 years ago. Many of the original models remain in active service, and most nozzle manufacturers continue to produce updated versions of these nozzles today.

Automatic Nozzles

While the advent of constant flow volume nozzles solved some important problems facing firefighters and pump operators, other issues prevent these nozzles from being foolproof. One of the most significant issues is that if the pump operator fails to supply the rated nozzle pressure to the nozzle, the flow volume will be less than the nozzle rating and the stream pattern's quality may

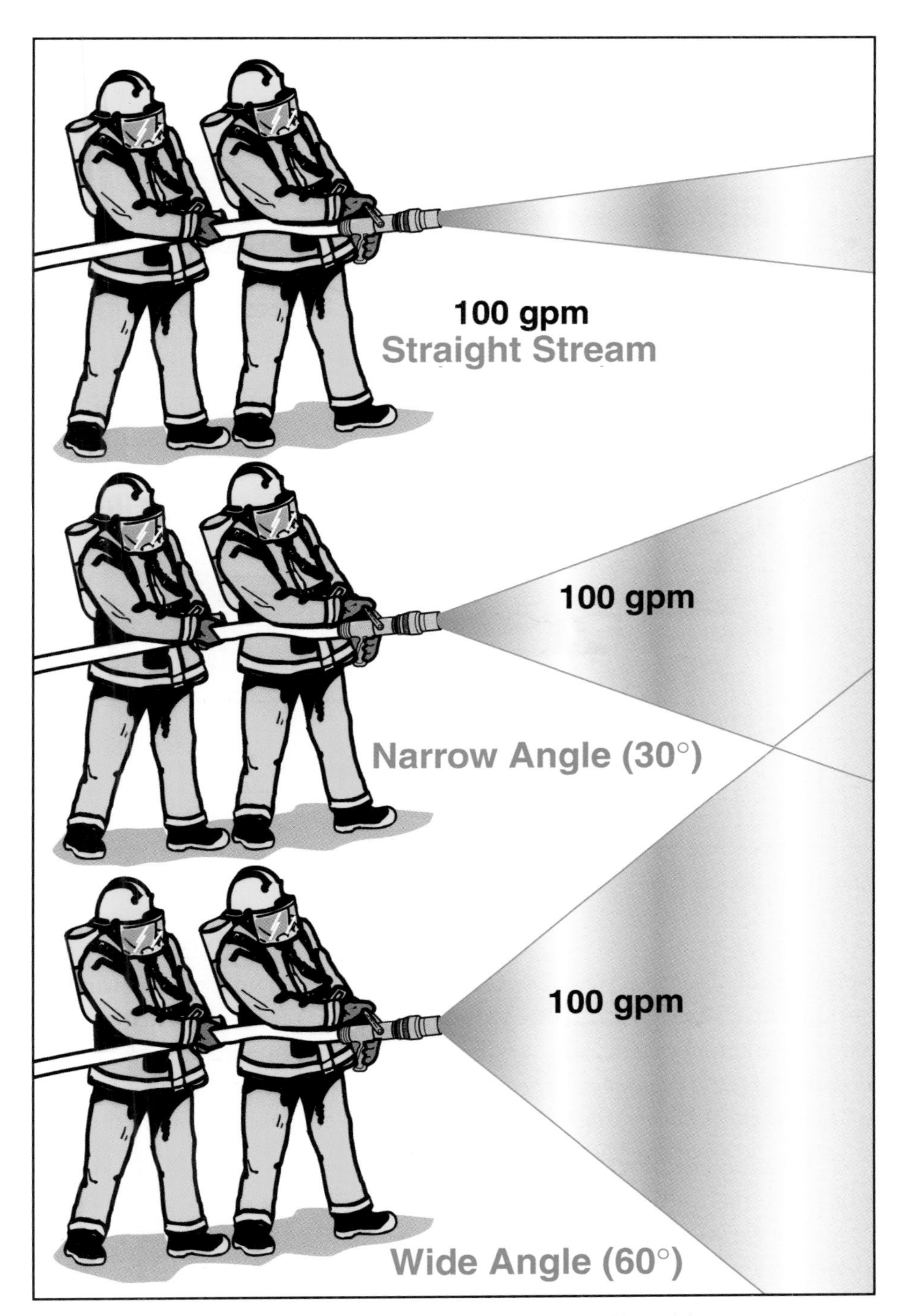

Figure 11.9 A constant flow nozzle delivers the same flow regardless of the stream pattern setting.

be unsatisfactory. On the other hand, if the pump operator supplies too much pressure to the nozzle, the nozzle becomes unwieldy and poses a physical hazard to the firefighters attempting to handle it.

The solution to these problems was the development of a *constant pressure* nozzle as opposed to a *constant flow volume* nozzle. Today's fire service more commonly refers to constant pressure nozzles as *automatic nozzles* (**Figure 11.10, p. 244**). An automatic nozzle maintains a constant nozzle pressure of

Figure 11.10 An automatic nozzle.

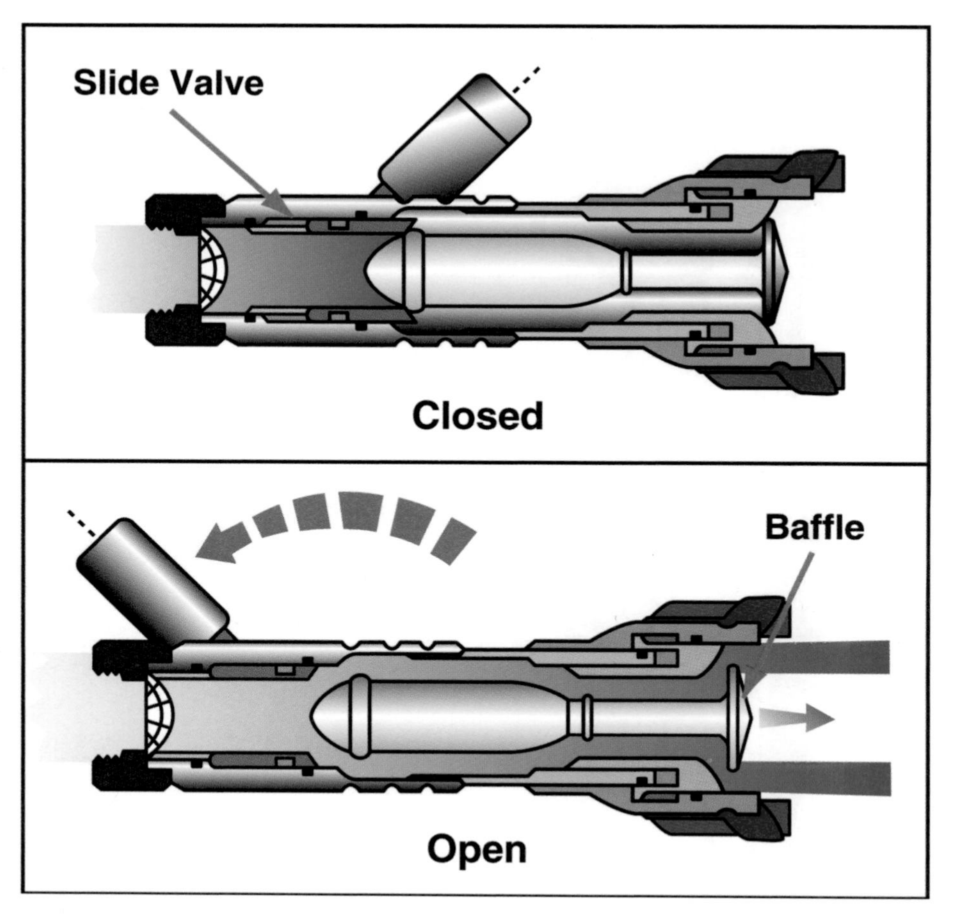

Figure 11.11 One common design for an automatic nozzle uses a slide barrel control inside the nozzle.

If the operator supplies too much pressure to an automatic nozzle, the nozzle pressure remains at 100 psi and the nozzle reaction remains constant.

approximately 100 psi (unless it is a low-pressure nozzle) no matter how much the pump discharge pressure is above this figure. This is accomplished by a baffle that moves automatically, varying the spacing between itself and the throat **(Figure 11.11)**. As pump discharge pressure increases, the nozzle automatically enlarges its effective opening size—within the nozzle's range—to match the flow.

While automatic nozzles provide a constant pressure when they are supplied at or above their design pressure (usually 100 psi), they do not also provide a constant flow volume under all conditions. Unlike the older, variable flow nozzles, at a constant pressure automatic nozzles will maintain the same flow regardless of the stream pattern. However, if the pressure (and corresponding volume) supplied to the automatic nozzle increases or decreases and nozzle pressure remains at 100 psi, the nozzle's flow volume will change accordingly. In this sense automatic nozzles technically may be considered variable flow nozzles, though in quite a different sense than the older variable flow nozzle designs.

Automatic nozzles allow pump operators somewhat safely to err on the high side in terms of supplying hose streams. If the operator supplies too much pressure to a constant flow volume nozzle, the nozzle pressure and resultant reaction can increase to dangerously unmanageable levels for the hose team. In contrast, if the operator supplies too much pressure to an automatic nozzle, the nozzle pressure remains at 100 psi and the nozzle reaction remains constant. As well, the flow volume being discharged may actually increase. This is a much safer scenario for the firefighters on the hose stream.

However, the automatic nozzles also have a downside. If the pump operator is not supplying the minimum rated pressure to the automatic nozzle, the stream pattern can look good but in reality may not be discharging sufficient water for fire extinguishment or firefighter protection.

Another potential pitfall of automatic nozzles is that they may allow the pump operator to supply too much water to the nozzle and actually exceed the incoming water supply. This is of particular concern when supplying large-volume master streams equipped with automatic nozzles. If the pump approaches the point where it is trying to discharge more water than it is taking in, the pump operator should notice that the engine is starting to speed up or become erratic or that the intake supply hose is becoming soft or beginning to collapse. The intake-pressure gauge also may be at or close to 0 psi. At this point the pump operator should back off on the engine throttle until the intake pressure gauge reads at least 10, and preferably 20, psi. Otherwise the pump may go into a damaging cavitation condition.

BROKEN STREAM NOZZLES

The third major category of fire hose nozzles is broken stream nozzles. Though in some cases the stream from a broken stream nozzle appears similar to that from a fog nozzle, in reality the two streams operate in significantly different ways. While fog stream nozzles deflect the column of water coming from the hose or use impinging streams to create the fog pattern, broken stream nozzles force the water through a series of small holes on the nozzle's discharge end. In general, broken streams produce larger droplets of water than do fog streams. This tends to give them better reach and penetrating power, which is what makes them useful for special applications.

Broken streams generally produce larger droplets of water than do fog streams, which gives them longer reach and more penetrating power.

In today's fire service, broken stream nozzles are almost exclusively special use nozzles for particular fire fighting operations. Some fire situations, such as chimney fires, basement fires, and fires in inaccessible spaces are more easily handled using a special use broken stream nozzle than a standard solid or fog stream nozzle. The following section describes some of the more common types of special use, broken stream nozzles found in the fire service. None of these special nozzles are required to be carried on fire department pumpers according to NFPA® 1901, *Standard for Automotive Fire Apparatus*. The decision to carry any of them is typically based on the experiences of the particular jurisdiction and the types of fire situation it commonly encounters. All of these nozzles can be used to discharge either plain water or Class A foam solution.

Cellar Nozzles

Cellar nozzles are specially designed to be used on basement fires that are otherwise inaccessible for standard fire fighting operations **(Figure 11.12, p. 246)**. Another common name for cellar nozzles is distributors. Most cellar nozzles consist of a playpipe that contains a 90° bend. They have a swivel inlet

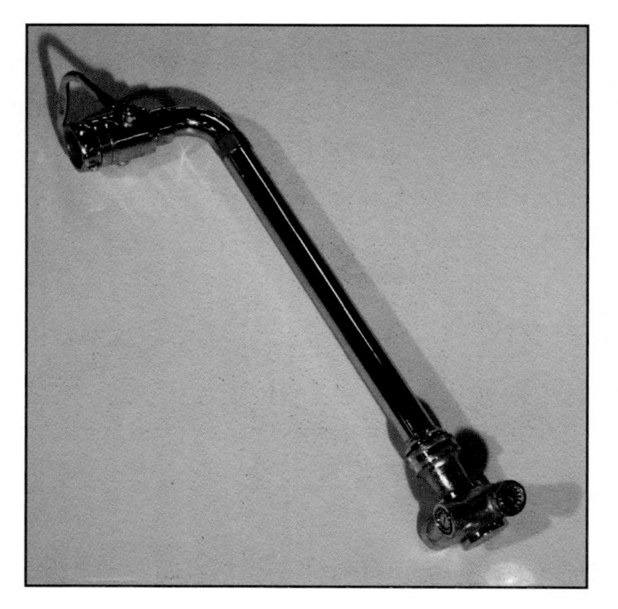

Figure 11.12 A typical cellar nozzle.

Figure 11.13 The cellar nozzle is inserted through a hole that is made in the floor.

at one end and a circular or spinning nozzle head at the other end. The nozzle head produces a cone of water that extends out from all sides and the top of the nozzle. Cellar nozzles may or may not have a shutoff valve on the play pipe. If the nozzle does not have an integral shutoff valve, an in-line shutoff valve should be placed at a convenient location back from the nozzle.

These nozzles are used primarily by lowering them into a fire area through holes cut in the floor or through other suitable openings **(Figure 11.13)**. Prior to lowering the nozzle through the hole, the firefighters should use a pike pole or similar tool to probe around the opening and ensure that nothing will obstruct the water stream.

Cellar nozzles also may be used to attack attic fires. The operation is similar to fighting a cellar fire, except that the nozzle is pushed through a hole in the ceiling to attack the fire above. Once again, when using this technique the firefighter first should probe through the hole in the ceiling with a pike pole or similar tool to ensure that no insulation or similar materials are hindering fire the stream.

Piercing Nozzles

Piercing nozzles commonly are used to apply water to areas that otherwise are inaccessible to water streams. They differ from cellar nozzles in that they eliminate the step of first creating a hole for the nozzle. Piercing nozzles create their own hole and then discharge the water onto the fire.

The piercing nozzle is generally a 3- to 6-foot-long hollow steel rod 1½ inches in diameter (**Figure 11.14;** *Courtesy of Superior Flamefighter, Inc.*). Its discharge end usually is a hardened steel point. This makes the tip suitable for driving through most materials, including masonry, metal, plywood, sheet rock, and other types of wall or partition assemblies. The point contains a series of small holes through which water or foam solution is flowed in all directions. Most piercing nozzles are designed for attachment to 1½- or 1¾-inch hose and can deliver about 100 gpm of water. Opposite the pointed nozzle end of the piercing nozzle is the driving end. This end of the nozzle is driven with a sledgehammer to force the point through the obstruction **(Figure 11.15)**.

In recent years it has become increasingly common for fire departments to equip one preconnected attack line on a pumper with a piercing nozzle. This allows the firefighters to deploy the piercing nozzle quickly to attack such fires as engine compartment fires on automobiles and structural fires in metal walled buildings.

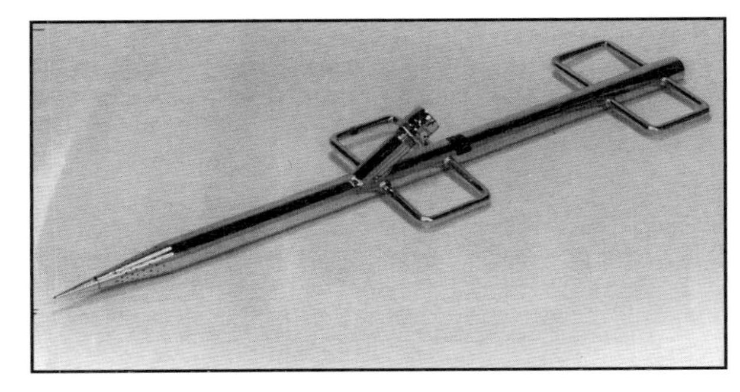

Figure 11.14 Piercing nozzles have a blunt end that is struck with a sledge to drive the pointed end through a barrier. *Courtesy of Superior Flamefighter Inc.*

Chimney Nozzles

Chimney nozzles are used to attack chimney flue fires **(Figure 11.16)**. Chimney flue fires occur when creosote that has built up on the inside of the chimney flue ignites. The result is a very hot fire that can crack the chimney's flue liner and extend into the structure. The nozzle is a solid piece of brass or steel with numerous, very small impinging holes around its circumference. The chimney nozzle is designed to be placed on the end of a booster hose. The hose and nozzle are lowered down the entire length of the chimney and then quickly pulled out **(Figure 11.17, p. 248)**.

At a nozzle pressure of 100 psi, a chimney nozzle generally produces only 1.5 to 3 gpm of water in a very fine, misty fog cone. The mist from the nozzle immediately turns to steam and chokes the flue fire as well as loosening the soot on the inside of the chimney. Because water is used in such small quantities and converts to steam so quickly, the flue liner is not damaged by the thermal shock of sudden cooling. This process may damage booster hose over time. Jurisdictions that frequently use chimney nozzles often connect an old section of booster hose to the end of the regular section when using a chimney nozzle.

Figure 11.15 At least two firefighters are needed to place a piercing nozzle in service.

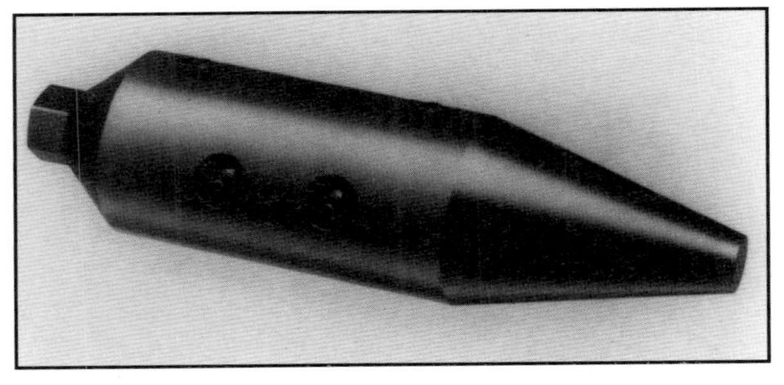

Figure 11.16 A typical chimney nozzle. *Courtesy of Jaffrey Fire Protection.*

Water Curtain Nozzles

Water curtain nozzles produce a fan-shaped stream that acts as a water curtain between a fire and a combustible material or firefighters **(Figure 11.18, p. 248)**. Some water curtain nozzles are designed for in-line use; thus they have both an intake fitting and a discharge fitting. This allows several water curtain nozzles to be placed at intervals along the line. The top of the water curtain nozzle contains a semicircular head with small holes around its circumference. These holes create the fan-shaped stream of water.

Figure 11.17 The chimney nozzle is lowered down the flue to extinguish the fire.

Figure 11.18 Water curtain nozzles were never proven to be overly effective and are rarely found in use today.

The modern North American fire service rarely uses water curtain nozzles today. One reason for this is that a water curtain is effective in absorbing only convected heat from a fire. Radiated heat is transmitted through the water curtain and can be reduced only by placing water on the exposure. When this is done, the water absorbs the radiated heat as it is produced on the exposed surface. However, water curtain nozzles are still commonly used in Europe and other areas of the world.

Thus, a water curtain between the fire and combustible material is not as effective as the same amount of water flowing over the surface of the combustible material. It is better, therefore, to direct fire streams onto exposed surfaces. This may be accomplished angling water curtain nozzles so that the water cascades down the side of the exposure being protected rather than simply spraying the water into the air.

NOZZLE PRESSURE AND NOZZLE REACTION

One of the most famous laws of physics is Newton's Third Law of Motion, which states that for every action there is an equal and opposite reaction. Newton's Law is certainly not lost on fire stream science. Anyone who has ever handled a fire hose understands that as water is discharged from a nozzle at a given pressure, a force pushes back on the firefighters handling the hoseline. This counterforce is known to the fire service as *nozzle reaction.*

Newton's Law makes it clear that the greater the nozzle discharge pressure, the greater the resulting nozzle reaction will be. This is what ultimately sets practical limits on the discharge pressure for both handline and portable master streams. If the firefighters or portable master stream device cannot handle the nozzle reaction from an excessive discharge pressure, the likely result will be a loose nozzle and hose that could injure or kill personnel in its proximity.

If firefighters or portable master stream devices cannot handle nozzle reaction from an excessive discharge pressure, the likely result will be a loose nozzle and hose that could injure or kill personnel in its proximity.

The fire service commonly assumes that the practical working limits for the velocity of fire streams are from 60 to 120 feet per second. Because we typically do not know the velocity of water moving through a hose and nozzle, we must equate this to the more common fire service measurement of pressure in pounds per square inch (psi). Most nozzles, regardless of their type or design, operate within these velocity parameters when their discharge pressures range from 25 to 100 psi.

The nozzle pressure of 100 psi at which traditional fog nozzles (handlines or master streams) are designed to operate obviously is at the upper limit of the manageability scale according to Newton's Law. That is why some jurisdictions now use low-pressure nozzles, which were originally intended solely for high-rise fire fighting, on all incidents. Because these nozzles are designed to flow their rated capacity at 75 psi, firefighters can flow the same amount of water as they would with a standard nozzle, but the reduced nozzle reaction will make the hose easier to handle.

Fog-nozzle pressure may be as much as 65 psi without making a handline unmanageable.

Fog nozzles can handle higher nozzle pressures because their inner workings dissipate the nozzle reaction to a certain extent. This is not the case with solid stream nozzles as they are essentially tapered, straight pipes with no internal obstructions of the waterway. Thus, solid stream nozzles produce more nozzle reaction than fog nozzles. Although the rule of thumb for handline solid stream nozzles is 50 psi, the nozzle pressure may be raised to 65 psi without becoming unmanageable if greater reach and volume are needed. Above this point, solid streams become increasingly difficult to handle.

In comparison, portable master stream devices equipped with solid stream nozzles should not be operated above the previously quoted 80 psi figure unless the manufacturer of the device specifically approves it. More likely, some manufacturers will specify a lower nozzle pressure, particularly for larger nozzle tips. Respecting the maximum rated nozzle pressure is especially crucial on aerial devices as exceeding it could cause a nozzle reaction that damages the aerial device.

Calculating Nozzle Reaction for Solid Stream Nozzles

Realistically and practically, determining the nozzle reaction from a given hose layout is certainly not something that a driver/operator or other fire protection professional would ever do on the fire scene. Calculations of this type would almost always be limited to preincident planning and the development of standard operating procedures. For example, you might want to determine the amount of nozzle reaction that would be created when determining hose and nozzle configurations for preconnected attack lines on an apparatus. By using the equations that follow, you can determine whether the given number of crew members realistically can handle the proposed hose lay.

The following equation is used to calculate the nozzle reaction from hoselines or appliances utilizing solid stream nozzles:

Equation 11.2

$NR = 1.57 \text{ x } d^2 \text{ x } NP$

Given:

NR = Nozzle reaction in pounds

1.57 = A constant

d = Nozzle diameter in inches

NP = Nozzle pressure in psi

Example 11.4

Determine the nozzle reaction from a 2½-inch handline equipped with a 1 ½-inch tip operating at a nozzle pressure of 65 psi.

Solution:

$NR = (1.57)(d)^2(NP)$

$NR = (1.57)(1.5)^2(65)$

$NR = (1.57)(2.25)(65)$

NR = 229.6 pounds

Example 11.5

Determine the nozzle reaction from a detachable ladder pipe that is attached to an aerial ladder and is equipped with a 2-inch solid stream tip operating at 80 psi.

Solution:

$NR = (1.57)(d)^2(NP)$

$NR = (1.57)(2)^2(80)$

$NR = (1.57)(4)(80)$

NR = 502.4 pounds

In the rare event that calculating nozzle reaction for a solid stream nozzle on the emergency scene would be necessary a simple rule-of-thumb formula gives an approximate figure. While the results will not be as accurate as those using Equation 11.2, they will be close enough to be safe.

Equation 11.3

$NR = Q/3$

Given:

NR = Nozzle reaction in pounds

Q = Total flow of water in gpm

3 = A constant

Example 11.6

Using the rule-of-thumb method, determine the nozzle reaction from a solid stream h andline nozzle flowing 328 gpm.

Solution:

$NR = Q/3$

$NR = 328/3$

NR = 109.3 pounds

Example 11.7

Using the rule-of-thumb method, determine the nozzle reaction from a solid stream portable master stream device equipped with a 1¾ inch tip at a nozzle pressure of 80 psi.

Solution:

$gpm = (29.7)(d)^2(\sqrt{NP})$

$gpm = (29.7)(1.75)^2(\sqrt{80})$

$gpm = (29.7)(3.0625)(8.94)$

$gpm = 814$

$NR = Q/3$

$NR = 814/3$

NR = 271.3 pounds

Calculating Nozzle Reaction for Fog Stream Nozzles

Fire protection professionals also can calculate the approximate nozzle reaction from a fog stream nozzle, whether it is a handline or a master stream appliance. It is important to note, however, that the calculation for fog stream nozzle reaction will be, at best, only an approximation. The exact nozzle reaction will vary depending on the nozzle's design. To calculate the nozzle reaction from hoselines or appliances utilizing fog stream nozzles, use the following equation:

Equation 11.4

$NR = 0.0505 \times Q \times \sqrt{NP}$

Given:

NR = Nozzle reaction in pounds

0.0505 = A constant

Q = Total flow through the nozzle in gpm

NP = Nozzle pressure in psi

Example 11.8

Find the nozzle reaction of a 1¾-inch hoseline with a high-rise fog nozzle flowing 175 gpm at 75 psi.

Solution:

$NR = (0.0505)(Q)(\sqrt{NP})$

$NR = (0.0505)(175)(\sqrt{75})$

$NR = (0.0505)(175)(8.66)$

NR = 76.5 pounds

Example 11.9

Find the nozzle reaction of a 1¾-inch hoseline with a standard fog nozzle flowing 175 gpm at 100 psi.

Solution:

$NR = (0.0505)(Q)(\sqrt{NP})$

$NR = (0.0505)(175)(\sqrt{100})$

$NR = (0.0505)(175)(10)$

NR = 88.4 pounds

As with solid stream nozzles, a simple rule-of-thumb formula can be used to calculate nozzle reaction for a fog stream nozzle on the emergency scene. Again, although the rule-of-thumb formula's results will not be as accurate as those of Equation 11.4, they will be close enough to be safe.

Equation 11.5

$NR = Q/2$

Given:

NR = Nozzle reaction in pounds

Q = Total flow of water in gpm

2 = A constant

Example 11.10

Using the rule-of-thumb method, determine the nozzle reaction from a fog stream handline nozzle flowing 95 gpm.

Solution:

$NR = Q/2$

$NR = 95/2$

NR = 47.5 pounds

SUMMARY

Applying the appropriate type of fire stream, flowing the necessary amount of water, remains one of the most basic yet important principles of manual fire fighting operations. Any fire protection professional who hopes to positively influence a fire fighting situation must thoroughly understand the principles of fire stream selection and application. In the previous chapter we detailed the three basic types of fire streams available for fire fighting operations: solid, fog or spray, and broken streams. In this chapter we reviewed the basic types of nozzles used to develop these streams. We also examined methods for determining the flow and nozzle reaction from these nozzles.

The information in this chapter serves as a prerequisite to the hydraulic calculations that we will address in the remainder of this text. Calculating friction loss and pump discharge pressure is not possible without first knowing the desired nozzle pressure and flows. In the remaining chapters we will incorporate all of this information.

REVIEW QUESTIONS

1. What does IFSTA generally recognize as the dividing flow line between handline and master stream nozzles?

___

2. Historically, what has the rule-of-thumb been for the ratio of the diameter of a solid stream nozzle tip to the diameter of the hoseline to which it attached?

___

3. What are the standard nozzle operating pressures for solid stream handline and master stream nozzles?

   Solid stream handline: ___

   Master stream: ___

4. Determine the flow volume from a 2¼-inch solid stream nozzle tip operating at 80 psi.

___

5. Determine the flow volume from a 5/8-inch solid stream nozzle tip operating at 50 psi.

___

6. On which operating principle do most modern fog stream nozzles operate?

___

7. What is the nozzle operating pressure for most standard fog nozzles?

___

8. State Newton's Third Law of Motion.

___

___

9. What are the fire service's commonly accepted practical working limits for velocity of fire streams?

___

10. Using Equation 11.2, p. 250, determine the nozzle reaction from a 2½-inch handline equipped with a 1⅝-inch tip operating at a nozzle pressure of 50 psi.

___

___

11. Determine the nozzle reaction from a turret nozzle that is attached to a fire boat and is equipped with a 3-inch solid stream tip operating at 100 psi.

___

___

12. Using the rule-of-thumb method, determine the nozzle reaction from a solid stream handline nozzle flowing 800 gpm.

___

___

13. Using the rule-of-thumb method, determine the nozzle reaction from a solid stream portable master stream device equipped with a 2-inch tip at a nozzle pressure of 80 psi.

____

____

14. Using Equation 11.4, p. 252, find the nozzle reaction of a 1½-inch hoseline with a fog nozzle flowing 125 gpm at 100 psi.

____

____

15. Using the rule-of-thumb method, determine the nozzle reaction from a fog stream handline nozzle flowing 250 gpm.

____

____

Answers found on page 431.

Part 5

Photo courtesty Peter Mathews

Fire Service Pressure Calculations

When asked what part of a fire service academic program the majority of students least look forward to, most educators will respond that it is the topic of fire service hydraulics. More specifically, the computation of pressure loss, pump discharge pressures, and other similar mathematical procedures are intimidating to students as well as to their teachers and to many experienced fire service practitioners. This should come as no surprise as the fear of error in mathematics is not limited to the fire profession. While mathematics is essential to most areas of science, technology, and academia, it ranks among the most unpopular topics in all courses of study.

As we will see in the section that follows, the fear of the mathematics involved in fire service hydraulics is largely unfounded. Much effort has been made over the years to simplify the calculations required of fire service professional. Most fire service computations can be performed using simple formulas that anyone with a high school education should be capable of performing. While computer programs and specially-designed electronic calculators can perform these functions, a true professional understands the theory behind them and can perform them mentally or with a simple calculator.

Chapter 12: Examines the calculation of pressure loss in various hose layouts.

Chapter 13: Explores the required pump discharge pressure to supply simple hose layouts.

Chapter 14: Discusses the methods for establishing relay pumping operations.

Chapter 15: Examines the procedures to support sprinkler and standpipe systems in buildings.

All of these tasks can be performed using relatively simple calculations. Firefighters, fire apparatus driver/operators, or other fire service personnel are not expected to perform many, or for that matter, any of these tasks on the emergency scene. Most of these calculations' value is for activities such as preincident planning, establishing set discharge pressures for preconnected hose lines, and training exercises. Working through the information in the remaining chapters of this text should substantially reduce the age-old fear of hydraulic calculations.

FESHE Course Objectives

The information in this chapter is intended to meet some of the objectives outlined for the Fire Protection Hydraulics and Water Supply course by the United States Fire Administration's National Model Core Curriculum as developed by the Fire and Emergency Services Higher Education (FESHE) initiative. The information in this chapter also allows the student to meet the requirements of NFPA® 1002, *Standard for Fire Apparatus Driver/Operator Professional Qualifications* (2009 edition) listed below.

FESHE Course Objectives

1. Calculate friction loss in hose using the historical (Underwriter's Formula) method.
2. Calculate pressure loss in single and multiple hoselines using the modern $FL = CQ^2L$ method.
3. Explain the procedure for calculating the friction loss coefficient for hose used by your fire department.
4. Describe the situations in which including appliance pressure loss is important and list the appropriate appliance loss figures.
5. Determine pressure loss or gain due to changes in elevation between the pump and the nozzle.
6. Calculate total pressure loss for simple and complex hose layouts.
7. Describe the various rule of thumb and other methods for performing field calculations of pressure loss.

NFPA® 1002 Requirements

5.2.1 (A) and 5.2.2 (A) Requisite Knowledge. Hydraulic calculations for friction loss and flow using both written formulas and estimation methods.

Chapter 12

Principles of Fire Service Pressure Loss Calculations

INTRODUCTION

Other than delivering the apparatus safely to the scene of an emergency, the most basic requirement of a pumping fire apparatus driver/operator is to deliver water at an adequate pressure and volume to the hoselines or master stream appliances that will be used to control the incident **(Figure 12.1)**. This is accomplished by making sure that a water supply has been established and that the fire pump is set to discharge water at an appropriate pressure for the hose lay or master stream device it is to supply. This necessary pump pressure is more commonly referred to as the *required pump discharge pressure*. Final calculation of the required pump discharge pressure will be covered in Chapter 13.

Figure 12.1 Supplying fire streams is one of the driver/operator's two primary functions. *Courtesy of Ron Jeffers*

Before the driver/operator can determine the required pump discharge pressure, he or she must first determine all the potential pressure losses in the hose system to be supplied. To deliver the required pump discharge pressure, these pressure losses must be overcome. Their sources include friction between the water and the inner surface of the hose or piping through which it is traveling,

Before the driver/operator can determine the required pump discharge pressure, he or she must first determine all the potential pressure losses in the hose system to be supplied.

pressure losses as the water travels through hose appliances, such as wyes or water thieves, and pressure that is lost (or gained) because of a difference in elevation between the pump and the discharge point.

This chapter begins by examining some of the older formulas and methods for calculating pressure loss in hose systems. While these older formulas are no longer pertinent to current fire hydraulics, they do provide a perspective on how the study of hydraulics has evolved, and they will give you an appreciation of how mathematical hydraulic calculations have been simplified in recent years.

The chapter next provides detailed information and examples of how to calculate pressure losses using formulas and rule-of-thumb methods. Depending on the situation (preincident planning or on the emergency scene), the pump operator or fire protection professional may have to use either of these methods and therefore must know both.

The formulas and information in this chapter are the most up-to-date methods of pressure loss calculation and information available. Despite all of the advances, however, ***mathematical and rule-of-thumb calculation of pressure loss remains, at best, an inexact science***. Do not be discouraged by the fact that given the same hose layout, you are likely to arrive at different answers using different calculation methods. Furthermore, if you then used properly calibrated gauges or flowmeters to actually measure the pressure loss through that same hose system, you might arrive at yet another figure.

None of the formulas or methods used to determine friction loss is guaranteed, or for that matter even expected, to determine the exact pressure loss. The actual figures will vary for many reasons, which are detailed in the appropriate sections of this chapter. The methods described in this chapter provide rough approximations of the pressure loss that can be expected in the given hose arrangements. This information is suitable for tactical planning hose evolutions and strategic preincident planning, although in most cases the calculations will result in somewhat higher water supply figures than those found in actual use. That provides a margin of safety, as operating at the calculated pressures will likely supply a little more water than is needed rather than a little less.

NOTE: The examples used in this chapter and the answers provided to review questions are all rounded up or down to the nearest one (1) gpm or psi. On occasion you will find that after calculating a particular problem several times your answer differs from the answer provided in the book by 1 or 2 psi or gpm. This is often caused by the manner in which particular calculators or computers round numbers during calculations. If you arrive at the same answer after calculating a problem two or three times and it remains slightly different than the one provided by this text, assume that you have performed the task successfully and write off the variance to differences in calculators.

Also, the calculations in this chapter assume that plain water is being flowed through the hoses and appliances. Any additive introduced into the water, such as Class A foam concentrates or other wetting agents, will make water "slipperier" and reduce the friction loss in the system. Tests conducted in the 1970s revealed that these types of additives could reduce friction loss as much as 40 percent. These tests, often referred to as the "wet water tests," were conducted to see if the additives could reduce friction loss while increasing the volume of water flowed through hoses of the same size. To date, no mathematical methods have been developed for calculating pressure loss or required pump discharge pressures when water additives are in use. The best recommendation is to use the methods described in this chapter and perhaps reduce the final answer by about 25 percent or so to provide a safe margin of error.

HISTORICAL METHOD OF FRICTION LOSS CALCULATIONS

To give you an appreciation of how far fire service hydraulic calculations have advanced, this chapter begins by looking at some of the older methods of computing pressure loss and required pump discharge pressure. The methods and formulas, often referred to as the Underwriter's Formulas, were among the earliest available to the fire service. Prior to their development, pump operators basically were limited to adjusting the discharge pressure on the pump until the water coming out the nozzle "seemed about right." This made preincident planning difficult or almost impossible. It also led to more frequent overpressurization of hoselines and the dangers that those situations posed to firefighters and equipment alike.

The advent of the formulas discussed in this section substantially eliminated many of these problems, and they served the fire service well for many years. Their use was prevalent even into the mid 1980s, when the more modern formulas discussed later in this chapter came into use. Some fire departments and instructors still prefer the Underwriter's Formulas and continue to use them. This is not advisable for a number of reasons including:

Although some fire departments still prefer the Underwriter's Formulas and continue to use them, a number of reasons admonish against doing so.

1. These formulas were based on hose technology of the 1930s and 1940s. Hose of that era had rubber linings that were not as refined as those in hose today. The rougher linings in that earlier hose increased friction loss. Thus, calculating pressure loss using the old formulas will produce figures significantly higher than the friction loss that actually occurs in modern hose.
2. Different formulas were required based on the amount of water being flowed. The newer formulas discussed later in the chapter are good for any flow.

Regardless of which formula you use, to produce effective fire streams you must know the amount of friction loss in the fire hose between the pumper and the nozzle. Friction loss is caused by a number of factors, such as hose and coupling condition, kinks, and bends. The primary determinant, however, is the volume of water flowing per minute, taking into account the length and diameter of the hoseline. The Underwriter's friction loss formula was the first to take into account all of these factors.

While the Underwriter's Formula would serve the fire service well for many years, it was based on a number of premises that now limit its usefulness. First, when the formula was developed, the fire service basically used only two sizes and types of fire hose: ¾- or 1-inch noncollapsible booster hose and 2½-inch double-jacketed, rubber-lined fire hose. Pump operators rarely needed to calculate friction loss in booster hose because a set amount was always stated on the reel; therefore, the friction loss formula was developed based on tests conducted on the 2½-inch hose. During the late 1950s and early 1960s other sizes of hose, such as 1½-inch attack lines and 3-inch supply lines, became more prevalent in the fire service. As we will see later in this section, those other hose sizes would require additional calculations.

The second premise of the Underwriters Formulas was that since the amount of hose used between the pumper and the nozzle would not always be the same, it would be most convenient to consider friction loss in pounds per square inch (psi) per 100 feet of hose. Thus if more than 100 feet of 2½-inch hose were used, the driver/operator would have to multiply the final answer derived using the formula by the number of hundreds of feet of hose in use. For example, if a 250 foot hose lay was in use, the formula calculated the friction loss in 100 feet of hose at that flow volume, and then the driver/operator had to multiply the answer by 2.5 (250 ÷ 100) in order to determine the total friction loss for the length of hose.

Because of the limitations of technology at the time, the developers of the Underwriter's Formula deemed it impossible to arrive at a single formula that would be accurate for all flows of water through the 2½-inch hose. Thus, as we will see later in this section, different formulas were required depending on whether the flow of water was greater or less than 100 gpm.

Calculating Friction Loss for a Single 2½-Inch Hose – Flows of 100 GPM or Greater

The base friction loss formula using the old calculation methods was the formula for determining friction loss in 100 feet of 2½-inch, double-jacketed, rubber-lined fire hose when flowing 100 gpm or more:

Equation 12.1

$$FL = 2Q^2 + Q$$

Given:

FL = Friction loss per 100 feet of 2½-inch hose

Q = Hundreds of gpm flowing (Total flow ÷ 100)

Example 12.1

Using Equation 12.1, calculate the friction loss in 100 feet of 2½-inch hose flowing 300 gpm through a fog nozzle.

$$FL = 2Q^2 + Q$$

$$FL = (2)(300/100)^2 + (300/100)$$

$$FL = (2)(3)^2 + 3$$

$$FL = (2)(9) + 3$$

$$FL = 18 + 3$$

$$\mathbf{FL = 21\ psi\ per\ 100\ feet}$$

Example 12.2

Using Equation 12.1, calculate the friction loss in 100 feet of 2½-inch hose supplying a handheld solid stream nozzle equipped with a 1¼-inch tip and a nozzle pressure of 50 psi.

$Q = (29.7)(d)^2(\sqrt{NP})$

$Q = (29.7)(1.25)^2(\sqrt{50})$

$Q = (29.7)(1.5625)(7.07)$

$\mathbf{Q = 328\ gpm}$

$FL = 2Q^2 + Q$

$FL = (2)(328/100)^2 + (328/100)$

$FL = (2)(3.28)^2 + 3.28$

$FL = (2)(10.76) + 3.28$

FL = 24.8, or 25 psi per 100 feet

The previous two examples are based on a 100-foot length of 2½-inch hose being used. Actual fireground situations may require shorter or longer lengths of hose. The following examples show the calculations necessary to compute the friction loss in 2½-inch hose that are shorter or longer than 100 feet.

Example 12.3

A portable master stream device equipped with a 250 gpm fog nozzle is being supplied by a single 50-foot section of 2½-inch hose. Calculate the friction loss in this section of hose using Equation 12.1.

$FL = 2Q^2 + Q$

$$F = (2)\left(\frac{250}{100}\right)^2 + \left(\frac{250}{100}\right)$$

$FL = (2)(2.5)^2 + 2.5$

$FL = (2)(6.25) + 2.5$

$FL = 12.5 + 2.5$

$FL = 15$ psi per 100 feet

$FL_{50\ feet} = (\text{Total length in feet}/100)(\text{FL per 100 feet})$

$$FL_{50\ feet} = \left(\frac{50}{100}\right)(15\ psi)$$

$FL_{50\ feet} = (0.50)(15\ psi)$

$\mathbf{FL_{50\ feet} = 7.5\ psi}$

Example 12.4

Using Equation 12.1, calculate the friction loss in a 2½-inch attack line that is 250 feet long and is equipped with a 1⅛-inch tip discharging at 50 psi.

$Q = (29.7)(d)^2(\sqrt{NP})$

$Q = (29.7)(1.125)^2(\sqrt{50})$

$Q = (29.7)(1.2656)(7.07)$

$\mathbf{Q = 266\ gpm}$

$FL = 2Q^2 + Q$

$$FL = (2)\left(\frac{266}{100}\right)^2 + \left(\frac{266}{100}\right)$$

$FL = (2)(2.66)^2 + 2.66$

$FL = (2)(7.08) + 2.66$

$FL = 14.16 + 2.66$

FL = 16.8, or 17 psi per 100 feet

$FL_{250\ feet}$ = (Total length in feet/100)(FL per 100 feet)

$$FL_{250\ feet} = \left(\frac{250}{100}\right)(17\ psi)$$

$FL_{250\ feet} = (2.5)(17\ psi)$

$\mathbf{FL_{250\ feet} = 42.5\ psi}$

Calculating Friction Loss for a Single 2½-Inch Hose – Flows of Less than 100 GPM

When the Underwriter's Formula was being developed, actual tests of friction loss in various hose arrangements showed that the formula was not particularly accurate for water flows less than 100 gpm. The friction loss for flows less than 100 gpm in 2½-inch hose was consistently somewhat less than calculated using Equation 12.1. The developers theorized that this reduction in friction loss occurred because such low flows were probably more laminar in form than turbulent. Laminar flows have less friction than turbulent flows. Further testing showed that the original formula could simply be adjusted slightly to produce a more accurate result for small flows:

Equation 12.2

$FL = 2Q^2 + \frac{1}{2}Q$

Given:

FL = Friction loss per 100 feet of 2½-inch hose

Q = Hundreds of gpm flowing (Total flow ÷ 100)

Example 12.5

Using Equation 12.2, calculate the friction loss in a 2½-inch hose that is 400 feet long and flowing 90 gpm.

$FL = 2Q^2 + ½ Q$

$FL = 2(0.9)^2 + (0.5)(0.9)$

$FL = 2(0.81) + 0.45$

$FL = 1.62 + 0.45$

FL = 2.07, or 2 psi per 100 feet of 2½-inch hose

$FL_{400\ feet} = \left(\frac{400}{100}\right)(2\ psi)$

$FL_{400\ feet} = (4)(2\ psi)$

$\mathbf{FL_{400\ feet} = 8\ psi}$

In reality, rarely would a 2½-inch hose ever be used to flow less than 100 gpm. Equation 12.2 saw little application prior to the advent of smaller attack lines. However, as we will see in the next section, hoselines other than 2½-inch hose require additional calculations as well.

Calculating Friction Loss for Hose Other than 2½-Inch Hose

The usefulness of the Underwriter's friction loss formula came into question as the fire service gradually gravitated to hose sizes other than 2½-inch. This included smaller diameter hoses for attack lines and larger diameter hoses for more efficient water supply operations. Obviously, it would have been impractical to develop different equations for every possible hose size at flows over and under a certain amount.

A variety of experiments revealed that the easiest approach would be to calculate the friction loss for the given flow rate in 2½-inch hose and then convert it to the flow rate for the size of hose actually in use. For example, to find the friction loss for 3-inch hose flowing 200 gpm, you would first find the friction loss for 2½-inch hose flowing 200 gpm. This process is often referred to as converting the friction loss to equivalent lengths of 2½-inch hose. Then you would either multiply or divide the 2½-inch hose friction loss by a predetermined conversion factor for the size hose being used **(Table 12.1, p. 266)**. Whether you multiply or divide is depends on your personal preference.

Example 12.6

Determine the friction loss per 100 feet of 1½-inch hose flowing 95 gpm.

Because the flow is less than 100 gpm, we will use Equation 12.2:

$FL = 2Q^2 + 1/2\ Q$

$FL = 2(0.95)^2 + (0.5)(0.95)$

$FL = (2)(0.9025) + 0.475$

$FL = 0.72 + 0.30$

FL = 2.28, or 2 psi per 100 feet of 2½-inch hose

Table 12.1
Conversion Factors for Finding Friction Loss

Single Lines

Diameter (in inches) and Type	Divide by	Multiply by
3/4 booster	0.0029	344.00
1 booster	0.011	91.00
1 1/4 rubber lined	0.025	40.00
1 1/2 rubber lined	0.074	13.50
1 3/4 rubber lined	0.16	5.95
2 rubber lined	2.60	0.385
3 rubber lined	2.60	0.385
3 rubber lined with 2 1/2 inche couplings	2.50	0.40
3 1/2 rubber lined	5.80	0.172
4 rubber lined	11.00	0.09
4 1/2 rubber lined	19.50	0.051
5 rubber lined	32.00	0.031
6 rubber lined	83.33	0.012
1 1/4 unlined linen	0.0157	63.60
1 1/2 unlined linen	0.039	25.60
2 unlined linen	0.16	6.25
2 1/2 unlined linen	0.47	2.13

Siamesed Lines of Equal Length

Diameter (in inches) and Type	Divide by	Multiply by
Two 2 1/2-inch	3.60	0.28
Three 2 1/2-inch	7.75	0.129
Two 3-inch	9.35	0.107
One 3-inch, One 2 1/2-inch	6.10	0.164
Two 2 1/2-inch, One 3-inch	11.50	0.087
Two 3-inch, One 2 1/2-inch	15.00	0.067

Sand Pipes (Diameter)

Diameter (in inches) and Type	Divide by	Multiply by
4-inches	7.50	0.133
5-inches	22.00	0.045
6-inches	52.00	0.019

To convert this figure to friction loss in 1½-inch hose, locate the appropriate conversion factors in Table 12.1. For 1½-inch hose, we have the option of multiplying the above figure by 13.5 or dividing it by 0.074 (the reciprocal of 13.5):

$FL_{1½\text{-inch}} = (2)\ (13.5) = 27$ psi per 100 feet of 1½-inch hose

—or—

$FL_{1½\text{-inch}} = 2 \div 0.074 = 27$ psi per 100 feet of 1½-inch hose

Example 12.7

Determine the total friction loss in a 400-foot length of 3½-inch hose supplying a pumper with 500 gpm.

Because the flow is greater than 100 gpm, Equation 12.1 applies:

$FL = 2Q^2 + Q$

$FL = 2(5)^2 + 5$

$FL = (2)(25) + 5$

$FL = 50 + 5$

$FL = 55$ psi per 100 feet of 2½-inch hose

$$FL_{400\ feet} = \left(\frac{400}{100}\right)(55\ psi)$$

$FL_{400\ feet} = (4)(55\ psi)$

$FL_{400\ feet} = 220$ psi

To convert this figure to friction loss in 3½-inch hose, locate the appropriate conversion factors in Table 12.1. For 3½-inch hose we have the option of multiplying the above figure by 0.172 or dividing it by 5.80 (the reciprocal of 0.172):

$FL_{3½\text{-inch}} = (220)\ (0.172) = 38$ psi for 400 feet of 3½-inch hose

—or—

$FL_{3½\text{-inch}} = 220 \div 5.80 = 38$ psi for 400 feet of 3½-inch hose

Up to this point, all of our friction loss calculations have involved single hose lines. Many fire scenes require the use of multiple hose lines, often of varying lengths and sizes. When multiple hoselines supply different nozzles or apparatus, it is necessary to calculate the friction loss for each of the individual hoselines and set the pump discharge pressure for the hose that requires the most pressure. Hoses needing less pressure simply have their individual pump discharges partially closed (often referred to as gated or gating down) until they are discharging an appropriate amount of pressure. (Chapter 13 of this text describes these principles in more detail).

When multiple hoselines supply different nozzles or apparatus, the pump operator must calculate the friction loss for each of the individual hoselines and set the pump discharge pressure for the hose that requires the most pressure.

On occasion, fire fighting operations require personnel to lay two or more parallel hose lines to supply an adequate amount of water to another pumper or a master stream device. This process of laying multiple hoses to another pumper is commonly referred to as laying "duals" (two lines) or "trips" (three lines). Two or more hoselines laid to supply a single master stream device are commonly called "siamesed lines."

Fire personnel using the old formula to calculate friction loss in multiple hose lines of equal length had a number of options available. If the hose lines were all of equal diameter, personnel could assume one-half the flow was going through each hose if there were two lines, or one-third the flow if there were three lines; then they could simply use Equation 12.1 to determine the friction loss for the flow through that one line. The other option would be to calculate the total friction loss through both or all three lines using the method described above for hose sizes other than 2½-inch. Table 12.1 provides some conversion factors for common lays of multiple hose lines.

Example 12.8

Determine the total friction loss in the hose lay when dual 2½-inch hose lines, each 150 feet long, are laid to supply a portable master stream device equipped with a 1½-inch tip discharging water at 80 psi.

$Q = (29.7)(d)^2(\sqrt{NP})$

$Q = (29.7)(1.5)^2(\sqrt{80})$

$Q = (29.7)(2.25)(8.94)$

$\mathbf{Q = 598\ gpm}$

Assume that each hoseline is carrying one-half of the total flow. Thus the flow in one of the 2½-inch hoselines would be:

$Q_{one\ line} = Q/2$

$Q_{one\ line} = 598/2$

$\mathbf{Q_{one\ line} = 299\ gpm}$

Using the flow through one line, calculate the friction loss in each hose line using Equation 12.1:

$FL = 2Q^2 + Q$

$FL = 2(2.99)^2 + 2.99$

$FL = (2)(8.94) + 2.99$

$FL = 17.9 + 2.99$

$FL = 20.89$, or 21 psi per 100 feet of 2½-inch hose

$FL_{150\ feet} = (150/100)(21\ psi)$

$FL_{150\ feet} = (1.5)(21\ psi)$

$\mathbf{FL_{150\ feet} = 31.5}$**, or 32 psi lost in the siamesed hose lay**

Example 12.9

Determine the total friction loss in the hose lay when two 3-inch hoselines and one 2½-inch hose line, each 550 feet long, are laid to supply an attack pumper with 800 gpm.

Using Equation 12.1, determine what the friction loss would be if the above flow were pumped through a single 2½-inch hoseline of the same length:

$FL = 2Q^2 + Q$

$FL = 2(800/100)^2 + (800/100)$

$FL = 2(8)^2 + 8$

$FL = (2)(64) + 8$

$FL = 128 + 8$

FL = 136 psi per 100 feet of 2½-inch hose

Next, multiply the equivalent 2½-inch hose figure by the actual distance of the hose lay:

$FL_{550\ feet} = (550/100)(136\ psi)$

$FL_{550\ feet} = (5.5)(136\ psi)$

$FL_{550\ feet}$ = 748 psi lost in a 550 foot length of 2½-inch hose flowing 800 gpm

Last, to convert this figure to friction loss in the three parallel hoselines, locate the appropriate conversion factors in Table 12.1. For two 3-inch hoselines and one 2½-inch hose line we have the option of multiplying the above figure by 0.067 or dividing it by 15 (the reciprocal of 0.067):

$FL_{siamesed\ lines}$ = (748)(0.067) = 50.1, or 50 psi

— or —

$FL_{siamesed\ lines}$ = 748 ÷ 15 = 49.9, or 50 psi

Many other examples of calculations using the old formulas could be cited; however, because this formula is no longer widely used, to do so would be pointless. The various examples above clearly demonstrate that even the simplest of friction loss calculations using the old formulas could involve multiple steps. This should give you a true appreciation of the modern methods described next.

THE MODERN FRICTION LOSS FORMULA

The fire service used the formula $2Q^2 + Q$ as the basis for mathematical friction loss calculations well into the 1980s. As hose design technology began to improve in the early 1980s, many fire professionals who conducted actual friction loss testing on hose noticed that the friction loss was substantially lower than in hose of older design technology. Correspondingly, they determined that the old friction loss formula gave significantly higher friction loss results than the field tests. Clearly, adjustments to the old formula or a completely new formula were needed to keep pace with technology.

Adjusting the old formula really was not the best answer. It was based solely on 2½-inch hose, but by the 1980s a wide variety of hose sizes from 1- to 6-inch and larger was being used. It would not be possible to easily convert the $2Q^2 + Q$ to a more versatile formula. Even if it were, the new formula still would require the additional steps of converting to equivalent lengths of 2½-inch hose and then adjusting for the length of the hose lay. A newer, simpler formula was needed.

The new formula, $FL = CQ^2L$ quietly appeared in the "Fire Streams" chapter of the NFPA® *Fire Protection Handbook,* 17th edition, in the mid-1980s. (In all my years of research, I have never conclusively identified its exact origin.) At that time, it was presented as an alternative to the $2Q^2 + Q$, and its appearance did not translate into wide acceptance by the fire service. However, fire hydraulics specialists and fire instructors quickly saw the merits of the new formula.

Almost certainly, the most influential factor in $FL = CQ^2L$ becoming the modern, primary method of mathematical friction loss calculation was IFSTA's adoption of the formula in its seventh edition of the *Fire Streams Practices* manual in 1989. IFSTA manuals were and remain the most influential fire service training manuals. All previous editions of *Fire Streams Practices* since the first in the late 1930s had propagated the use of the $2Q^2 + Q$. The members of the IFSTA *Fire Streams* seventh edition validation committee recognized the limitations of continuing to propagate the old $2Q^2 + Q$ formula and saw the new formula's advantages for fire service personnel. Thus, they made the switch to the $FL = CQ^2L$ formula. IFSTA members and the staff at their home office, Fire Protection Publications at Oklahoma State University, report that this change caused a considerable stir in the fire service training world for the first couple of years. However, once the instructors and students took the time to learn the new formula, virtually everyone agreed that it was much easier to use than the old formula.

Calculating Friction Loss in Single Hoselines

The modern friction loss formula, $FL = CQ^2L$, simplifies friction loss calculation by taking into account the size of the hose and the length of the lay in the original calculation. This eliminates the multistep calculation necessary with the old formula when anything other than 2½-inch hose was in use. The new formula is also accurate over the entire range of possible flows, eliminating the need to use different formulas for different flows.

The merits of the modern friction loss formula are easily demonstrated:

Equation 12.3

$$FL = CQ^2L$$

Given:

FL = Friction loss in psi

C = Friction loss coefficient from **Table 12.2**

Q = Flow rate in hundreds of gpm (flow/100)

L = Hose length in hundreds of feet (length/100)

The coefficients listed in Table 12.2 are those commonly used for the various sizes of hose in use today. Both IFSTA and the NFPA® use them. They are only approximations of the friction loss for each size hose. The actual coefficient for any particular piece of hose varies with the condition of the hose and the manufacturer. The coefficients provided in this manual provide results that reflect a worst-case situation. In other words, the results are probably slightly higher

Table 12.2
Friction Loss Coefficients

Single Hoselines

Hose Diameter (in inches)	Coefficient
3/4 booster	1,100
1 booster	150
1 1/4 booster	80
1 1/2	24
1 3/4	15.5
2	8
2 1/2	2
3 with 2 1/2- inch couplings	0.8
3 with 3-inch couplings	0.677
3 1/2	0.34
4	0.2
4 1/2	0.1
5	0.08
6	0.05

Sandpipes

Diameter (in inches)	Coefficient
4	0.374
5	0.126
6	0.052

than the actual friction loss. Departments that require more exact friction loss calculations should obtain coefficients from the manufacturer of their fire hose or through the calculation covered later in this chapter.

The steps for determining friction loss in a single hoseline using Equation 12.3 are:

Step 1: Obtain the friction loss coefficient for the hose being used from Table 12.2.

Step 2: Determine the number of hundreds of gallons of water per minute flowing (Q) through the hose by using the equation Q = gpm/100.

Step 3: Determine the number of hundreds of feet of hose (L) by using the equation L = feet/100.

Step 4: Substitute the numbers from Steps 1, 2, and 3 in Equation 12.2 to determine the total friction loss.

Example 12.10

Determine the friction loss in a 200-foot preconnected 1¾-inch attack line that is flowing 150 gpm.

$$FL = CQ^2L$$

$$FL = (15.5)\left(\frac{150}{100}\right)^2\left(\frac{200}{100}\right)$$

$$FL = (15.5)(1.5)^2(2)$$

$$FL = (15.5)(2.25)(2)$$

$$\mathbf{FL = 69.75, or\ 70\ psi}$$

Example 12.11

Determine the friction loss in a 1,000-foot lay of 5-inch supply hose that is flowing 500 gpm.

$$FL = CQ^2L$$

$$FL = (0.08)\left(\frac{150}{100}\right)^2\left(\frac{1,000}{100}\right)$$

$$FL = (0.08)(5)^2(10)$$

$$FL = (0.08)(25)(10)$$

$$\mathbf{FL = 20\ psi}$$

Example 12.12

A 2½-inch attack line that is 150 feet long has a reducer at its discharge end to which 100 feet of 1½-inch hose is attached (Figure 12.2). The nozzle is flowing 95 gpm. Determine the total friction loss in the hose lay. (**Note:** The pressure loss created by the reducer is not considered in this example; this will be added in later in this chapter.)

$$FL_{2½} = CQ^2L$$

$$FL_{2½} = (2)\left(\frac{95}{100}\right)^2\left(\frac{150}{100}\right)$$

$$FL_{2½} = (2)(0.95)^2(1.5)$$

$$FL_{2½} = (2)(0.9025)(1.5)$$

$$\mathbf{FL_{2½} = 2.7, or\ 3\ psi}$$

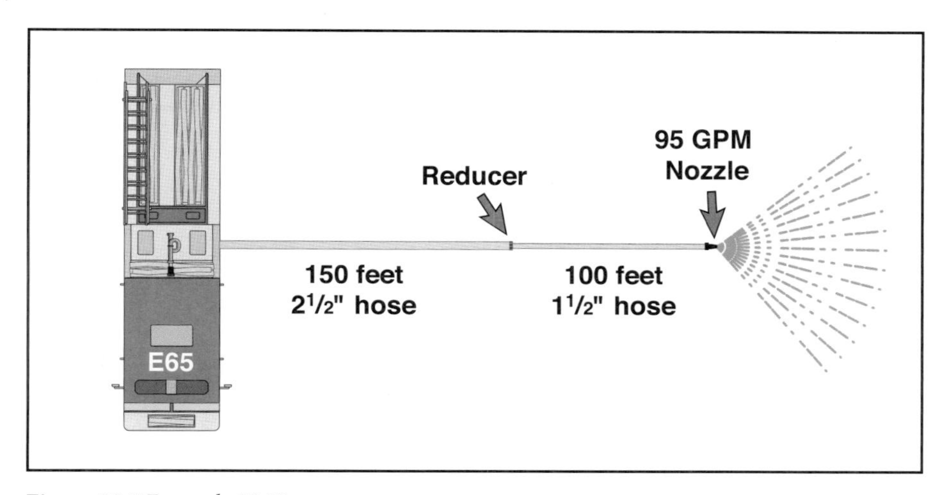

Figure 12.2 Example 12.12.

$$FL_{1½} = CQ^2L$$

$$FL_{1½} = (24)\left(\frac{95}{100}\right)^2\left(\frac{100}{100}\right)$$

$$FL_{1½} = (24)(0.95)^2(1)$$

$$FL_{1½} = (24)(0.9025)(1)$$

$$FL_{1½} = 21.7, \text{ or } 22 \text{ psi}$$

$$FL_{Total} = FL_{2½} + FL_{1½}$$

$$FL_{Total} = 3 + 22$$

$$\mathbf{FL_{Total} = 25 \text{ psi}}$$

Calculating Friction Loss in Siamesed Hoselines (Equal Length)

To increase the flow rate through a given hose, additional pressure is needed to overcome friction in the hose. When conditions require great volumes of water or when hose lays are long, the friction loss created by the desired flow of water may become so excessive that the pump and hose cannot supply that volume of water within their design limitations. In the absence of hose with a larger diameter, firefighters may keep friction loss within reasonable limits by laying two or more parallel hoselines to supply the needed quantity of water. Fire personnel have three options at the discharge end of these parallel lines:

1. They may connect the lines into a multiple inlet master stream device.
2. They may connect the lines into a siamese appliance that in turn discharges a single hose line.
3. They may connect the lines into the intakes of one or more fire apparatus equipped with pumps.

Of those three possibilities, only the second meets the true definition of siamesed hoselines. However, from a hydraulic calculation standpoint the three options are identical. The same formula that is used for single hoselines can also be used for parallel or siamesed hoselines. When calculating friction loss in siamesed lines, however, it is necessary to use a different set of coefficients (C) than for single hoselines. These coefficients are found in **Table 12.3, p. 274.**

Table 12.3
Friction Loss Coefficients

Siamese or Parallel Hoselines

Number of Hoses and Their Diameters (Inches)	Coefficient
Two 2½-inch	0.5
Three 2½-inch	0.22
Two 3-inch with 2½-inch couplings	0.2
One 3-inch with 2½-inch couplings, one 2½-inch	0.3
One 3-inch with 3-inch couplings, one 2½-inch	0.27
Two 2½-inch, one 3-inch couplings, with 2½-inch couplings	8
Two 2½-inch, with 2½-inch couplings, one 2½-inch	0.12

The steps for determining friction loss in siamesed lines are:

Step 1: Compute the total number of hundreds of gpm flowing by using the equation:

$$Q = \frac{\text{gpm flowing}}{100}$$

Step 2: (If applicable) Determine the friction loss in the attack line using Equation 12.3 ($FL = CQ^2L$). Use Table 12.1 for the coefficient in this step.

Step 3: Determine the amount of friction loss in the siamesed lines using Equation 12.3 ($FL=CQ^2L$). Use Table 12.2 to obtain the coefficient in this step.

Step 4: (If applicable) Add the friction loss from the siamesed lines, attack line, and 10 psi for the siamese appliance if flow is greater than 350 gpm to determine the total friction loss.

The advantage of laying parallel hoselines to reduce friction loss is readily apparent in comparisons of actual test results or of the coefficients in Tables 12.1 and 12.2. Actual tests show that when two hoselines of equal lengths are siamesed to supply a fire stream, the friction loss is approximately 25 percent of that in a single hoseline at the same flow and nozzle pressure. When three hoselines of equal length are siamesed, the friction loss is approximately 10 percent of that in a single line at the same flow and nozzle pressure.

Similarly, Table 12.1 gives a coefficient of 2 for a single 2½-inch hoseline, while Table 12.2 gives a coefficient of only 0.5 for two 2½-inch hoselines. Mathematically, 0.5 is 25 percent of 2. Further, Table 12.2 gives a coefficient of just

0.22 for three 2½-inch hoselines. Mathematically, 0.22 is roughly 10 percent (actually 11 percent) of 2. Thus the coefficients used with the modern friction loss formula emulate the test information described in the previous paragraph.

Example 12.13

A pumper attached to a fire hydrant is supplying 600 gpm through 400 feet of dual 3-inch hoses equipped with 2½-inch couplings to a pumper at the fire scene. What is the friction loss in the hose lay?

$$FL = CQ^2L$$

$$FL = (0.2)\left(\frac{600}{100}\right)^2\left(\frac{400}{100}\right)$$

$$FL = (0.2)(6)^2(4)$$

$$FL = (0.2)(36)(4)$$

$$\mathbf{FL = 28.8 \text{ or } 29 \text{ psi}}$$

Example 12.14

A pumper is supplying a portable master stream device equipped with a 1 ⅜-inch solid stream nozzle discharging at 80 psi. A single 100-foot length of 3-inch hose with 3-inch couplings is supplying the master stream device. This hose is in turn being supplied by siamesed lines consisting of one 3-inch hose with 3-inch couplings and one 2½-inch hose, each 300 feet long **(Figure 12.3)**. What is the friction loss in the hose lay?

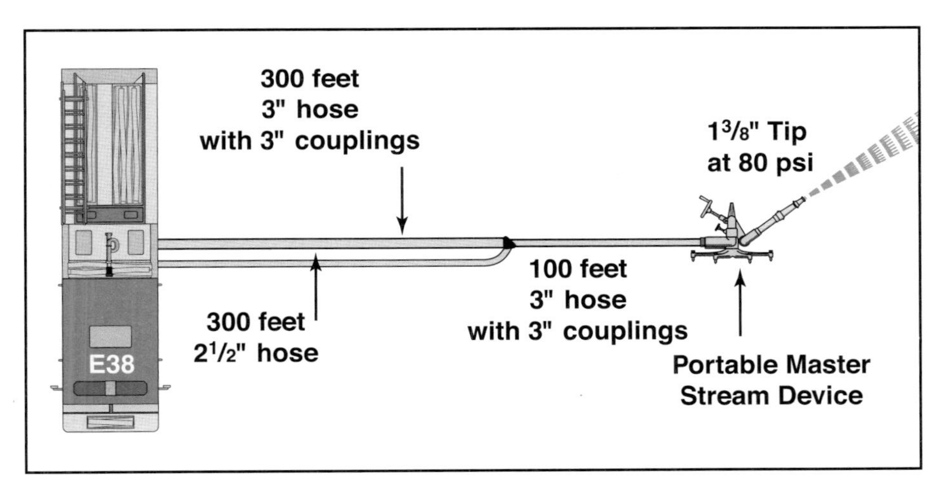

Figure 12.3 Example 12.14.

$$GPM = 29.7\,(d)^2(\sqrt{NP})$$

$$GPM = (29.7)(1.375)^2(\sqrt{80})$$

$$GPM = (29.7)(1.89)(8.94)$$

$$\mathbf{GPM = 502}$$

Note: Because the quantity exceeds 350 gpm, appliance friction loss for the siamese device and the master stream nozzle will need to be factored in later.

$$FL_{3\text{-inch}} = CQ^2L$$
$$FL_{3\text{-inch}} = (0.677)\left(\frac{502}{100}\right)^2\left(\frac{100}{100}\right)$$
$$FL_{3\text{-inch}} = (0.677)(5.02)^2(1)$$
$$FL_{3\text{-inch}} = (0.677)(25.2)(1)$$
$$\mathbf{FL_{3\text{-inch}} = 17.1 \text{ or } 17 \text{ psi}}$$

$$FL_{\text{siamesed lines}} = CQ^2L$$
$$FL_{\text{siamesed lines}} = (0.27)\left(\frac{502}{100}\right)^2\left(\frac{300}{100}\right)$$
$$FL_{\text{siamesed lines}} = (0.27)(5.02)^2(3)$$
$$FL_{\text{siamesed lines}} = (0.27)(25.2)(3)$$
$$\mathbf{FL_{\text{siamesed lines}} = 20.4, \text{ or } 20 \text{ psi}}$$

DETERMINING YOUR OWN FRICTION LOSS COEFFICIENTS

The coefficients in Tables 12.1 and 12.2 represent commonly accepted averages for particular sizes of hose. In real life, the actual coefficient that should be applied to a specific hose varies with a number of factors, including the:

- Age of the hose
- Manufacturer of the hose
- Design of the hose (e.g., rubber liner versus synthetic liner)
- How well the hose has been maintained
- Number of bends, cracks, and creases in the hose

Departments that wish to know the exact coefficients of the hose they have in service can get this information in two ways:

- They can request the specific friction coefficient for a particular hose from the manufacturer when it is purchased.
- They can test their own hose to determine the actual coefficients.

If a department has the equipment, the procedure for determining friction loss and calculating the coefficient for a particular hose is rather simple. Before performing these tests, however, several basic principles must be considered. Getting results that accurately indicate averages that can be expected on the fireground requires using the same hose that will be used on the fireground. It is best to conduct tests on hose that is in service, not hose that is sitting on a storage rack or that has never been put into service (unless new hose is about to be placed into service).

When conducting tests on a particular type of hose, it is best to test only one type at a time. For example, if the department is testing 3-inch cotton/polyester, double-jacket hose, do not use any hose of different size or construction along with test hose. Different hose have varying amounts of friction loss due to differences in construction, fabrics, rubber liners, couplings, and wear. If the department regularly mixes different kinds of hose, it may be very difficult to determine a friction loss coefficient suitable for all situations.

Test results are only as accurate as the equipment used to obtain them. Thus, it is important that all measuring devices (pitot tubes, in-line gauges, flowmeters, and so on) be properly calibrated for optimum results. The following is a list of the equipment needed to conduct these tests:

- Pitot tube or flowmeter
- Two in-line gauges, preferably calibrated in increments of 5 psi or less
- Hose to be tested
- Smoothbore nozzle if using pitot tube
- Any type nozzle if using flowmeter
- Worksheet 12.1

Test results are only as accurate as the equipment used to obtain them.

The following is a step-by-step procedure for determining friction loss in any size hose **(Figure 12.4)**:

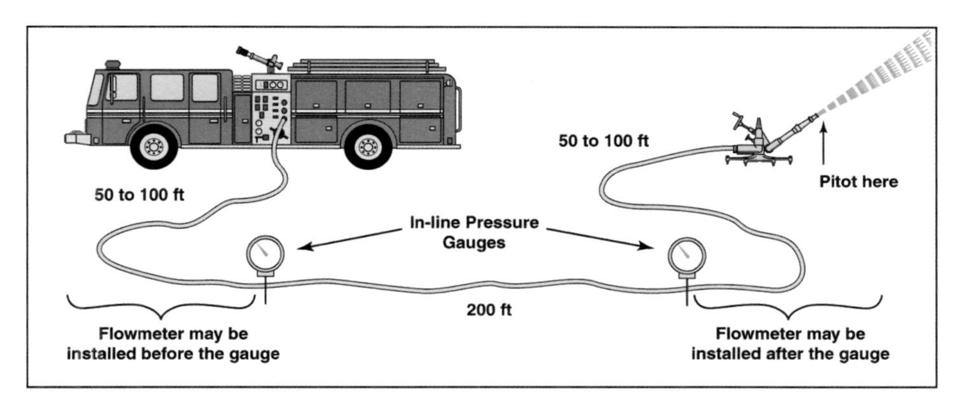

Figure 12.4 This layout can be used to determine a friction loss coefficient for any hose.

Step 1: Lay out on a level surface the lengths of hose to be tested. Lay out 300 feet if the hose is in lengths of 50 feet or 400 feet if it is in lengths of 100 feet.

Step 2: Connect one end of the hoseline to a discharge on the pumper being used to carry out the tests. Connect a nozzle to the opposite end of the hoseline. If using a pitot tube to determine the nozzle pressure and corresponding flow of water, use a smoothbore nozzle. If using a flowmeter, any nozzle is suitable (Figure 8.6).

Step 3: Insert gauge 1 in the hoseline at the connection between the first and second sections of hose away from the discharge. Make this connection 50 feet from the pumper if the hose being tested is in 50-foot lengths or 100 feet from the pumper if it is in 100-foot lengths.

Step 4: Insert gauge 2 at a distance of 200 feet from gauge 1, regardless of the length of the hose sections. You should have 50 feet of hose between gauge 2 and the nozzle if you are using 50-foot hose lengths or 100 feet of hose between gauge 2 and the nozzle if you are using 100-foot hose lengths. If using a portable flowmeter, **DO NOT** insert it anywhere between the gauges. It may be inserted anywhere in the hoseline except between the test gauges.

Step 5: When all appliances have been inserted into the hoseline, begin the tests. Supply water to the hoseline at a constant pump discharge pressure for the duration of each test run. Make three or four test runs for each size hose; you may use different pump discharge pressures for different runs as long as they remain constant within the same test run. Use sufficient pump discharge pressure to produce a satisfactory fire stream from the nozzle.

Step 6: Once water is flowing, record the pump discharge pressure, the readings from gauge 1 and gauge 2 and from the flowmeter or pitot gauge in the appropriate spaces of **Worksheet 12.1**.

Step 7: Complete Worksheet 12.1 as instructed on the table. This provides the friction loss coefficient for your particular hose.

Worksheet 12.2 represents the test results from a fire department that chose to determine its own coefficient for its 2½-inch hose. The department tested the hose at three different pump discharge pressures. By completing the work sheet, the department determined that the actual coefficient for its 2½-inch hose was 1.6. This represents a 20 percent decrease from the standard coefficient for 2½-inch hose used in this manual (C = 2).

The actual testing and calculation of specific coefficients will give the department a more accurate picture of the friction loss characteristics in their hose *at that point in time.* As hose ages and is used repeatedly for emergency operations, its internal condition will eventually deteriorate and cause more friction loss. This means that the friction coefficient will also change over time. In most cases the coefficient will become larger due to increased friction loss in the hose. The only way to know accurate, hose-specific friction loss coefficients for particular hose over its lifetime is to test it periodically and adjust the coefficient as results dictate.

The only way to know accurate, hose-specific friction loss coefficients for particular hose over its lifetime is to test it periodically and adjust the coefficient as results dictate.

DETERMINING ELEVATION PRESSURE

To this point we have focused solely on determining the pressure lost due to friction inside the hose. Friction causes only part of the total pressure loss within a hose lay. Another cause of pressure loss (or gain) in a hose lay is a change in elevation between the pump and nozzle. Fireground operations often require the use of hoselines at varying elevations, such as up a ladder, on an upper story of a building, or in a basement. The effects of elevation on the pressure in the hose lay must be considered when determining total pressure loss.

Worksheet 12.1
Friction Loss Coefficient Determination Chart

Date:___/___/____ **Hose Size ________ Inches** **Hose Construction ______________**

Person Conduction Tests ____

Column 1 Test Run No.	Column 2 Pump Discharge Pressure psi	Column 3 Pressure@ Gauge 1 psi	Column 4 Pressure@ Gauge 2 psi	Column 5 Nozzle Pressure* psi	Column 6 Flow From Flowmeter or by Equation**	Column 7 $\left(\frac{GPM}{100}\right)^2$ or $\left(\frac{Col.6}{100}\right)^2$	Column 8 Friction Loss per 100 feet or $\left(\frac{Col. 3-Col. 4}{2}\right)$	Column 9 C $\left(\frac{Col. 8}{Col. 7}\right)$
1								
2								
3								
4								

Average C= $\frac{\text{Total of all Col. 9 Answers}}{\text{No. of Tests Conducted}}$

Average C= ☐

* Not necessary if flowmeter is used

$^{**}GPM = 29.7\ d^2 \sqrt{NP}$

Worksheet 12.2
Friction Loss Coefficient Determination Chart

Date: ___/___/____ **Hose Size** ________ **Inches** **Hose Construction** ____________

Person Conduction Tests ____________________________________

Column1 Test Run No.	Column2 Pump Discharge Pressure psi	Column 3 Pressure@ Gauge 1 psi	Column 4 Pressure@ Gauge 2 psi	Column 5 Nozzle Pressure* psi	Column 6 Flow From Flowmeter or by Equation**	Column 7 $\left(\frac{GPM}{100}\right)^2$ or $\left(\frac{Col.6}{100}\right)^2$	Column 8 Friction Loss per 100 feet or $\left(\frac{Col.\ 3-Col.\ 4}{0.2}\right)$	Column 9 C $\left(\frac{Col.\ 8}{Col.\ 7}\right)$
1	**130**	**122**	**107**	—	**220**	$\left(\frac{220}{100}\right)^2=$ **4.84**	**7.5**	**1.56**
2	**150**	**143**	**123**	—	**250**	$\left(\frac{220}{100}\right)^2=$ **6.25**	**10**	**1.6**
3	**170**	**157**	**131**	—	**280**	$\left(\frac{280}{100}\right)^2=$ **7.84**	**13**	**1.64**
4	—	—	—	—	—	—	—	—

Average C= $\dfrac{\text{Total of all Col. 9 Answers} \quad \mathbf{4.83}}{\text{No. of Tests Conducted}}$

Average C= **1.6**

* Not necessary if flowmeter is used

**GPM = $29.7\ d^2 \sqrt{NP}$

As discussed in Chapters 2 and 3, water exerts a pressure of 0.433 psi per foot of elevation. Another way of looking at the same principle is that a column of water must be 2.31 feet high to create 1 psi of pressure at its bottom. When a nozzle is operating at an elevation higher than the apparatus, this pressure is exerted back against the pump **(Figure 12.5)**. To compensate for this pressure "loss," the driver/operator must add the elevation pressure to the friction loss to determine total pressure loss. On the other hand, operating a nozzle lower than the pump results in pressure pushing against the nozzle **(Figure 12.6)**. Subtracting the elevation pressure from the total friction loss compensates for this "gain" in pressure, and the pump has to do less work to supply the proper discharge pressure to the nozzle.

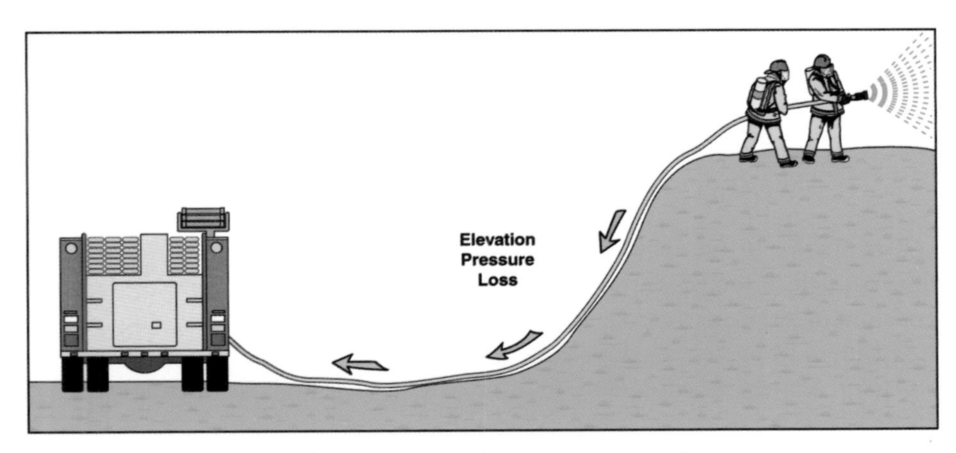

Figure 12.5 In this case, an elevation pressure loss would occur at the pump.

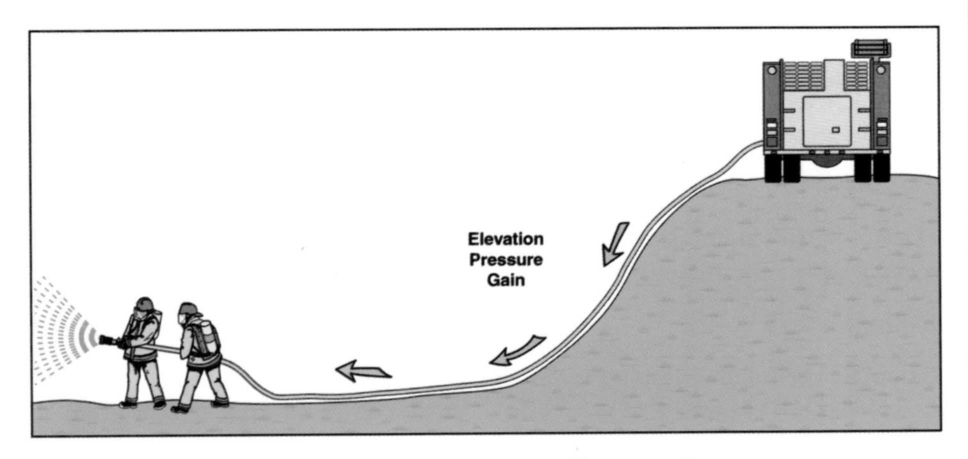

Figure 12.6 In this case an elevation pressure gain would occur at the pump.

In Chapter 2 we illustrated the calculation of elevation (head) pressure using Equation 2.3, Pressure Loss = (0.433)(height). Using Equation 2.3 on an emergency scene would be highly impractical without the assistance of an electronic calculator. To simplify elevation pressure calculations on the fireground, you may use either of the following formulas:

Equation 12.4

$EP = 0.5H$

Given:

EP = Elevation pressure (loss or gain) in psi

0.5 = A constant

H = Height in feet

— or—

Equation 12.5

$EP = H/2$

Given:

EP = Elevation pressure (loss or gain) in psi

2 = A constant

H = Height in feet

Using either of these equations will provide the pump operator with a workable figure for elevation pressure loss or gain. The difference between the scientific constant (0.433) and the rule of thumb constant (0.5) will provide a figure that is within ± 13 percent of the actual elevation pressure. This difference can be considered fairly inconsequential in the overall pressure loss calculation.

To estimate the pressure loss or gain in a multistory building, many fire professionals prefer to use the following equation based on the number of floors rather than feet:

Equation 12.6

EP = (5 psi)(number of stories -1)

This formula assumes that each story of the structure accounts for an elevation change of about 10 feet by using the rule-of-thumb constant (5 psi ¸ 0.5 psi/foot) or about 11.5 feet using the scientific constant (5 psi ¸ 0.433/foot). Both of these figures are most accurate for commercial and industrial structures but may be a little too high for multistory single-family dwellings.

Example 12.15

Calculate the total elevation pressure loss for a hoseline operating at the top of an 80-foot hill **(Figure 12.7)**.

$EP = 0.5H$

$EP = (0.5)(80)$

EP = 40 psi

— or —

$EP = H/2$

$EP = 80/2$

EP = 40 psi

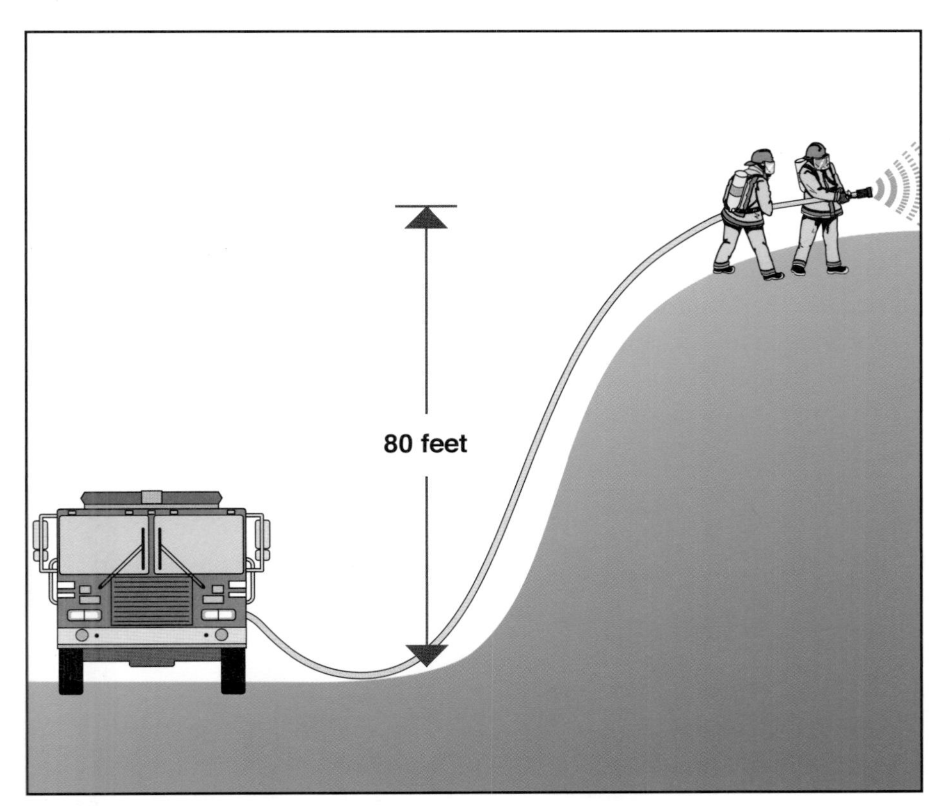

Figure 12.7 Example 12.15.

Example 12.16

A hoseline connected to a building's standpipe system is operated on a fire on the 13th floor. What is the total pressure loss due to elevation at the base of the standpipe system?

EP = (5 psi)(number of stories –1)

EP = (5 psi)(13 –1)

EP = (5)(12)

EP = 60 psi

HOSE LAYOUT APPLICATIONS

Because the modern friction loss and elevation pressure formulas are so simple, the preceding section has covered all the necessary basics for using them. However, anyone who has ever observed a modern fireground operation can attest to the fact that rarely are the hose lays as simple as those calculated in the previous section. Almost endless configurations of fire hose may be used during the course of emergency scene operations. The section that follows will detail some of the more common configurations and demonstrate how to calculate the total pressure loss, including friction loss, elevation pressure, and appliance loss, for these situations.

Appliance Pressure Loss

In addition to hose friction loss and elevation pressure, the various appliances used in fireground hose also cause pressure loss. To make the connections necessary for the situation, fireground operations often require the use of hoseline appliances including reducers, increasers, gates, wyes, siameses, manifolds, aerial apparatus, and standpipe systems **(Figure 12.8)**. As water flows from a hose into and through these devices, a certain amount of pressure or friction loss occurs. The amount of friction loss varies with each type of appliance. Appliance friction loss generally is considered significant only in cases where the total flow through the appliance is at least 350 gpm, so we will not include it in calculations for flows less than that.

Figure 12.8a Reducer.

Figure 12.8b Gated wye.

Figure 12.8c LDH manifold.

Figure 12.8d Aerial device piped waterway. *Courtesy of Harvey Eisner*

We will assume a 0 psi loss for flows less than 350 gpm and a 10 psi loss for *each appliance* (other than master stream devices) in a hose assembly when flowing 350 gpm or more. These include:

- Wyes—gated or not
- Siameses
- Reducers and increasers
- Hose manifolds, including LDH manifolds
- Fire department connections (FDC) to sprinkler and standpipe systems if the pressure loss in the system piping is calculated separately
- Hose discharges in standpipe systems if the pressure loss in the system piping is calculated separately

We will not consider friction loss caused by handline nozzles, as it is generally insignificant in the overall hose assembly.

We also will assume a friction loss of 25 psi for all flows through the following systems or devices:

- All master stream appliances regardless of the flow and whether they are fixed or portable, this includes detachable aerial ladder pipes
- Fixed aerial apparatus waterway systems, including the nozzle
- Standpipe systems, as a rule-of-thumb including the FDC, outlet, and piping

As with fire hose, the only sure way for a department to determine the exact friction loss of each appliance is to conduct its own friction loss tests.

Simple Hose Layouts

The majority of fire fighting operations are conducted with fairly simple hose layouts. These include single hoselines, equal length multiple hoselines, equal length wyed hoselines, and equal length siamesed hoselines. These layouts do not present great difficulty in determining total pressure loss; however, little nuances may be present and easy to overlook for inexperienced fire personnel.

Single Hoselines

The most commonly used hose lay is the single-hoseline layout. This hose lay, whether used as an attack line or supply line, requires the simplest friction loss calculations. The previous section used some very basic friction loss calculations for single hoselines to explain the modern friction loss formula. The following examples demonstrate how to determine the total pressure loss in a single-hoseline layout to include other factors, such as appliance loss and elevation pressure.

Example 12.17

A pumper is supplying a 300-foot hoseline with 125 gpm flowing. The hoseline is composed of 200 feet of 2½-inch hose reduced to 100 feet of 1¾-inch hose **(Figure 12.9)**. What is the pressure loss due to friction in the hose assembly?

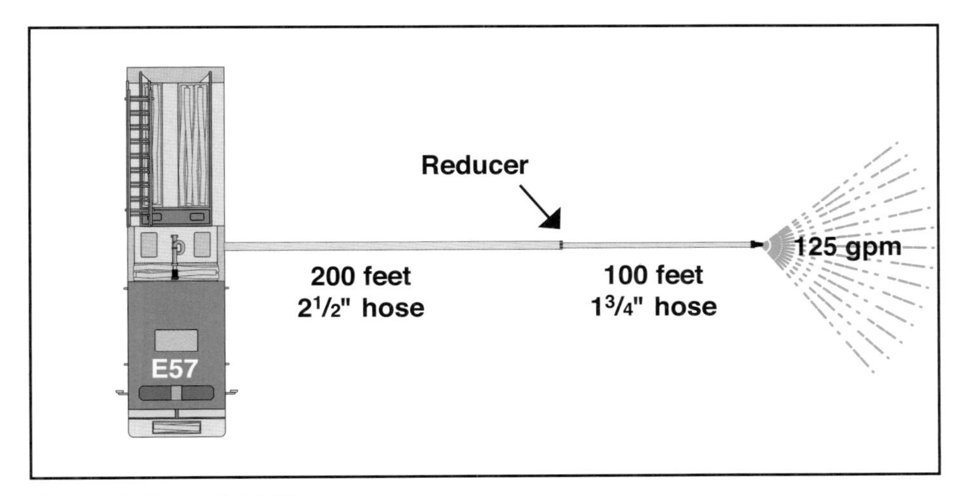

Figure 12.9 Example 12.17.

$FL_{2½} = CQ^2L$

$FL_{2½}(2)\left(\frac{125}{100}\right)^2\left(\frac{200}{100}\right)$

$FL_{2½} = (2)(1.25)^2(2)$

$FL_{2½} = (2)(1.5625)(2)$

$FL_{2½} = 6.25$, or 6 psi

Change the formula to $FL_{1¾}$:

$FL_{1¾} = (15.5)(1.25)^2(1)$

$FL_{1½} = (15.5)(1.5625)(1)$

$FL_{1½} = 24.22$, or 24 psi

$FL_{total} = FL_{2½} + FL_{1¾}$ + Appliance Loss (if greater than 350 gpm)

FL_{total} = 6 psi + 24 psi + 0 psi (flow < 350 gpm)

FL_{total} = 30 psi

Example 12.18

A fire is discovered on the eighth floor of a structure. An arriving engine company proceeds to the seventh floor of the structure and connects 150 feet of 2½-inch hose to the standpipe outlet and proceeds to the eighth floor to attack the fire. The hoseline is equipped with a 1⅛-inch tip solid stream nozzle that will be flowed at 50 psi. Determine the total pressure loss due to friction and elevation pressure at the standpipe system fire department connection. (Use the rule-of-thumb for friction loss in the standpipe system.)

$Q = 29.7(d)^2(\sqrt{NP})$

$Q = (29.7)(1.125)^2(\sqrt{50})$

$Q = (29.7)(1.2656)(7.07)$

$Q = 266$ gpm

$FL_{2½} = CQ^2L$

$FL_{2½}(2)\left(\frac{266}{100}\right)^2\left(\frac{150}{100}\right)$

$FL_{2½} = (2)(2.66)^2(1.5)$

$FL_{2½} = (2)(7.08)(1.5)$

$FL_{2½} = 21.2$, or 21 psi

EP = (5 psi)(number of stories – 1)

EP = (5 psi)(8 – 1)

EP = (5)(7)

EP = 35 psi

$FL_{total} = FL_{2½} + EP + \text{standpipe system loss}$

$FL_{total} = 21 \text{ psi} + 35 \text{ psi} + 25 \text{ psi}$

$\mathbf{FL_{total} = 81 \text{ psi}}$

Example 12.19

A pumper is supplying a portable master stream device equipped with a 500 gpm fog nozzle. The master stream device is being supplied by 300 feet of 4-inch hose and is located on top of a 20-foot embankment **(Figure 12.10)**. Determine the total pressure loss due to friction and elevation pressure at the pumper.

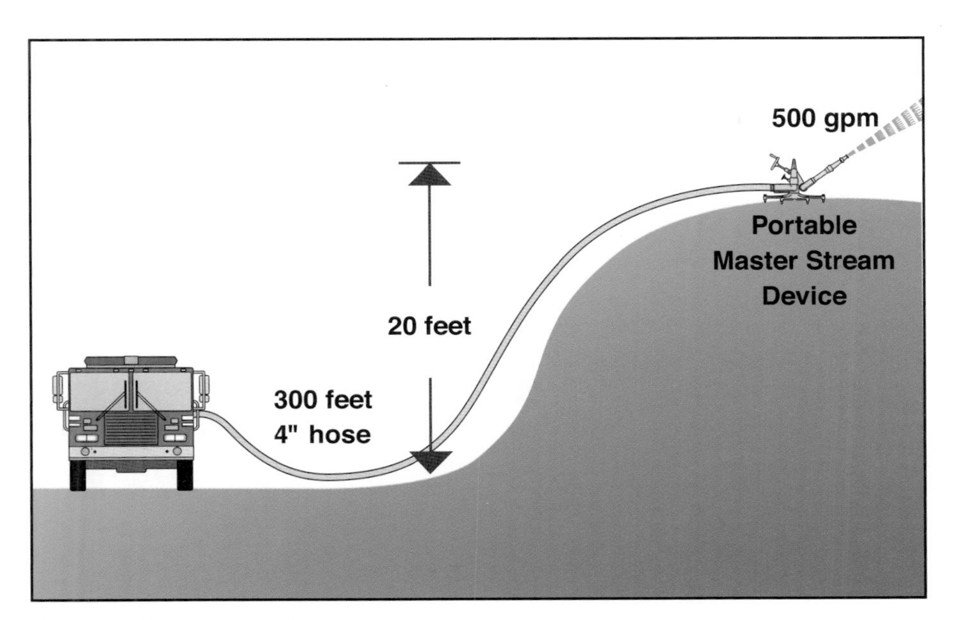

Figure 12.10 Example 12.19.

$FL_4 = CQ^2L$

$FL_4(0.2)\left(\frac{500}{100}\right)^2\left(\frac{300}{100}\right)$

$FL_4 = (0.2)(5)^2(3)$

$FL_4 = (0.2)(25)(3)$

$FL_4 = 15 \text{ psi}$

$EP = (0.5 \text{ psi})(H)$

$EP = (0.5 \text{ psi})(20 \text{ feet})$

$EP = 10 \text{ psi}$

$FL_{total} = FL_4 + EP + \text{Appliance Loss}$

$FL_{total} = 15 \text{ psi} + 10 \text{ psi} + 25 \text{ psi}$

$\mathbf{FL_{total} = 50 \text{ psi}}$

Multiple Hoselines (Equal Length)

Frequently, fireground operations involve the use of more than one hoseline from a pumper. When determining the friction loss in equal length multiple lines whose diameters are the same, the pump operator only needs to perform calculations for one line. This is because each of the other hoselines will have approximately the same friction loss. Conversely, when the diameters of the hoselines vary, the operator must calculate friction loss for each hoseline and then set the pump discharge pressure for the highest pressure. The corresponding valve on any hose requiring less pressure is partially closed to reduce the pressure from that discharge. Example 12.20 shows how to determine total pressure loss when multiple hoselines are used.

Example 12.20

A pumper is supplying two preconnected handlines being used to fight a garage fire. The attack line is 200 feet of 2-inch hose equipped with a 200 gpm fog nozzle. The exposure protection line is a 200-foot 1-inch booster line flowing 35 gpm **(Figure 12.11)**. What is the total pressure loss due to friction in each hoseline, and upon which line should the operator base the pump pressure?

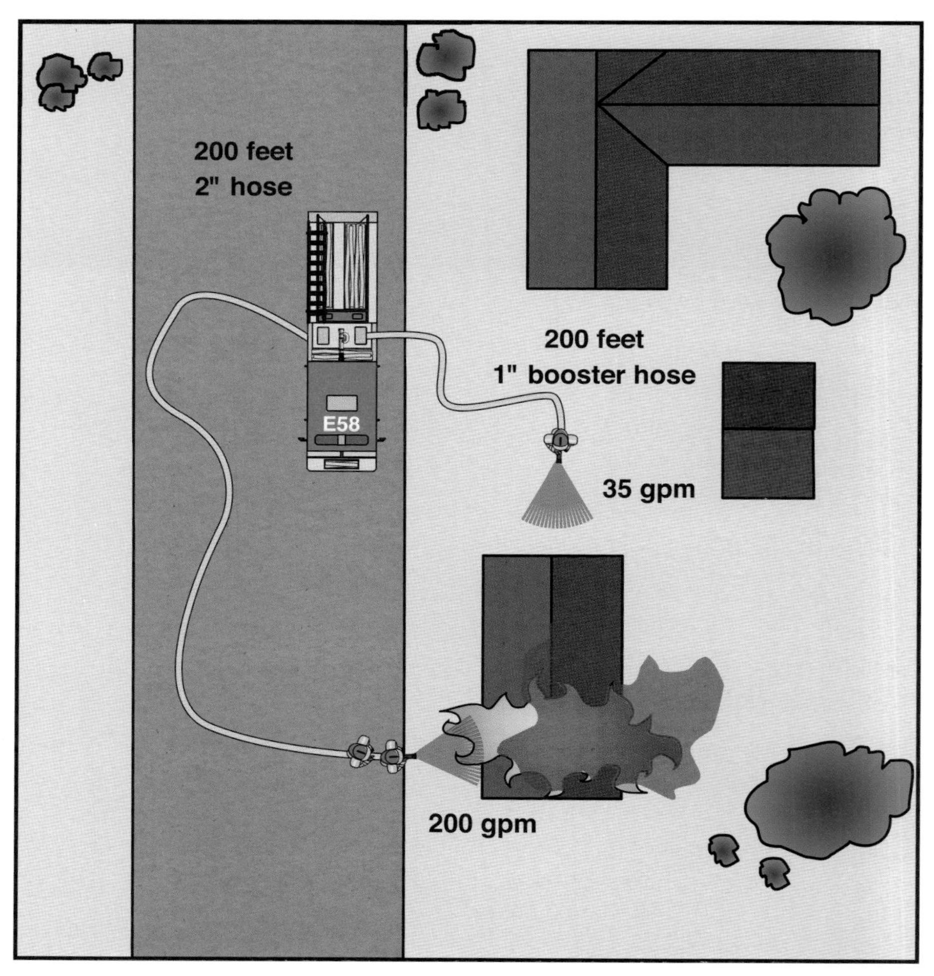

Figure 12.11 Example 12.20.

$$FL_2 = CQ^2L$$

$$FL_2 = (8)\left(\frac{200}{100}\right)^2\left(\frac{200}{100}\right)$$

$$FL_2 = (8)(2)^2(2)$$

$$FL_2 = (8)(4)(2)$$

$$FL_2 = 64 \text{ psi}$$

$$FL_1 = CQ^2L$$

$$FL_1 = (150)\left(\frac{35}{100}\right)^2\left(\frac{200}{100}\right)$$

$$FL_1 = (150)(0.35)^2(2)$$

$$FL_1 = (150)(0.1225)(2)$$

$$FL_1 = 36.75 \text{ or } 38 \text{ psi}$$

Answer: The pump pressure must be based on the 2-inch attack line, and the booster-line valve must be gated down.

Wyed Hoselines (Equal Length and Diameter)

Fire departments that routinely perform reverse hose lay operations or commonly encounter fires in structures such as garden apartment complexes, townhouses, or multiple-occupancy commercial buildings frequently have their pumping apparatus set up to quickly deploy wyed hoselines for a multiple-head fire attack. Wyed hoseline assemblies involve the use of one supply line, usually a 2½-, 3-, or 4-inch hose, wyed into two or more smaller attack lines. These attack lines may range in size from 1½ to 2½ inches and are generally equal in length. It is important that these wyed lines be the same length and diameter in order to avoid having two different nozzle pressures and an exceptionally difficult friction loss problem.

Determining friction loss in wyed hoselines of equal length and diameter is not a complex problem. When the nozzle pressure, hose length, and diameter are the same on both lines, the wye appliance equally splits the total water flowing. This requires the pump operator to consider only one of the wyed hoselines when computing the total pressure lost.

When wyed lines are used to fight fires at occupancies such as garden apartments, one line is often placed at a higher elevation than the other. For example, one hose from the wye might be stretched to the second floor while the other was used on the first floor. In these situations the pressure loss should be calculated on a "worst case" basis, which means the elevation pressure of the line on the upper floor should be factored in. It is unlikely that the elevation difference between these two lines would be so great that the lower line(s) would be dangerously overpressurized. If the difference were significant, partially closing the valve on the wye to the line requiring less pressure would compensate.

Use the following steps when calculating the friction loss in an equal-length wyed hoseline assembly.

Step 1: Compute the number of hundreds of gpm flowing in each wyed hoseline.

Step 2: Determine the friction loss in one of the wyed attack lines using Equation 12.3.

Step 3: Compute the total number of hundreds of gpm flowing through the supply line to the wye by using the following equation:

$$Q_{total} = \frac{(\text{gpm in attack line 1}) + (\text{gpm in attack line 2})}{100}$$

Step 4: Determine the friction loss in the supply line using Q_{total} as the figure for Q in Equation 12.3.

Step 5: Add the friction loss from the supply line, one of the attack lines, 10 psi for the wye appliance if the total flow exceeds 350 gpm, and the elevation pressure if applicable to determine the total pressure loss.

Example 12.21

Determine the pressure lost due to friction in a hose assembly in which two 1¾-inch hoselines, each 150 feet long and flowing 125 gpm, are wyed off 300 feet of 2½-inch hose **(Figure 12.12)**.

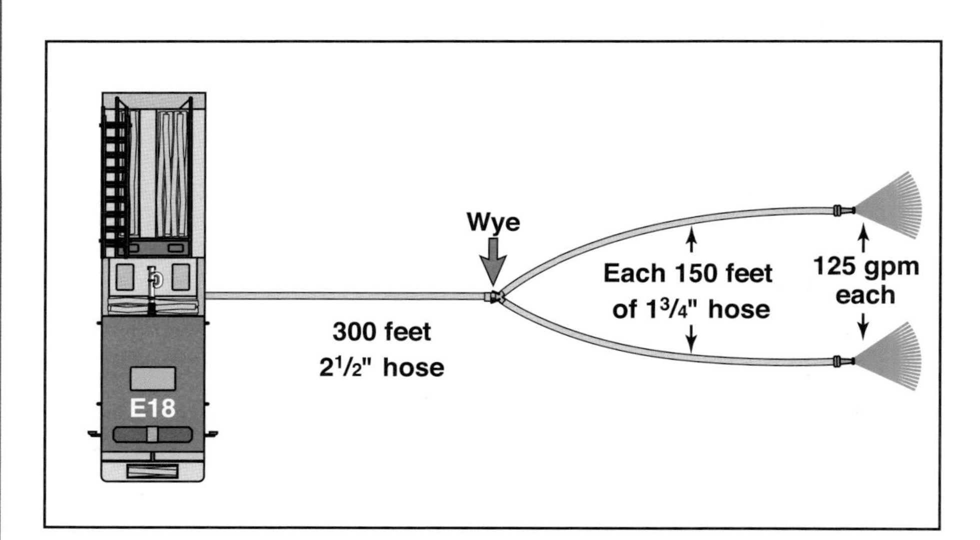

Figure 12.12 Example 12.21.

$$FL_{1¾} = CQ^2L$$

$$FL_{1½} = (15.5)\left(\frac{125}{100}\right)^2\left(\frac{150}{100}\right) \quad (FL_{1¾})$$

$$FL_{1¾} = (15.5)(1.25)^2(1.5)$$

$$FL_{1¾} = (15.5)(1.5625)(1.5)$$

$$FL_{1¾} = 36.33, \text{ or } 36 \text{ psi}$$

$FL_{2½} = C(Q_{total})^2 L$

$FL_{2½} = (2)\left(\frac{125 + 125}{100}\right)^2\left(\frac{300}{100}\right)$

$FL_{2½} = (2)(250/100)^2(3)$

$FL_{2½} = (2)(2.5)^2(3)$

$FL_{2½} = (2)(6.25)(3)$

$FL_{2½} = 37.5$ or 38 psi

$FL_{total} = FL_{2½} + FL_{1¾} +$ Appliance Loss (if greater than 350 gpm)

$FL_{total} = 36$ psi + 38 psi + 0 psi (flow < 350 gpm)

$\mathbf{FL_{total} = 74}$ **psi**

Example 12.22

Determine the pressure lost due to friction in a hose assembly in which two 2½-inch hoselines, each 200 feet long and equipped with a 1¼-inch tip solid stream nozzle to be operated at 50 psi, are wyed off 500 feet of 4-inch hose. One of the nozzles will be operated off the roof of a building that is 30 feet above street level.

$Q = 29.7(d)^2(\sqrt{NP})$

$Q = (29.7)(1.25)^2(\sqrt{50})$

$Q = (29.7)(1.5625)(7.07)$

$Q = 328$ gpm

$FL_{2½} = CQ^2L$

$FL_{2½} = (2)\left(\frac{328}{100}\right)^2\left(\frac{200}{100}\right)$

$FL_{2½} = (2)(3.28)^2(2)$

$FL_{2½} = (2)(10.76)(2)$

$FL_{2½} = 43$ psi

$EP = (0.5\ psi/ft)(H)$

$EP = (0.5\ psi/ft)(30\ feet)$

$EP = 15$ psi

$FL_4 = C(Q_{total})^2 L$

$FL_4 = (2)\left(\frac{328 + 328}{100}\right)^2\left(\frac{500}{100}\right)$

$FL_4 = (0.2)\left(\frac{656}{100}\right)^2(5)$

$FL_4 = (0.2)(6.56)^2(5)$

$$FL_4 = (0.2)(43)(5)$$

$$FL_4 = 43 \text{ psi}$$

$$FL_{total} = FL_{2½} + EP + FL_4 + \text{Appliance Loss (if greater than 350 gpm)}$$

$$FL_{total} = 43 \text{ psi} + 15 \text{ psi} + 43 \text{ psi} + 10 \text{ psi (flow > 350 gpm)}$$

$$\mathbf{FL_{total} = 111 \text{ psi}}$$

Siamesed Lines

Example 12.23 is typical of an actual fireground application for siamesed lines.

Example 12.23

A pumper is supplying an aerial ladder equipped with a detachable ladder pipe and a 500 gpm fog nozzle. The pumper is supplying two siamesed 3-inch hoses with 2½-inch couplings; each hose is 300 feet long. These lines are siamesed into a single 100-foot section of 3½-inch hose that is attached to the ladder pipe. The tip of the ladder is elevated approximately 60 feet **(Figure 12.13)**. Determine the total pressure loss in this arrangement.

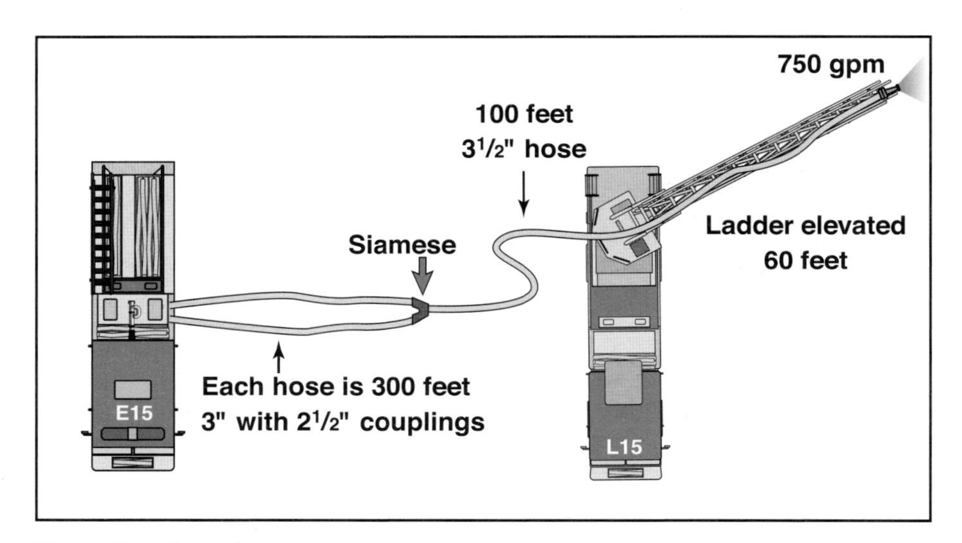

Figure 12.13 Example 12.23.

$$FL_{3½} = CQ^2L$$

$$FL_{3½} = (0.34)\left(\frac{500}{100}\right)^2\left(\frac{100}{100}\right)$$

$$FL_{3½} = (0.34)(5)^2(1)$$

$$FL_{3½} = (0.34)(25)(1)$$

$$FL_{3½} = 8.5, \text{ or } 9 \text{ psi}$$

$$FL_{\text{siamesed lines}} = CQ^2L$$

$$FL_{\text{siamesed lines}} = (0.2)\left(\frac{500}{100}\right)^2\left(\frac{300}{100}\right)$$

$$FL_{\text{siamesed lines}} = (0.2)(5)^2(3)$$

$FL_{siamesed\ lines} = (0.2)(25)(3)$

$FL_{siamesed\ lines} = 15$ psi

$EP = (0.5\ psi/ft)(H)$

$EP = (0.5\ psi/ft)(60\ feet)$

$EP = 30$ psi

$FL_{Total} = FL_{3\frac{1}{2}} + FL_{siamesed\ lines} + EP +$ appliance loss (if greater than 350 gpm)

$FL_{Total} = 9$ psi + 15 psi + 30 psi + 10 psi (siamese) + 25 psi (ladder pipe)

$FL_{Total} = 89$ psi

Complex Hose Layouts

On occasion, fire personnel will confront fireground situations that require hose lines of varying sizes and lengths. In these cases there will be no simple way to determine the friction loss in all of the lines together or the continuous hose layout. Instead the pump operator must calculate each line individually and then determine which part of the layout has the greatest amount of pressure loss. Ultimately, the operator will have to set the pump discharge pressure for the line that has the highest friction loss and gate down the other lines at the pump panel or at a gated wye or manifold so that they will not be overpressurized. (Determining pump discharge pressure is covered in Chapter 13.)

When hose layouts require lines of varying sizes and lengths, the pump operator must calculate each line individually, determine which line has the greatest pressure loss, and then set the pump discharge pressure for that line.

Hose lays for relay pumping operations and sprinkler and standpipe operations typically would be considered complex friction loss calculations as well. Those evolutions are covered in Chapters 14 and 15.

Multiple Hoselines (Unequal Length)

Frequently, the pump operator may be required to supply multiple hoselines of equal or unequal diameter that are not the same length. This can result from the addition of a new hoseline to a pumper, the addition of hose lengths to an existing line, or the pulling of multiple preconnected attack lines that differ in diameter and length. When unequal length and diameter hoselines are used, the amount of friction loss varies in each line, so the pump operator must calculate friction loss in each hoseline.

Example 12.24

A pumper is supplying four preconnected attack lines at a structure fire. Two of the lines are 200 feet of 1¾-inch hose equipped with a 3/4-inch tip solid stream nozzle operating at 50 psi. The third line is 150 feet of 1½-inch hose equipped with a 125 gpm fog nozzle. The fourth line is a 250-foot 3-inch attack line equipped with a 1¼-inch tip solid stream nozzle operating at 50 psi. Determine the total pressure loss due to friction in each hoseline and upon which line to base the pump discharge pressure.

Lines 1 and 2

$$GPM = 29.7\ d^2\sqrt{NP}$$

$$GPM = (29.7)(0.75)^2(\sqrt{50})$$

$$GPM = (29.7)(0.5625)(7.07)$$

$$GPM = 118\ gpm$$

$$FL_{1\frac{3}{4}} = CQ^2L$$

$$FL_{1\frac{3}{4}} = (15.5)\left(\frac{118}{100}\right)^2\left(\frac{200}{100}\right)$$

$$FL_{1\frac{3}{4}} = (15.5)(1.18)^2(2)$$

$$FL_{1\frac{3}{4}} = (15.5)(1.39)(2)$$

$$\mathbf{FL_{1\frac{3}{4}} = 43.2,\ or\ 43\ psi}$$

Line 3

$$FL_{1\frac{1}{2}} = CQ^2L$$

$$FL_{1\frac{1}{2}} = (24)\left(\frac{125}{100}\right)^2\left(\frac{150}{100}\right)$$

$$FL_{1\frac{1}{2}} = (24)(1.25)^2(1.5)$$

$$FL_{1\frac{1}{2}} = (24)(1.5625)(1.5)$$

$$\mathbf{FL_{1\frac{1}{2}} = 56.3,\ or\ 56\ psi}$$

Line 4

$$GPM = 29.7\ d^2\sqrt{NP}$$

$$GPM = (29.7)\ (1.25)^2(\sqrt{50})$$

$$GPM = (29.7)\ (1.5625)(7.07)$$

$$GPM = 328\ gpm$$

$$FL_3 = CQ^2L$$

$$FL_3 = (2)\left(\frac{328}{100}\right)^2\left(\frac{250}{100}\right)$$

$$FL_3 = (2)(3.28)^2(2.5)$$

$$FL_3 = (2)(10.76)(2.5)$$

$$\mathbf{FL_3 = 53.8,\ or\ 54\ psi}$$

Total Pressure Loss

The total pressure loss in the system is based on the line with the highest loss, which in this case would be **Line 3 at 56 psi**.

Example 12.25

A pumper is supplying two handlines that are being used to protect exposures and one large-diameter hose that is supplying an aerial master stream. One handline is a 250-foot length of 2-inch hose supplying a 175 gpm fog nozzle at ground level. The second handline is a 200-foot length of 1½-inch hose supplying a 95 gpm fog nozzle positioned on the roof of a 28-foot-high exposed garage. The aerial device is equipped with a piped waterway that has a 1,000 gpm fog nozzle and is raised 80 feet; it is being supplied by 300 feet of 5-inch hose. Determine the total pressure loss due to friction in each hoseline and upon which line to base pump discharge pressure.

2-Inch Handline

$$FL_2 = CQ^2L$$

$$FL_2 = (8)\left(\frac{175}{100}\right)^2\left(\frac{250}{100}\right)$$

$$FL_2 = (8)(1.75)^2(2.5)$$

$$FL_2 = (8)(3.06)(2.5)$$

$$\mathbf{FL_2 = 61.3, \text{ or } 61 \text{ psi}}$$

1½-Inch Handline

$$FL_{1½} = CQ^2L$$

$$FL_{1½} = (24)\left(\frac{95}{100}\right)^2\left(\frac{200}{100}\right)$$

$$FL_{1½} = (24)(0.95)^2(2)$$

$$FL_{1½} = (24)(0.9025)(2)$$

$$FL_{1½} = 43.3, \text{ or } 43 \text{ psi}$$

$$EP = (0.5 \text{ psi/ft})(H)$$

$$EP = (0.5 \text{ psi/ft})(28 \text{ feet})$$

$$EP = 14 \text{ psi}$$

$$FL_{total} = FL_{1½} + EP$$

$$FL_{total} = 43 + 14$$

$$\mathbf{FL_{total} = 57 \text{ psi}}$$

5-Inch Supply to Aerial Device

$$FL_5 = CQ^2L$$

$$FL_5 = (0.08)\left(\frac{1,000}{100}\right)^2\left(\frac{300}{100}\right)$$

$$FL_5 = (0.08)(10)^2(3)$$

$$FL_5 = (0.08)(100)(3)$$

$$FL_5 = 24 \text{ psi}$$

$EP = (0.5 \text{ psi/ft})(H)$

$EP = (0.5 \text{ psi/ft})(80 \text{ feet})$

$EP = 40 \text{ psi}$

$FL_{total} = FL_5 + EP + \text{appliance loss (if the flow exceeds 350 gpm)}$

$FL_{total} = 24 + 40 + 25$

$\mathbf{FL_{total} = 89 \text{ psi}}$

Total Pressure Loss

The total pressure loss in the system is based on the line with the highest loss, which in this case would be the line supplying the aerial device at 89 psi.

Wyed Hoselines (Unequal Length, Diameter, and/or Elevation) and Manifold Hoselines

We have seen that determining the friction loss for wyed lines of the same length and diameter is not particularly difficult because the flow through the two lines theoretically would be identical. That is not the case when wyed lines are of unequal length or diameter and/or they are operated at different elevations. These differences result in unequal water flows at the split at the wye appliance. Thus we have to determine friction for each of the unequal length wyed lines and then use the line with the greater pressure loss in our final calculation for the entire hose lay.

When we talk about using a wye, or wyed lines, we are typically talking about an appliance that converts one larger supply line into two equal or smaller attack lines. The intake on most wyes (except those used in wildland fire fighting) is equipped with a 2½-inch swivel female connection. The discharges are most commonly equipped with 1½-inch male connections. Of course, since 1½-, 1¾-, and 2-inch hose are all equipped with 1½ couplings, any combination of these hose could be connected to the wye.

Some wyes that have a 2½-inch intake and two 2½-inch discharges may also be found in service. Though a single 2½-inch supply may be used to supply two separate 2½-inch attack lines off this type of wye, it is more hydraulically efficient to use a 3- or 3½-inch hose equipped with 2½-inch couplings (or with a reducer) to supply this type of wye.

If fire personnel need to divide a supply hose into more than two attack lines, a number of other devices are available. The most common of these is the water thief. The water thief is equipped with a 2½-inch intake. On the discharge side it is equipped with one 2½-inch discharge and two 1½-inch discharges. Each discharge is equipped with a shutoff valve.

Since the introduction of large diameter supply hose (LDH) to the fire service, a variety of other manifolds have been designed to divide the water from a single LDH (4-inch or larger) supply line. These manifolds have one LDH intake and as many as five discharge connections. Typically the smallest discharge connections on this type of manifold would be 2½-inches. These manifolds may also have one or more LDH discharge connections.

Again, keep in mind that when hose lengths are unequal in length and/or diameter, the total pressure loss in the system is based on the highest pressure loss in any of the lines. In a real-life situation, the hoselines requiring less than the maximum pressure are gated down at the wye or manifold. Gating down the lines at the manifold to achieve the proper pressure in each line is, at best, a guessing game unless the wye or manifold is equipped with a pressure gauge on each discharge. The following steps can be used to calculate friction loss in unequal length wyed or manifold hoselines.

Step 1: Compute the number of hundreds of gpm flowing in each of the wyed hoselines by using the following equation:

$$Q = \frac{\text{discharge in gpm}}{100}$$

Step 2: Determine the friction loss in each of the wyed lines using Equation 12.3.

Step 3: Compute the total number of hundreds of gpm flowing in the supply line to the wye or manifold by adding the sum of the flows in the attack lines and dividing by 100.

Step 4: Determine the friction loss in the supply line using Equation 12.3 as follows:

$$FL = (C)(Q_{Total})^2L$$

Step 5: Add the friction loss from the supply line, the wye or manifold appliance if total flow is greater than 350 gpm, elevation loss, and the wyed line with the greatest amount of friction loss to determine the total pressure loss.

The following examples illustrate how to compute the total pressure loss in unequal wyed hose and manifold hose assemblies:

Example 12.26

Determine the total pressure loss due to friction and elevation pressure for a hoseline assembly in which one 300-foot, 3-inch hose with 2½-inch couplings is supplying two attack lines. The first attack line consists of 200 feet of 1¾-inch hose stretched through the front door of a structure. It is flowing 150 gpm. The second attack line is 250 feet of 2-inch hose that is carried up a ground ladder and stretched through a second-floor window. It is flowing 200 gpm.

Attack Line 1 (1¾-inch)

$$FL_{1¾} = CQ^2L$$

$$FL_{1¾} = (15.5)\left(\frac{150}{100}\right)^2\left(\frac{200}{100}\right)$$

$$FL_{1¾} = (15.5)(1.5)^2(2)$$

$$FL_{1¾} = (15.5)(2.25)(2)$$

$$\mathbf{FL_{1¾} = 69.75, \text{ or } 70 \text{ psi}}$$

Attack Line 2 (2-inch)

$$FL_2 = CQ^2 L$$
$$FL_2 = (8)\left(\frac{200}{100}\right)^2\left(\frac{250}{100}\right)$$
$$FL_2 = (8)(2)^2(2.5)$$
$$FL_2 = (8)(4)(2.5)$$
$$FL_2 = 80 \text{ psi}$$

$$EP = (5 \text{ psi})(\text{no. of floors} - 1)$$
$$EP = (5 \text{ psi})(2 - 1)$$
$$EP = (5 \text{ psi})(1)$$
$$EP = 5 \text{ psi}$$

$$FL_{2\ total} = FL_2 + EP$$
$$FL_{2\ total} = 80 + 5$$
$$\mathbf{FL_{2\ total} = 85\ psi}$$

Supply Line (3-inch with 2½-inch couplings)

$$FL_3 = C(Q_{total})^2 L$$
$$FL_3 = (0.8)\left(\frac{150+200}{100}\right)^2\left(\frac{300}{100}\right)$$
$$FL_3 = (0.8)\left(\frac{350}{100}\right)^2(3)$$
$$FL_3 = (0.8)(3.5)^2(3)$$
$$FL_3 = (0.8)(12.25)(3)$$
$$\mathbf{FL_3 = 29.4,\ or\ 29\ psi}$$

Total Pressure Loss

The total pressure loss in the system is based on the attack line with the higher loss, which in this case would be Line 2, plus the friction loss in the supply line. Because the flow rate is 350 gpm, adding in the appliance loss is also required.

$$FL_{total} = FL_2 + FL_3 + \text{Appliance loss}$$
$$FL_{total} = 85 + 29 + 10$$
$$\mathbf{FL_{total} = 124\ psi}$$

Example 12.27

Determine the total pressure loss due to friction for a hoseline assembly in which one 500-foot, 3-inch hose with 3-inch couplings is supplying three attack lines attached to a water thief. The first attack line consists of 150 feet of 1½-inch hose flowing 125 gpm. The second attack line is 100 feet of 1½-inch hose flowing 95 gpm. The third attack line is 150 feet of 2½-inch hose flowing 225 gpm (**Figure 12.14**).

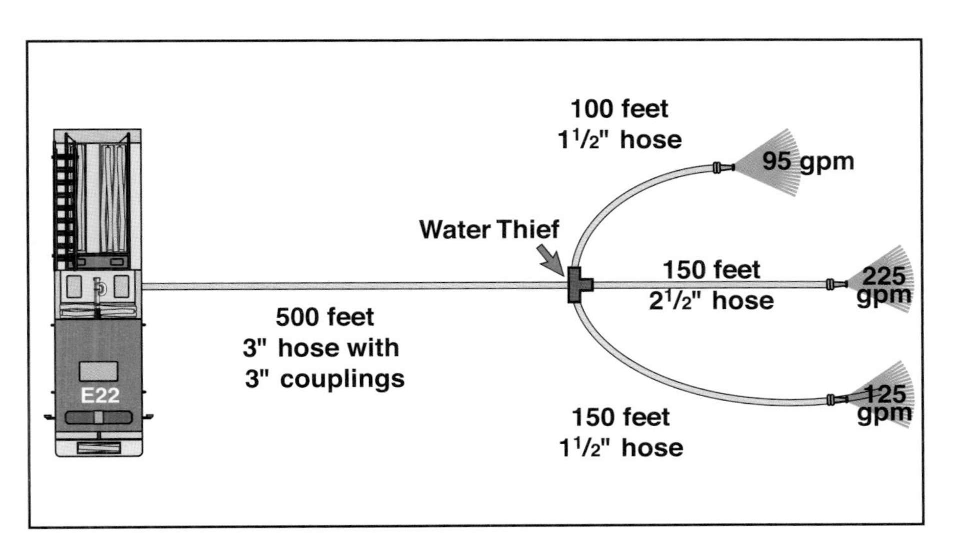

Figure 12.14 Example 12.27.

Attack Line 1 (1½-inch)

$$FL_{1½} = CQ^2L$$

$$FL_{1½} = (24)\left(\frac{125}{100}\right)^2\left(\frac{150}{100}\right)$$

$$FL_{1½} = (24)(1.25)^2(1.5)$$

$$FL_{1½} = (24)(1.5625)(1.5)$$

$$\mathbf{FL_{1½} = 56.3, \text{ or } 56 \text{ psi}}$$

Attack Line 2 (1½-inch)

$$FL_{1½} = CQ^2L$$

$$FL_{1½} = (24)\left(\frac{95}{100}\right)^2\left(\frac{100}{100}\right)$$

$$FL_{1½} = (24)(0.95)^2(1)$$

$$FL_{1½} = (24)(0.9025)(1)$$

$$\mathbf{FL_{1½} = 21.7, \text{ or } 22 \text{ psi}}$$

Attack Line 3 (2½-inch)

$$FL_{2½} = CQ^2L$$

$$FL_{2½} = (2)\left(\frac{225}{100}\right)^2\left(\frac{150}{100}\right)$$

$$FL_{2½} = (2)(2.25)^2(1.5)$$

$$FL_{2½} = (2)(5.0625)(1.5)$$

$$\mathbf{FL_{2½} = 15.2, \text{ or } 15 \text{ psi loss}}$$

Supply Line (3-inch with 3-inch couplings)

$$FL = (C)(Q_{Total})^2(L)$$

$$FL = (0.677)\left(\frac{125+95+225}{100}\right)^2\left(\frac{500}{100}\right)$$

$$FL = (0.677)(4.45)^2(5)$$

$FL = (0.677)(19.8025)(5)$

$\mathbf{FL = 67\ psi}$

Total Pressure Loss

The total pressure loss in the system is based on the attack line with the highest loss, which in this case would be Line 1, plus the friction loss in the supply line. Because the total flow rate exceeds 350 gpm, it is necessary to add the 10 psi appliance loss.

$FL_{total} = FL_{Line\ 1} + FL_3 + \text{Appliance loss}$

$FL_{total} = 56 + 67 + 10$

$\mathbf{FL_{total} = 133\ psi}$

Example 12.28

Determine the total pressure loss due to friction for a hoseline assembly in which one 600-foot, 5-inch hose is supplying three hoselines attached to a large-diameter hose manifold. One hoseline is 200 feet of 4-inch hose supplying a portable master stream device. The master stream is discharging 80 psi through a 1½-inch tip. The second hoseline is 150 feet of 2½-inch hose equipped with a 250 gpm fog nozzle. The third hoseline is 250 feet of 3-inch hose with 2½-inch couplings flowing 275 gpm through a fog nozzle.

Master Stream

$GPM = 29.7(d^2)(\sqrt{NP})$

$GPM = (29.7)\ (1.5)^2(\sqrt{80})$

$GPM = 598\ gpm$

$FL_4 = CQ^2L$

$FL_4 = (0.2)\left(\frac{598}{100}\right)^2\left(\frac{200}{100}\right)$

$FL_4 = (0.2)(5.98)^2(2)$

$FL_4 = (0.2)(35.7604)(2)$

$FL_4 = 14.3$, or 14 psi

$FL_{total} = FL_4 + \text{Appliance loss}$

$FL_{total} = 14 + 25$

$\mathbf{FL_{total} = 39\ psi}$

2½-inch Handline

$FL_{2½} = (C)(Q)^2(L)$

$FL_{2½} = (2)\left(\frac{250}{100}\right)^2\left(\frac{150}{100}\right)$

$FL_{2½} = (2)(2.5)^2(1.5)$

$FL_{2½} = (2)(6.25)(1.5)$

$\mathbf{FL_{2½} = 18.75}$, **or 19 psi**

3-inch Handline (with 2 1/2-inch couplings)

$FL_3 = (C)(Q)^2(L)$

$FL_3 = (0.8)\left(\frac{275}{100}\right)^2\left(\frac{250}{100}\right)$

$FL_3 = (0.8)(2.75)^2(2.5)$

$FL_3 = (0.8)(7.5625)(2.5)$

$\mathbf{FL_3 = 15.1, or\ 15\ psi}$

Supply Line (5-inch)

$FL = (C)(Q_{Total})^2(L)$

$FL_5 = (0.8)\left(\frac{598+250+275}{100}\right)^2\left(\frac{500}{100}\right)$

$FL = (0.08)(11.23)^2(6)$

$FL = (0.08)(126.1129)(6)$

$\mathbf{FL = 60.53\ or\ 61\ psi}$

Total Pressure Loss

The total pressure loss in the system is based on the attack line with the highest loss, which in this case is the 4-inch line supplying the master stream, plus the friction loss in the supply line. Because the total flow rate exceeds 350 gpm, it is necessary to add the 10 psi appliance loss.

$FL_{Total} = FL_4 + FL_5 + \text{appliance loss}$

$FL_{Total} = 39 + 61 + 10$

$\mathbf{FL_{Total} = 110\ psi}$

Master Streams

The basic principles for calculating friction loss associated with master streams are essentially the same as those for other fire streams. In fact, we have already calculated friction loss for hoses supplying master streams in several of our previous examples. The primary difference between calculating friction loss for master stream hose arrangements and handlines is the greater volume of water that master streams flow.

To supply the necessary volume of water to a master stream device, the nozzle often will be fed by two or three medium-diameter (2½- or 3-inch) hose lines (MDH) or one large-diameter hose line. Most master stream nozzles used by municipal fire departments are capable of flows up to 1,000 gpm. That amount of water can be safely supplied using three MDH hoses or a single LDH supply line. Some aerial devices may have nozzles that can flow as much as 2,000 gpm; achieving that flow may require two or more LDH supply hoses. Some industrial fire brigades, particularly those that protect petrochemical facilities, may have very large master stream nozzles capable of flows of up to 10,000 gpm **(Figure 12.15, p. 302)** or more. A nozzle discharging this quantity of water may need to be supplied by from six to eight 5-inch or larger supply hoses being pumped

Figure 12.15 This master stream device is supplied by exceptionally large diameter hose.
Courtesy of Ron Jeffers

from two or more large-capacity industrial pumpers. Some fire brigades use supply hose as large as 12 inches in diameter. Industrial pumpers are commonly constructed with fire pumps with a rated capacity of 3,000 gpm or more.

The hose lays used to supply master streams are essentially the same as those used for other fire streams. For this reason, the concepts used in determining friction loss are also the same. The only difference occurs if hoselines of unequal length or diameter are used to supply a master stream appliance.

If the hoselines supplying a master stream device are of unequal lengths but the same diameter and are being pumped from the same pumper, determine an average of the hose lengths for ease of calculation. To obtain this average, add the length of each hoseline and divide the sum by the number of hoselines being used. When possible, use the coefficients for wyed hoselines in Table 12.3 with the total flow through the nozzle to complete the friction loss calculations. If Table 12.3 does not give a coefficient for the particular combination of hoses in use, average the length of the lines and assume an equal amount of water is going through each hose.

If multiple lines of the same diameter and different lengths are being supplied to a single master stream device by multiple pumpers, calculate the friction loss in each of the respective hose lays so that the individual pumps are supplying the proper amount of pressure to each hose lay feeding the master stream device.

If multiple lengths of hose that are different lengths and diameters are being used to supply a single master stream, they cannot be combined into a single hydraulic calculation. Each hose must be calculated separately, and the pump discharge pressure will be based on the line requiring the highest amount of pressure.

Manual aerial devices with piped waterways are treated the same as master stream appliances, using a friction loss of 25 psi to include pressure losses created by the intake, internal piping, and nozzle. Elevation pressure loss is calculated separately. For exact figures, consult the aerial manufacturer for specific friction loss data or perform field tests to derive data. If a traditional detachable ladder pipe and hose assembly are used, the friction loss within the siamese (if used), hose, and ladder pipe must all be accounted for. Examples 12.29 and 12.30 show how to calculate the total pressure loss in a master stream hose layout.

Example 12.29

A 5,000-gpm master stream device is being set up to attack a petroleum storage tank fire. Two industrial pumpers will supply the master stream nozzle, and we will assume that each is supplying one-half of the required volume. Pumper 1 is pumping through two 6-inch hoselines that are each 700 feet long. Pumper 2 is supplying three 5-inch hoselines that are each 500 feet long (**Figure 12.16**). Determine the total pressure loss in each hose assembly.

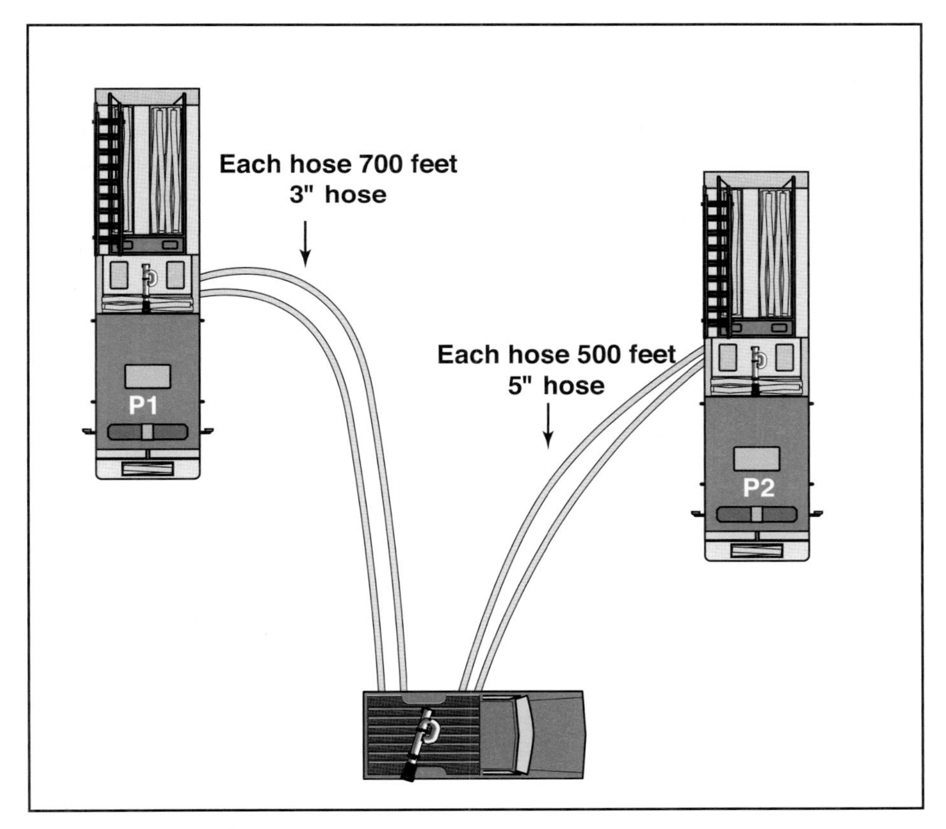

Figure 12.16 Example 12.29.

Pumper 1

We know that each pumper will be responsible for supplying 2,500 gpm. Since Pumper 1 is supplying two hose lines, we will assume that each 6-inch line is carrying 1,250 gpm. Thus,

$$FL_6 = CQ^2L$$

$$FL_6 = (0.05)\left(\frac{1{,}250}{100}\right)^2\left(\frac{700}{100}\right)$$

$$FL_6 = (0.05)(12.5)^2(7)$$

$$FL_6 = (0.05)(156.25)(7)$$

$$\mathbf{FL_6 = 56.7, \text{ or } 57 \text{ psi}}$$

Pumper 2

Since Pumper 2 is supplying three hose lines, we will assume that each 5-inch line is carrying 833 gpm. Thus,

$$FL_5 = CQ^2L$$

$$FL_3 = (0.08)\left(\frac{833}{100}\right)^2\left(\frac{500}{100}\right)$$

$$FL_5 = (0.08)(8.33)^2(5)$$

$$FL_5 = (0.08)(69.39)(5)$$

$$\mathbf{FL_5 = 27.75, \text{ or } 28 \text{ psi}}$$

Example 12.30

A pumper is supplying an aerial device equipped with a piped waterway. The aerial device is elevated 70 feet and is discharging 900 gpm. One 3½-inch hoseline is 450 feet long and the other is 500 feet long. Assume that each line is supplying half the volume. Determine the total pressure loss in this hose assembly. Since Table 12.3 does not list a coefficient for dual 3½-inch hose lines, calculate the friction loss using the flow through each line and the average of the two lengths of hose.

$$FL_{3½} = CQ^2L_{avg}$$

$$FL_{3½} = (0.34)\left(\frac{450}{100}\right)^2\left(\frac{[450 + 500]/2}{100}\right)$$

$$FL_{3½} = (0.34)(4.5)^2\left(\frac{475}{100}\right)^2$$

$$FL_{3½} = (0.34)(4.5)^2(4.75)$$

$$FL_{3½} = (0.34)(20.25)(4.75)$$

$$FL_{3½} = 32.7, \text{ or } 33 \text{ psi}$$

$$EP = (0.5 \text{ psi/ft})(H)$$

$$EP = (0.5)(70)$$

$$EP = 35 \text{ psi}$$

Total Pressure Loss

$$FL_{Total} = FL_{3½} + EP + \text{appliance loss}$$

$$FL_{Total} = 33 + 35 + 25$$

$$\mathbf{FL_{Total} = 93 \text{ psi}}$$

FIREGROUND HYDRAULIC CALCULATIONS

The formulas to this point provide useful background information on the hydraulics of fire fighting. However, the use of these mathematical calculations on the emergency scene is generally impractical. If calculations must be performed on the fireground, the driver/operator can rely on one or more of the following methods for determining pressure loss and required pump discharge pressure:

- Flowmeters
- Hydraulic calculators
- Pump charts
- Hand method
- Condensed "Q" formula
- GPM flowing

Flowmeters

The ultimate purpose of all fireground hydraulic calculations is to discharge an appropriate amount of water from the nozzle(s) being used to attack a fire. Chapter 8 discussed the two types of flowmeters available to the fire service: the paddlewheel type and the spring-probe type. Rather than measuring the pressure going through a discharge line, flowmeters measure the volume of water flow in gallons per minute. Unlike pressure, which is lost as water travels through a hose assembly, the volume of water flowing through a hose assembly is not reduced. Whatever amount of water is pumped into a hose must be discharged at the other end.

Flowmeters reduce the number of pressure calculations required of the driver/operator. The number displayed on the flowmeter requires no further calculation because it reflects how much water is moving through the discharge valve and consequently the nozzle **(Figure 12.17)**. This quantity of water diminishes before it reaches the nozzle only if there is a leak or break in the hoseline. Flowmeters can make it possible for driver/operators to pump (within the limits of the pump) the correct volume of water to nozzles without having to know the length of hoseline, the amount of friction loss, or the elevation of the nozzles. Thus, if a driver/operator is supplying a 1½-inch hoseline equipped with a 95-gpm nozzle, the driver/operator does not have to calculate the friction loss and add in the nozzle pressure to determine the appropriate pump discharge pressure. If both the nozzle and the pump discharge are open, the driver/operator needs only to increase the engine throttle until the flowmeter shows that 95 gpm is flowing through the line. This will indicate that the pump is at the correct pressure to supply the necessary flow.

Figure 12.17 Flowmeters provide a digital readout.

Flowmeters are particularly advantageous when supplying hoselines or master stream devices equipped with automatic nozzles. As discussed in Chapter 11, automatic nozzles automatically maintain nozzle pressure at a predetermined level. This may lead driver/operators to supply insufficient discharge pressures. These pressures may create what appears to be an acceptable fire stream but in reality has insufficient volume to extinguish the fire. If the discharge is equipped with a flowmeter, it is easier for the driver/operator to ensure an adequate flow rate.

Flow meters are particularly advantageous when supplying hoselines or master stream devices equipped with automatic nozzles.

NFPA® 1901, *Standard for Automotive Fire Apparatus*, allows the use of flowmeters instead of pressure gauges on all discharges 1½ to 3 inches in diameter. Discharges that are 3½ inches or larger may be equipped with flowmeters, but they must also have an accompanying pressure gauge. The flowmeter must provide a readout that is in increments no larger than 10 gpm.

Flowmeter Applications

Regardless of which type of flowmeter is used on a fire apparatus, properly calibrated flowmeters can reduce the difficulty of certain conditions faced by the driver/operator in a number of situations.

Diagnosing Waterflow Problems

The flowmeter can be used as a diagnostic tool to identify waterflow problems. If the flow does not increase when the driver/operator increases pressure, several problems are likely, such as:

- The hose is kinked
- A midline valve (such as a gated wye) may be partially closed
- A vehicle tire is parked on top of the hoseline

A scary and dangerous situation for firefighters involved in an interior attack is the loss of water volume and pressure at the nozzle. If firefighters on a line advise the pump operator or Incident Commander of a loss of water at the nozzle, but the flowmeter shows no reduction or perhaps even an increase in flow, personnel can assume that a hose has burst and will need to be replaced.

Relay Pumping

Use of a flowmeter during relay pumping makes it possible to feed a supply line without having to know the number of gallons flowing from the pumper receiving the water. This is done by monitoring the master discharge gauge and the flowmeter as the throttle is increased during the setup stage of the operation. As engine speed (rpm) increases so do the discharge and the gpm reading from the flowmeter. Increasing the engine speed until the flowmeter reading no longer increases sets the pump at the correct discharge pressure to supply an adequate flow to the receiving pumper. It also provides the driver/operator with a reading of the water volume being used by the receiving pumper. Although watching the flowmeter is helpful, pump operators also should watch the master intake pressure gauge and not allow the incoming pressure to drop much below 20 psi. Chapter 14 includes more information on relay pumping.

Standpipe Operations

One of the more challenging operations for pump operators is supplying standpipe systems that are supporting attack lines in large or multistory structures. Often the pump operator does not have accurate information on the length and diameter of the hoselines being used, the types of nozzles with which they are equipped, or the exact floors they are operating on. This lack of information makes calculating the friction loss and elevation pressure in these situations nearly impossible.

While flowmeters can be very beneficial in supporting standpipe systems, they are by no means a miracle cure. So that the operator can set up the pump to supply the necessary volume and pressure of water into the system, firefighters still need to relay some basic information including:

- The number of hose lines to be supplied
- The rated or desired volume to be flowed by each nozzle

This information allows the pump operator to add all of the flows from the nozzles and then pump that volume of water into the system. As long as the flowmeter shows that flow, the pump operator is supplying the system with adequate volume and pressure. Of course, the pump operator must realize that once the hoseline(s) is (are) charged, there is no flow through the system until a nozzle(s) is (are) opened. This requires the pump operator to set the pump for a discharge pressure relatively close to what will be required when the nozzle(s) is (are) flowing. Once the nozzle(s) is (are) fully opened and flowing, the driver/operator can adjust the discharge pressure until the appropriate amount of water is flowing.

During standpipe operations, the pump operator must be in communication with the firefighters on the nozzles to ensure that nozzle pressures (and nozzle reactions) are correct. This is important because nozzles placed several floors apart may receive pressures somewhat greater or less than optimum. This is a problem no matter which method is used to set the pressure, and communication is the best way to make adjustments and correct it. Chapter 15 includes more detailed information on supplying standpipes and sprinkler systems.

During standpipe operations, the pump operator must be in communication with the firefighters on the nozzles to ensure that nozzle pressures are correct.

Hydraulic Calculators

Some driver/operators use hydraulic calculators to determine friction loss and the pump discharge pressure required to supply a hose layout. This eliminates the need to do these calculations mentally or on paper, both of which are very difficult during emergency operations. Three basic types of hydraulic calculators are available to the fire service:

- Manual
- Mechanical
- Electronic

Manual and mechanical calculators have a slide or dial that indicates the water flow, size of hose, and length of the hose lay. By moving the slide or dial to line up each of these components properly, the driver/operator can then read the amount of friction loss or the required pump discharge pressure. These calculators are most commonly supplied to fire departments by hose and nozzle manufacturers. Contact your local fire equipment dealer for more information.

Some apparatus may be equipped with electronic hydraulic calculators. These specially programmed devices allow the driver/operator to input the known information: water flow, size of hose, length of hose lay, and any elevation changes. Through preprogrammed formulas from either the factory or other available software, these calculators compute the pressure loss and the required pump discharge pressure. They may be portable or mounted near the pump panel. Inexpensive electronic programmable calculators can be preprogrammed for fireground calculations and carried on apparatus.

Some departments are now also experimenting with hydraulic software that is installed in handheld computers (PDAs) and the mobile data or computer terminals mounted in the fire apparatus cab. As computer technology continues to advance, more types of electronic aids will become available to the fire service.

Pump Charts

To reduce the need for calculations on the emergency scene some fire departments use pump charts, also known as "cheat sheets." Pump charts contain the required pump discharge pressures for the various hose lays and assemblies the department uses. They may be placed on laminated sheets carried on the apparatus or on plates affixed to the pump panel. Departments may develop their own pump charts or use those supplied by fire hose or nozzle manufacturers.

It is highly recommended that departments take the time to develop pump charts for the specific types of hose lays that they commonly use. Pump charts supplied by manufacturers may be somewhat helpful but typically contain a lot of information that will not apply to the particular jurisdiction. To develop and use a pump chart, it is important to understand the column headings. The nozzle heading should include only those nozzles and devices used by the department. In addition to containing information on standard hose lays for supply and attack lines, the chart should include other applications such as sprinkler system support and relay pumping operations. The flow rate (gpm) column indicates the flow being provided to the particular nozzle or layout. The NP column indicates the nozzle pressure being produced. The columns headed 100, 200, and so on indicate the number of feet of hose being used to supply a given nozzle or layout. The reliability of the chart depends entirely on the accuracy of the calculations.

Fire departments should take the time to develop pump charts for the specific types of hose lays they commonly use.

The first step in developing a pump chart is to identify all nozzles, devices, and layouts used by the department and enter them in the nozzle column. Then enter the rated or desired flow in gallons per minute and the desired nozzle pressure for each item in the appropriate columns. If, for instance, a master stream device may be supplied by either 2½- or 3-inch hose, make a separate entry for each layout in the nozzle column, as indicated in the sample chart.

The second step is to calculate the required pump discharge pressures for each of the listed layouts using the formulas and tables given earlier in this manual or figures derived from field testing. Observe the following rules when making these calculations:

- Be certain to include the pressure loss in master stream appliances, aerial devices, and hose appliances when flowing in excess of 350 gpm.
- For wyed hoselines, the length of the layout numbers indicates the number of feet of hose between the pumper and the wye. If the attack lines are kept preconnected to the wye, they should be calculated into the overall figure, as they will be constant.

- When a master stream nozzle may be supplied by a different number or size of hoselines, indicate these on the chart.
- Because few apparatus gauges read to one psi, round pump discharge pressure answers on the chart to the nearest 5 psi.
- Do not list pump discharge pressures that exceed the test pressure used for the size of hose concerned.
- When calculating pump discharge pressures for relay operations, provide established departmental residual pressures at the intake of the pumper being supplied. This residual pressure may be indicated on the chart as a nozzle pressure.

To use a pump chart, locate the nozzle column and the nozzle or layout being used on the chart. Follow that line across to the vertical column headed by the number of feet (meters) of hose in that layout. The figure in the block where the two columns intersect is the required pump discharge pressure.

Example 12.31

Using **Table 12.4**, determine the pump discharge pressure for a master stream device that is equipped with a 1¾-inch solid stream nozzle tip and is supplied by two 2½-inch hoselines, each 200 feet long.

Locate the master stream device with a 1¾-inch nozzle tip in the left column of the chart. Choose the one that shows the device being supplied by two 2½-inch lines. Follow this line to the column for a 200-foot hose lay. You should land on the figure **169 psi**. This is the required pump discharge pressure.

Hand or Counting Fingers Method

Over the years many driver/operators have used the hand, or "counting fingers," method for determining friction loss in 2½-inch hose. Starting with the thumb of the left hand, as illustrated in **Figure 12.18, p. 310**, each finger is numbered at the base in terms of hundreds of gallons per minute. Returning to the thumb and again moving from left to right, the tip of each finger is given a successive even number, beginning with two. Because nozzle capacities vary in gpm, the nearest half-hundred can be used with slight variations. The numbers 3, 5, 7, and 9 can be used for flows of 150, 250, 350, and 450 gpm, respectively. These half-hundred figures can be assigned to the spaces between the fingers. To determine the friction loss for 100 feet of 2½-inch hose at a desired flow, select the finger to which that flow has been assigned and multiply the number at the tip of the finger by the first digit at the base of the finger. Thus, the friction loss for a flow of 500 gpm can be found by using the numbers assigned to the little finger, or (5)(10) = 50 psi friction loss per 100 feet of 2½-inch hose. Likewise, the friction loss for a flow of 200 gpm is found using the numbers assigned to the index finger. Thus, (2)(4) = 8 psi friction loss per 100 feet of 2½-inch hose.

This method gives a reasonable estimate of the expected friction loss in any particular hoseline. If more accurate figures are required, one of the other methods must be employed. (**NOTE:** This method has no conversion for the metric system of measurement.)

Table 12.4 (US)
Pump Chart

Nozzles	GPM	NP	100	200	300	400	500	600	700	800	900	1,000	1,100	1,200
			Length of Lay in Feet											
Booster 1"	23	100		115										
Preconnect 1¾"	150	100		170										
Wyed Line:														
200' of 1½"on 2½" skid	190	100	150	157	164	172	179	186	193	201	208	215	222	229
2½" Fog	250	100	112	125	138	150	162	175	187	200	212	225	237	250
Master Stream:														
1¾" (Two 2½" Lines)	800	80	137	169	201	233								
1¾" (Three 2½" Lines)	800	80	119	133	147	161	175	189	203					
1¾" (Two 3" Lines)	800	80	118	131	144	157	170	183	196	209				
Relay:														
One 3" Line	250	20	25	30	35	40	45	50	55	60	65	70	75	80
One 3" Line	500	20	40	60	80	100	120	140	160	180	200			
Two 3" Lines	750	20	31	43	54	65	76	88	99	110	121	133	144	155
Two 3" Lines	1,000	20	100	180										
Sprinklers	Maintain 150 psi													
Elevation	Add 1/2 psi per foot or 5 psi per floor													

NOTE: All pressures rounded to the nearest whole number for this table only. Computations rounded to the nearest 5 psi are acceptable.

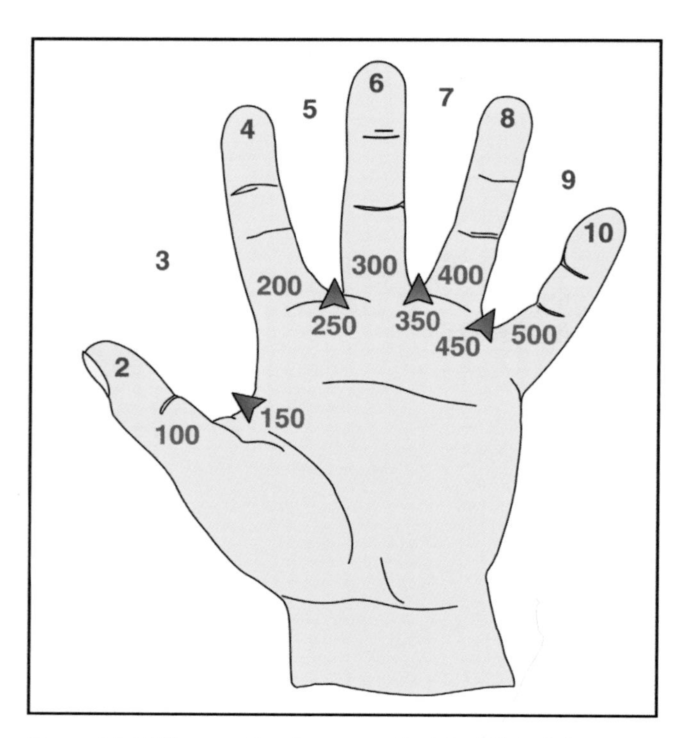

Figure 12.18 The counting fingers method for 2½-inch hose.
Courtesy of the Maryland Fire & Rescue Institute.

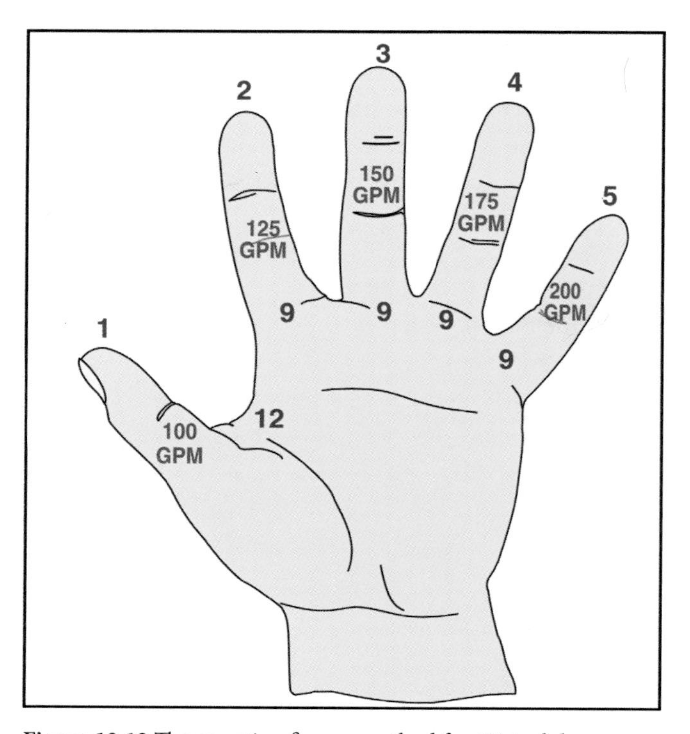

Figure 12.19 The counting fingers method for 1¾-inch hose.

Example 12.32

Using the hand method, determine the total pressure loss due to friction in 200 feet of 2½-inch hose when a fog nozzle is flowing 200 gpm at 100 psi.

(2)(4) = 8 psi loss/100 feet

(8)(200/100) = **16 psi loss in the hose assembly**

Example 12.33

Using the hand method, calculate the total pressure loss due to friction in a 2½-inch hoseline 300 feet long with 350 gpm flowing through the line.

(3.5)(7) = 24.5 psi loss/100 feet

(24.5)(3) = **73.5 psi loss in the hose assembly**

In recent years the counting fingers method has also been applied to 1¾-inch hoselines by numbering the fingers as shown in **Figure 12.19**. In this case, you can calculate the friction loss in 100 feet of 1¾-inch hose by going to the finger that corresponds to the flow you are using and multiplying the number at the tip of the finger by the number at the base of the same finger. For example, if you are going to flow 150 gpm, multiply the number at the tip of your middle finger (3) with the number at the base (9). Thus the friction loss in 100 feet of 1¾-inch hose flowing 150 gpm is 3 times 9, or 27 psi.

Example 12.34

Using the hand method, what is the total pressure loss due to friction in a 1¾-inch hoseline 200 feet long with 175 gpm flowing through the line?

(4)(9) = 36 psi loss/100 feet

(36)(2) = **72 psi loss in the hose assembly**

Condensed "Q" Formula

Once the fire service began using hoses larger than 2½-inches, pump operators needed quick, reliable fireground methods of calculating pressure loss in these hoselines. One such method, the Condensed "Q" formula, with minor variations, can be used to determine friction loss in 3-, 4-, and 5-inch hose.

3-Inch Hose

The basic, unaltered Condensed Q formula is for calculating friction loss in 3-inch hose. It does not matter whether the hose has 2½- or 3-inch couplings. This formula is not as accurate as the equation $FL = CQ^2L$, but it is faster and sufficiently precise for fireground operations. The amount of friction loss calculated using the Condensed Q formula will be 20 percent greater than if the same situation is calculated using $FL = CQ^2L$. This could mean a difference as high as 50 psi in a 1,000-foot, 3-inch hose lay. With this knowledge in hand, the pump operator may adjust the calculated pump pressure accordingly.

Equation 12.7

$FL = Q^2$

Given:

FL = Friction loss per 100 feet of 3-inch hose

Q = Number of hundreds of gpm flowing

Example 12.35

What will be the total pressure loss due to friction when 300 gpm is being discharged from a nozzle attached to 200 feet of 3-inch hose?

$FL = Q^2$

$FL = (300/100)^2$

$FL = (3)^2$

FL = 9 psi per 100 feet

$FL_{Total} = (\text{FL per 100 feet})(L)$

$FL_{Total} = (9)(200/100)$

$FL_{Total} = (9)(2)$

FL_{Total} = 18 psi

4-Inch Hose

With a simple modification, the Condensed Q formula can also be used to calculate friction loss in 4-inch supply hose.

Equation 12.8

$$\text{FL per 100 feet} = \frac{Q^2}{5}$$

Given:

FL = Friction loss in 100 feet of 4-inch hose

Q = Number of hundreds of gallons per minute

5 = A constant

Example 12.36

What will be the total pressure loss in 600 feet of 4-inch hose flowing 800 gpm?

$$FL = \frac{Q^2}{5}$$

$$FL = \frac{(800/100)^2}{5}$$

$$FL = \frac{8^2}{5}$$

$$FL = \frac{64}{5}$$

FL = 12.8, or 13 psi per 100 feet

$FL_{Total} = (FL \text{ per } 100 \text{ feet})(L)$

$FL_{Total} = (13)(600/100)$

$FL_{Total} = (13)(6)$

$\mathbf{FL_{Total} = 78 \text{ psi}}$

5-Inch Hose

With yet another simple modification, the Condensed Q formula can also be used to calculate friction loss in 5-inch supply hose.

Equation 12.9

$$FL = \frac{Q^2}{15}$$

Given:

FL = Friction loss in 100 feet of 4-inch hose

Q = Number of hundreds of gallons per minute

15 = A constant

Example 12.37

What will be the total pressure loss in 500 feet of 5-inch hose flowing 1,000 gpm

$$FL = \frac{(1{,}000/100)^2}{15}$$

$$FL = \frac{10^2}{15}$$

$$FL = \frac{100}{15}$$

FL = 6.67, or 7 psi per 100 feet

$FL_{Total} = (FL \text{ per } 100 \text{ feet})(L)$

$$FL_{Total} = (7)\left(\frac{500}{100}\right)$$

$FL_{Total} = (7)(5)$

$\mathbf{FL_{Total} = 35 \text{ psi}}$

SUMMARY

In this chapter we have reviewed both the historical and modern methods of mathematically calculating pressure loss in hose systems. We also have looked at rule-of-thumb, or fireground, methods of making these same calculations. Upon completing this chapter, it should be apparent that great strides have been made over the years in simplifying the process of performing hydraulic calculations. The ability to perform these calculations and apply them to an actual pumping situation is what makes the difference between a professional and a "lever puller." Anyone can pull levers on a pump panel and make water come out of the pump. The true professional performs this task in a controlled manner that leads to the correct volume of water being supplied to the nozzle at a useable pressure.

However, the calculations in this chapter are really only one-half of the entire process. Once pump operators know the pressure losses they face, they must determine the appropriate pressure at which to set the pump. Thus, in the next chapter we will factor in this chapter's pressure loss calculations with the other necessary factors and determine how to calculate the pump discharge pressure.

REVIEW QUESTIONS

1. Why do the formulas in this chapter not apply when Class A foam concentrates or wetting agents are being used?

2. List two reasons why the $2Q^2+Q$ method of friction loss determination is no longer preferred.

   1.

   2.

3. Using Equation 12.1, calculate the friction loss in 300 feet of 2 1/2-inch hose flowing 200 gpm through a fog nozzle.

4. Using Equation 12.1, determine the friction loss in 150 feet of 1¾-inch hose flowing 150 gpm.

5. Determine the friction loss in an 800-foot lay of 4-inch supply hose that is flowing 600 gpm.

6. A 3-inch attack line with 2½-inch couplings that is 200 feet long has another 150 feet of 2½-inch hose attached to its end. Attached to the end of the 2½-inch hose is a nozzle that is flowing 225 gpm. Determine the total friction loss in the hose lay.

7. A pumper attached to a fire hydrant is supplying 500 gpm through 600 feet of dual 2½-inch hoses to a pumper at the fire scene. What is the friction loss in the hose lay?

8. A pumper is supplying a portable master stream device that is equipped with a 1½-inch solid stream nozzle discharging at 80 psi. The master stream device is being supplied by a single 50-foot-long 3-inch hose with 3-inch couplings. This hose in turn is being supplied by siamesed lines consisting of dual 3-inch hose, each 400 feet long with 2½-inch couplings. What is the friction loss in the hose lay?

9. Calculate the total pressure loss due to elevation pressure for a hoseline operating on the roof of a 40-foot tall warehouse.

10. A hoseline is connected to a building's standpipe system and is operated on a fire on the ninth floor. What is the total pressure loss due to elevation at the base of the standpipe system?

____

____

11. A fire is discovered on the twelfth floor of a structure. An arriving engine company proceeds to the 11th floor of the structure and connects 200 feet of 2-inch hose to the standpipe outlet and continues to the 12th floor to attack the fire. The hoseline is equipped with a 1-inch tip solid stream nozzle that will be flowed at 50 psi. Determine the total pressure loss due to friction and elevation pressure at the standpipe system fire department connection. (Use the rule-of-thumb for friction loss in the standpipe system.)

____

____

12. Determine the total pressure loss due to friction and elevation pressure for a hoseline assembly in which one 200-foot long 2½-hose is supplying two attack lines. The first attack line consists of 200 feet of 1¾-inch hose that is stretched through the front door of a structure. The first attack line is flowing 175 gpm. The second attack line is 200 feet of 2-inch hose that is carried up a ground ladder and stretched through a third-floor window. The second line is flowing 225 gpm.

____

____

13. A pumper is supplying an aerial device equipped with a piped waterway. The aerial device is elevated 90 feet and is discharging 1,500 gpm. One 4-inch hoseline is 350 feet long, and the other is 400 feet long. Assume that each line is supplying half the volume. Determine the total pressure loss in this hose assembly.

____

____

14. When viewing the readout on a flowmeter, if the flow does not increase when the driver/operator increases pressure, what conditions are likely to exist?

____

____

____

15. Using Table 12.4, determine the pump discharge pressure for wyed lines of 200 feet of 1¾-inch hose on a 2½-inch skid if the 2½-inch hose is 600 feet long?

____

____

16. Using the hand method, determine the total pressure loss due to friction in 400 feet of 2½-inch hose when a fog nozzle is flowing 300 gpm at 100 psi.

____

____

Answers found on page 432.

FESHE Course Objectives

The information in this chapter is intended to meet some of the objectives outlined for the Fire Protection Hydraulics and Water Supply course by the United States Fire Administration's National Model Core Curriculum as developed by the Fire and Emergency Services Higher Education (FESHE) initiative. The information in this chapter also allows the student to meet the requirements of NFPA® 1002, *Standard for Fire Apparatus Driver/Operator Professional Qualifications* (2009 edition) listed below.

FESHE Course Objectives

1. Determine the required pump discharge pressure for simple hose layouts.
2. Determine the required pump discharge pressure for complex hose layouts.
3. Determine the required pump discharge pressure when supplying aerial master streams.
4. Determine net pump discharge pressure when operating from pressurized and static water supply sources.

NFPA® 1002 Requirements

5.2.1 (A) and 5.2.2 (A) Requisite Knowledge. Hydraulic calculations for friction loss and flow using both written formulas and estimation methods.

Chapter 13

Determining Pump Discharge Pressure

INTRODUCTION

Chapter 12 explained the ways in which pressure is lost (or in some cases gained) between the water supply source and the point of discharge. It also illustrated numerous methods for calculating those losses. Determining the pressure lost in a hose system, however, is only the first step in figuring the actual discharge pressure at which a fire pump must be set to supply an effective fire stream. In addition to understanding and calculating pressure losses, the driver/operator must also understand the concepts of pump discharge pressure and net pump discharge pressure and their relation to nozzle pressure.

To deliver the necessary water flow to the fire location, the pump discharge pressure at the apparatus must be able to overcome the sum of all pressure losses.

To deliver the necessary water flow to the fire location, the pump discharge pressure at the apparatus must be able to overcome the sum of all pressure losses.

These pressure losses *combined with the required nozzle pressure* are used to determine the pump discharge pressure. Pump discharge pressure (PDP) can be calculated using the following equation:

Equation 13.1

$$PDP = NP + TPL$$

Given:

PDP = pump discharge pressure in psi

NP = nozzle pressure in psi

TPL = total pressure loss in psi (appliance, friction, and elevation losses)

All nozzles are designed for optimum operations at a particular nozzle discharge pressure. In general these pressures are:

- Solid stream handline nozzles: 50 psi
- Solid stream master stream nozzles: 80 psi
- Low-pressure fog nozzles: 75 psi or manufacturer's recommendation
- Standard fog nozzles: 100 psi, handlines and master streams

Unless otherwise specified, the pump operator should base pump discharge pressure on these nozzle pressures.

If the hose line being supplied by the pumper is being used to supply another pumper instead of a nozzle, the desired intake pressure at the pumper being supplied should replace the nozzle pressure in Equation 13.1. For example, in relay pumping (as we will see in chapter 14), experts commonly advise that the intake pressure at a receiving pumper should never be below 20 psi. Thus, when determining the pump discharge pressure in a relay pumping situation, we would substitute the 20 psi minimum intake pressure for nozzle pressure in Equation 13.1. Departments that require or desire a different intake pressure can adjust the figure accordingly.

A pumper supplying multiple hose lines, wyed hose lines, or manifold hose lines likely will have to provide different pump discharge pressures for each attack line. This may be due to different amounts of pressure loss in each line and/or different required nozzle pressure at the end of each line. Since the pump can be set to only one overall pump discharge pressure, the pump operator must use another method to compensate for individual pressure requirements. When this occurs, the operator must set the pump discharge pressure for the hose line with the greatest pressure demand. If multiple hose lines are attached to the pump, the operator must gate back the valves for the remaining hose lines at the discharge outlets, using their individual discharge pressure gauges as a guide. If the pump is supplying wyed or manifold hose lines that have different attack line pressures, the operator should use the ball valve on the appliance to gate back the hose lines until obtaining the desired pressure on each line. Unless the appliance has individual pressure gauges on each discharge, adjusting the lower demand attack lines will depend on trial and error.

This chapter outlines the procedures for determining the required pump discharge pressure for simple and complex hose lays as well as for supplying aerial master stream devices. The final part of the chapter explains the concepts and calculation of net pump discharge pressure.

SIMPLE HOSE LAYOUTS

The vast majority of fire fighting situations require a single handline or two or more individual hose lines. Calculating the required pump discharge pressure for these situations is not particularly complex. The first step is to determine the total amount of pressure loss in the hose lay. The second is to identify the required nozzle pressure for the nozzle(s) being used. Having determined these two figures, the driver/operator may use Equation 13.1 to determine the required pump discharge pressure. These principles are illustrated by examples 13.1 through 13.4.

Example 13.1

Determine the required pump discharge pressure to supply a 200-foot, 1-inch booster line at ground level. The line is equipped with a 30-gpm fog nozzle.

Step 1: Determine the total pressure loss.

$$FL = CQ^2L$$

$$FL = (150)\left(\frac{30}{100}\right)^2\left(\frac{200}{100}\right)$$

$$FL = (150)(0.30)^2(2)$$

$$FL = (150)(0.09)(2)$$

$$FL = 27 \text{ psi}$$

Step 2: Determine the nozzle pressure.

The nozzle pressure for a handline fog stream nozzle is 100 psi.

Step 3: Determine the pump discharge pressure.

PDP = NP + TPL

PDP = 100 + 27

PDP = 127 psi

Example 13.2

Determine the required pump discharge pressure to supply a 100-foot, 1½-inch preconnected "trash" line at ground level. The line is equipped with a 95-gpm fog nozzle.

Step 1: Determine the total pressure loss.

$$FL = CQ^2L$$

$$FL = (24)\left(\frac{95}{100}\right)^2\left(\frac{100}{100}\right)$$

$$FL = (24)(0.95)^2(1)$$

$$FL = (24)(0.9025)(1)$$

FL = 21 .66, or 22 psi

Step 2: Determine the nozzle pressure.

The nozzle pressure for a handline fog stream nozzle is 100 psi.

Step 3: Determine the pump discharge pressure.

PDP = NP + TPL

PDP = 100 + 22

PDP = 122 psi

Many departments have chosen to replace the booster reels on their apparatus with short preconnected 1½-inch or larger hose lines to handle the small jobs for which booster lines were previously used. Examples 13.1 and 13.2 show that a short 1½-inch preconnect (called a trash line in many jurisdictions) will actually require a lower pump discharge pressure than the booster line while providing more than three times the volume of water. In almost every fire fighting situation, more water volume is desirable.

Example 13.3

A pumper is supplying two preconnected attack lines at a small dwelling fire. One line is a 150-foot 1½ -inch hose equipped with a 125-gpm fog nozzle. The other line is a 200-foot 1¾ -inch hose equipped with an automatic nozzle flowing 150 gpm. Determine the required pump discharge pressure for this pumper.

Step 1: Determine the total pressure loss in each line.

$$FL_{1½} = CQ^2L$$
$$FL_{1½} = (24)\left(\frac{125}{100}\right)^2\left(\frac{150}{100}\right)$$
$$FL_{1½} = (24)(1.25)^2(1.5)$$
$$FL_{1½} = (24)(1.5625)(1.5)$$
$$FL_{1½} = 56.25\text{, or } 56 \text{ psi}$$

$$FL_{1¾} = CQ^2L$$
$$FL_{1¾} = (15.5)\left(\frac{150}{100}\right)^2\left(\frac{200}{100}\right)$$
$$FL_{1¾} = (15.5)(1.5)^2(2)$$
$$FL_{1¾} = (15.5)(2.25)(2)$$
$$FL_{1¾} = 69.75 \text{ or } 68 \text{ psi}$$

Step 2: Determine the nozzle pressures.

The nozzle pressure for both handline fog stream nozzles is 100 psi.

Step 3: Determine the pump discharge pressure for each line.

$$PDP_{1½} = NP + TPL$$
$$PDP_{1½} = 100 + 56$$
$$PDP_{1½} = 156 \text{ psi}$$

$$PDP_{1¾} = NP + TPL$$
$$PDP_{1¾} = 100 + 68$$
$$\mathbf{PDP_{1¾} = 168 \text{ psi}}$$

The pump discharge pressure must be set for the higher of the two lines, which in this case is the 1¾-inch hose line at 168 psi.

Example 13.4

A pumper is supplying multiple handlines to attack a commercial fire (**Figure 13.1**). One discharge is supplying 150 feet of 2½-inch hose that is wyed into two 1½-inch attack lines, each 100 feet long and flowing 100 gpm through an automatic fog nozzle. The second discharge is equipped with 200 feet of 3-inch hose that is supplying a handline solid stream nozzle with a 1⅜-inch tip. Determine the required pump discharge pressure for this pumper.

Step 1: Determine the total pressure loss in each line.

For the wyed lines:

$$FL_{1½} = CQ^2L$$
$$FL_{1½} = (24)\left(\frac{100}{100}\right)^2\left(\frac{100}{100}\right)$$
$$FL_{1½} = (24)(1)^2(1)$$

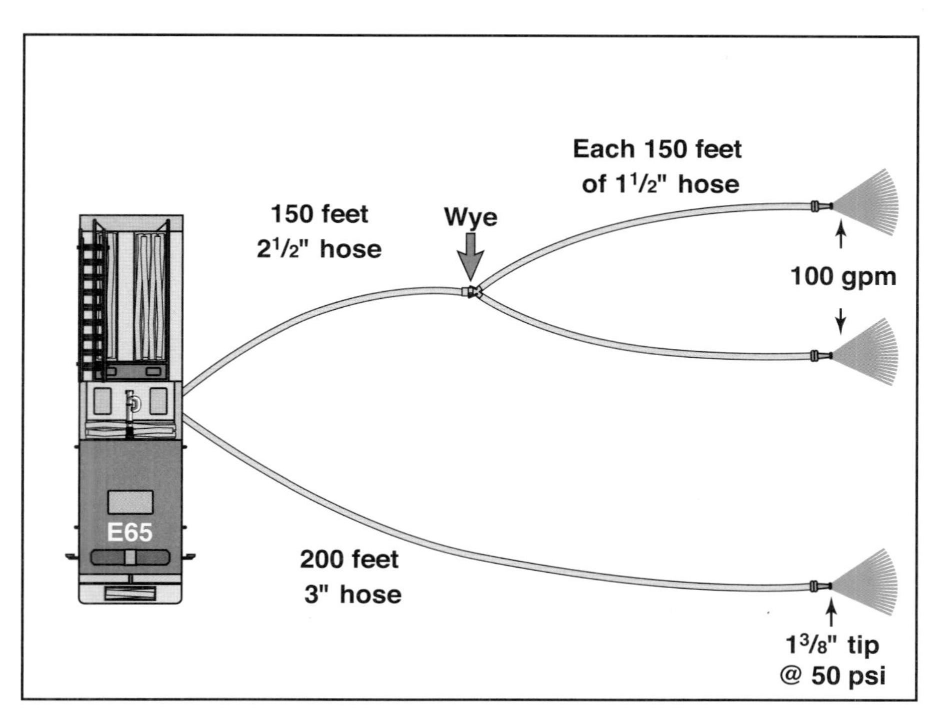

Figure 13.1 Example 13.4.

$FL_{1½} = (24)(1)(1)$

$FL_{1½} = 24 \text{ psi}$

$FL_{2½} = C(Q_{total})^2L$

$$FL_{2½} = (2)\left(\frac{100+100}{100}\right)^2\left(\frac{150}{100}\right)$$

$$FL_{2½} = (2)\left(\frac{200}{100}\right)^2(1.5)$$

$FL_{2½} = (2)(2)^2(1.5)$

$FL_{2½} = (2)(4)(1.5)$

$FL_{2½} = 12 \text{ psi}$

$TPL_{\text{wyed line}} = FL_{1½} + FL_{2½}$

$TPL_{\text{wyed line}} = 24 + 12$

Note: Appliance loss is not applicable because the flow is less than 350 psi.

$TPL_{\text{wyed line}} = 36 \text{ psi}$

For the 3-inch attack line:

$GPM = (29.7)(d)^2(\sqrt{NP})$

$GPM = (29.7)(1.375)^2(\sqrt{50})$

$$GPM = (29.7)(1.89)(7.07)$$

$$GPM = 397 \text{ gpm}$$

$$FL_3 = CQ^2L$$
$$FL_3 = (0.8)\left(\frac{397}{100}\right)^2\left(\frac{200}{100}\right)$$
$$FL_3 = (0.8)(3.97)^2(2)$$
$$FL_3 = (0.8)(15.76)(2)$$
$$FL_3 = 25.2, \text{ or } 25 \text{ psi}$$

Step 2: Determine the nozzle pressures.

The nozzle pressure for the wyed line handline fog stream nozzles is 100 psi.

The nozzle pressure for the 3-inch handline solid stream nozzle hose is 50 psi.

Step 3: Determine the pump discharge pressure for each line.

PDP wyed lines = NP + TPL wyed lines

PDP wyed lines = 100 + 36

PDP wyed lines = 136 psi

PDP 3 = NP + TPL3

PDP 3 = 50 + 25

PDP 3 = 75 psi

The pump discharge pressure must be set for the higher of the two lines, which in this case are the wyed lines at 136 psi.

COMPLEX HOSE LAYOUTS

The pump discharge pressure for complex hose layouts must be determined methodically, in much the same manner as determining total pressure loss. These concepts are illustrated by Examples 13.5 and 13.6.

Example 13.5

A pumper is supplying multiple handlines to attack a commercial fire **(Figure 13.2)**. One discharge is supplying 200 feet of 3-inch hose that is wyed into two attack lines. One attack line is 150 feet of 1½-inch hose equipped with a 95-gpm fog nozzle. This line is deployed to the roof of a 20-foot building for exposure protection. The other attack line connected to the wye is 200 feet of 2-inch hose equipped with a 200-gpm fog nozzle. The pumper is also supplying two 3-inch hoses equipped with 2½-inch couplings, each 300 feet long and supplying a portable master stream device with a 1¾-inch solid-stream tip. Determine the required pump discharge pressure.

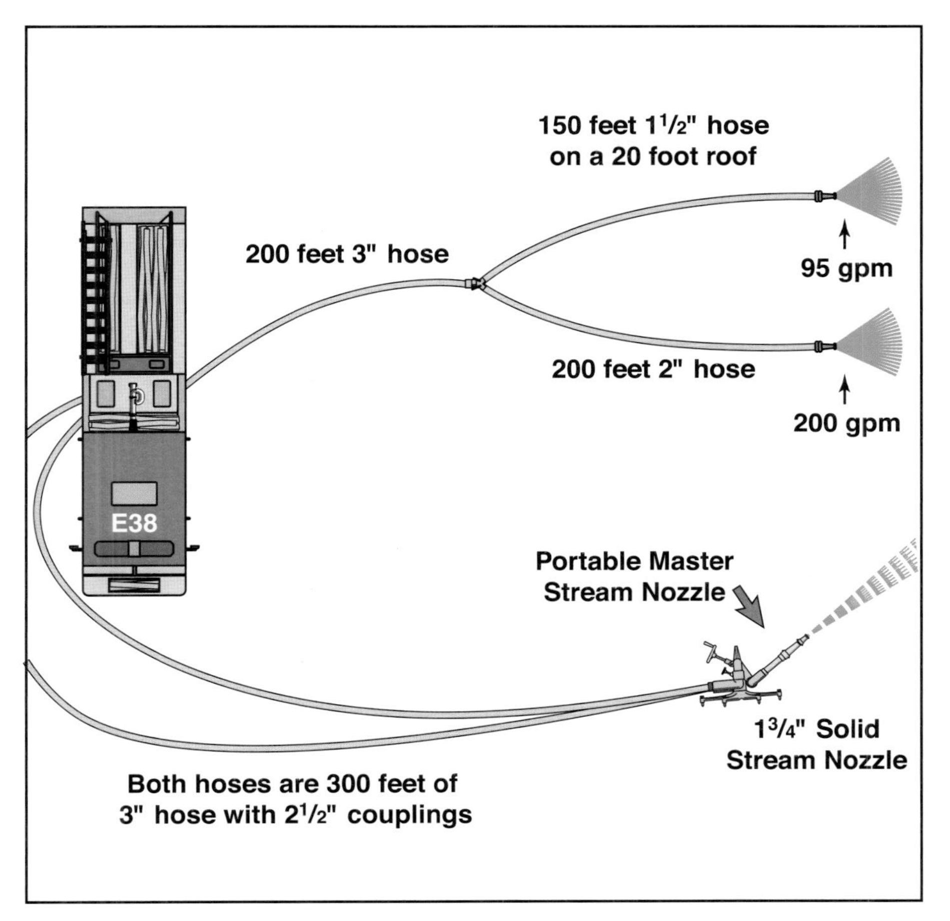

Figure 13.2 Example 13.5.

Step 1: Determine the total pressure loss in each line.

For the wyed lines:

$FL_{1½} = CQ^2L$

$FL_{1½} = (24)\left(\frac{95}{100}\right)^2\left(\frac{150}{100}\right)$

$FL_{1½} = (24)(0.95)^2(1.5)$

$FL_{1½} = (24)(0.9025)(1.5)$

$FL_{1½} = 32.49$, or 32 psi

$EP_{1½} = (0.5 \text{ psi/ft})(H)$

$EP_{1½} = (0.5 \text{ psi/ft})(20 \text{ feet})$

$EP_{1½} = 10$ psi

$TPL_{1½} = FL_1 + EP_1$

$TPL_{1½} = 32 + 10$

$TPL_{1½} = 42$ psi

$$FL_2 = CQ^2L$$

$$FL_2 = (8)\left(\frac{200}{100}\right)^2(2)$$

$$FL_2 = (8)(2)^2(2)$$

$$FL_2 = (8)(4)(2)$$

$$FL_2 = 64 \text{ psi}$$

Note: Since there is no elevation change, this is also the TPL for this line.

$$FL_3 = C(Q_{total})^2L$$

$$FL_3 = (0.8)\left(\frac{95+200}{100}\right)^2\left(\frac{200}{100}\right)$$

$$FL_3 = (0.8)\left(\frac{295}{100}\right)^2\left(\frac{200}{100}\right)$$

$$FL_3 = (0.8)(2.95)^2(2)$$

$$FL_3 = (0.8)(8.7)(2)$$

$$FL_3 = 13.9 \text{ or } 14 \text{ psi}$$

The greater pressure loss in the two attack lines is in the 2-inch line, so that figure is used to determine the total pressure loss in the wyed line assembly:

$$TPL_{\text{wyed line}} = FL_3 + FL_2$$

$$TPL_{\text{wyed line}} = 14 + 64$$

Note: Appliance loss is not applicable because the flow is less than 350 psi.

$$TPL_{\text{wyed line}} = 78 \text{ psi}$$

For the portable master stream device:

$$GPM = (29.7)(d)^2(\sqrt{NP})$$

$$GPM = (29.7)(1.75)^2\ (\sqrt{80})$$

$$GPM = (29.7)(3.0625)(8.94)$$

$$GPM = 814 \text{ gpm}$$

$$FL_{\text{dual 3-inch lines}} = CQ^2L$$

$$FL_{\text{dual 3-inch}} = (0.2)\left(\frac{814}{100}\right)^2\left(\frac{300}{100}\right)$$

$$FL_{\text{dual 3-inch lines}} = (0.2)(8.14)^2(3)$$

$$FL_{\text{dual 3-inch lines}} = (0.2)(66.3)(3)$$

$$FL_{\text{dual 3-inch lines}} = 39.76, \text{ or } 40 \text{ psi}$$

Step 2: Determine the nozzle pressures.

The nozzle pressure for the wyed line handline fog stream nozzles is 100 psi.

The nozzle pressure for the master stream solid stream nozzle hose is 80 psi.

Step 3: Determine the pump discharge pressure for each line.

PDP wyed lines = NP + TPL wyed lines

PDP wyed lines = 100 + 78

PDP wyed lines = 178 psi

PDP dual 3-inch lines = NP + TPL3

PDP dual 3-inch lines = 80 + 40

PDP dual 3-inch lines = 120 psi

The pump discharge pressure must be set for the higher of the two lines, which in this case are the wyed lines at 178 psi.

Example 13.6

Determine the pump discharge pressure in a hose line assembly in which one 300-foot, 3-inch hose with 3-inch couplings is supplying three attack lines attached to a water thief. The first attack line consists of 100 feet of 1½-inch hose flowing 125 gpm. The second attack line is 200 feet of 1½-inch hose flowing 95 gpm. The third attack line is 150 feet of 2½-inch hose flowing 250 gpm. All nozzles are fog streams.

Step 1: Determine the total pressure loss in each line.

$FL_{\text{Line 1}} = CQ^2L$

$FL_{\text{Line 1}} = (24)\left(\frac{125}{100}\right)^2\left(\frac{100}{100}\right)$

$FL_{\text{Line 1}} = (24)(1.25)^2(1)$

$FL_{\text{Line 1}} = (24)(1.5625)(1)$

$FL_{\text{Line 1}} = 37.5$ or 38 psi

$FL_{\text{Line 2}} = CQ^2L$

$FL_{\text{Line 2}} = (24)\left(\frac{95}{100}\right)^2\left(\frac{200}{100}\right)$

$FL_{\text{Line 2}} = (24)(0.95)^2(2)$

$FL_{\text{Line 2}} = (24)(0.9025)(2)$

$FL_{\text{Line 2}} = 43.3$ or 43 psi

$FL_{\text{Line 3}} = CQ^2L$

$FL_{\text{Line 3}} = (2)\left(\frac{250}{100}\right)^2\left(\frac{150}{100}\right)$

$FL_{\text{Line 3}} = (2)(2.5)^2(1.5)$

$FL_{\text{Line 3}} = (2)(6.25)(1.5)$

$FL_{\text{Line 3}} = 18.8$ or 19 psi

$$FL_{3\text{-inch supply hose}} = (C)(Q_{total})^2(L)$$

$$FL_{3\text{-inch supply hose}} = (0.677)\left(\frac{125+95+250}{100}\right)^2\left(\frac{300}{100}\right)$$

$$FL_{3\text{-inch supply hose}} = (0.677)\left(\frac{470}{100}\right)^2\left(\frac{300}{100}\right)$$

$$FL_{3\text{-inch supply hose}} = (0.677)\ (4.7)^2(3)$$

$$FL_{3\text{-inch supply hose}} = (0.677)(22.09)(3)$$

$$FL_{3\text{-inch supply hose}} = 44.9\text{, or } 45 \text{ psi}$$

The total pressure loss in the system is based on the highest loss of the three attack lines, which in this case would be Line 2, plus the friction loss in the supply line. Because the total flow rate exceeds 350 gpm, it also is necessary to add the 10 psi appliance loss.

$$\text{TPL wyed lines} = \text{FL Line}_2 + FL_{3\text{-inch supply hose}} + \text{appliance loss}$$
$$\text{TPL wyed lines} = 43 + 44 + 10$$
$$\text{TPL wyed lines} = 97 \text{ psi}$$

Step 2: Determine the nozzle pressures.

The nozzle pressure for all nozzles is 100 psi.

Step 3: Determine the pump discharge pressure.

$$\text{PDP wyed lines} = \text{NP} + \text{TPL wyed lines}$$
$$\text{PDP wyed lines} = 100 + 97$$
$$\textbf{PDP wyed lines = 197 psi}$$

It will be necessary to gate back the discharge outlets on the water thief for Line 1 and 3 until their desired hose line pressures are obtained.

Aerial Master Streams

Supplying aerial master stream devices is among the pump operator's most critical functions **(Figure 13.3, p. 328)**. This is so for a number of reasons. First, if an incident has evolved to the point where controlling it requires aerial master streams, it goes without saying that it is a large fire that must be contained as quickly as possible to avoid a potential conflagration. This requires the pump operator to ensure that the aerial master stream receives an adequate water supply. Second, the pump operator needs to know the operational limits of the aerial master stream nozzle and water supply system in order to avoid supplying too much pressure to the nozzle. Excessive pressure could result in failure of the water supply equipment or in placing excessive force on the aerial device. Either scenario could endanger the firefighters operating in the area.

Supplying aerial master stream devices is among the pump operator's most critical functions.

It is crucial that the pump operator who must supply an aerial master stream device be given all of the pertinent information before charging the hose line(s). This includes the type of aerial device being supplied, the type of nozzle and its height, and the length of the hose lay. If any of these factors change during the operation, the pump operator should be notified immediately. For example, if the height of the nozzle were lowered from 90 feet to 30 feet, the required pump discharge pressure would drop 30 psi. Failure to adjust the discharge pressure could result in overpressurization at the lower height.

It is crucial that the pump operator who must supply an aerial master stream device be given all of the pertinent information before charging the hoseline(s).

Figure 13.3 Determining the proper pump discharge pressure is crucial to the safety of elevated master stream operations. *Courtesy of Chris Mickal, New Orleans Fire Department*

Examples 13.7 through 13.9 show some common aerial master stream operations and calculations.

Example 13.7

A large-capacity industrial fire brigade aerial master stream device is protecting exposures at a large fire. The fixed waterway on this device is equipped with a 3,000 gpm fog nozzle that is elevated 90 feet. A 3,000 gpm industrial pumper will supply this device with dual 6-inch hoses, each 500 feet long. Determine the pump discharge pressure for this pumper.

Step 1: Determine the total pressure loss.

Since no coefficient for dual 6-inch hoses is available, assume that each hose is supplying one-half of the total water supply or 1,500 gpm. Then calculate the pressure loss in one of the hoses. Remember that the rule-of-thumb for pressure loss in aerial device fixed waterway systems is 25 psi.

$$FL_6 = CQ^2L$$

$$FL_6 = (0.05)\left(\frac{1{,}500}{100}\right)^2\left(\frac{500}{100}\right)$$

$$FL_6 = (0.05)(15)^2(5)$$

$$FL_6 = (0.05)(225)(5)$$

$$FL_6 = 56.25\text{, or } 56 \text{ psi}$$

$$EP = (0.5 \text{ psi/ft})(H)$$

$$EP = (0.5 \text{ psi/ft})(90)$$

$$EP = 45 \text{ psi}$$

$TPL = FL_6 + EP + \text{appliance loss}$

$TPL = 56 + 45 + 25$

$TPL = 126 \text{ psi}$

Step 2: Determine the nozzle pressure.

The nozzle pressure for an aerial master stream fog nozzle is 100 psi.

Step 3: Determine the pump discharge pressure.

$PDP = NP + TPL$

$PDP = 100 + 126$

$\mathbf{PDP = 226 \text{ psi}}$

From a practical point of view, a single 3,000 gpm industrial pumper probably would not be able to supply 3,000 gpm of water at a pressure of 226 psi, since most pumps are rated to flow their capacity at 150 psi. Adding a hose line or two from the pumper would do little to help meet the pressure requirement, as 170 psi of the 226 psi requirement would be fixed by the nozzle pressure, height, and system pressure loss. Thus, if the pumper must flow 3,000 gpm from a height of 90 feet, the situation will call for two 3,000 gpm pumpers, each flowing 1,500 gpm. A 3,000 gpm pump can flow half of its rated capacity at 250 psi. Of course, this assumes that the 6-inch hose is rated for such high pressures.

Example 13.8

Determine the pump discharge pressure in the hose assembly when a fire department pumper is supplying two 3-inch hose lines with 2½-inch couplings, each 200 feet long. These hoses are connected to a siamese appliance that is in turn supplying 100 feet of 3-inch hose with 3-inch couplings attached to a detachable ladder pipe. The ladder pipe is elevated 60 feet and is discharging through a 1½-inch diameter solid stream nozzle at 80 psi.

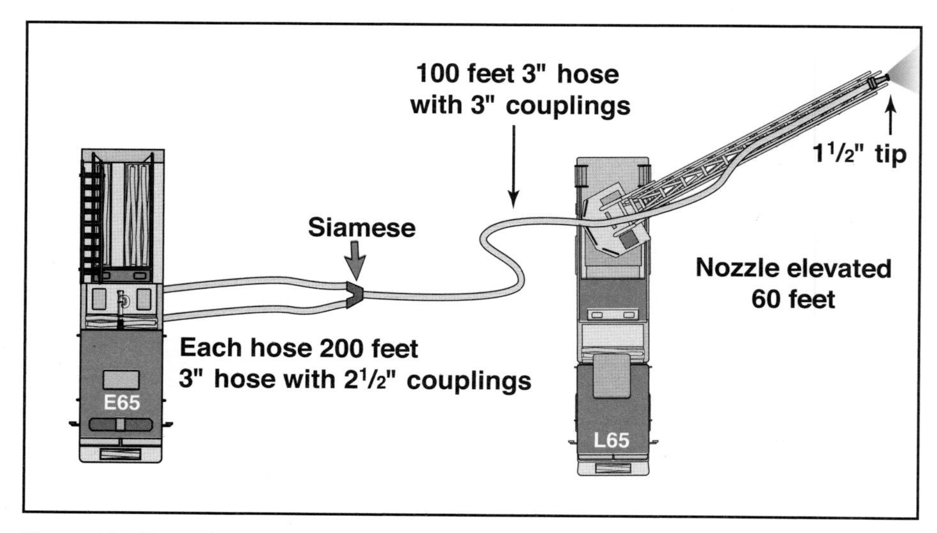

Figure 13.4 Example 13.8.

Step 1: Determine the total pressure loss.

$GPM = 29.7\ d^2\ (\sqrt{NP})$

$GPM = (29.7)\ (1.5)^2\ (\sqrt{80})$

$GPM = (29.7)\ (2.25)^2\ (8.94)$

$GPM = 598\ gpm$

$FL_{dual\ 3\text{-}inch} = CQ^2L$

$FL_{dual\ 3\text{-}inch} = (0.2)\left(\frac{598}{100}\right)^2\left(\frac{200}{100}\right)$

$FL_{dual\ 3\text{-}inch} = (0.2)(5.98)^2(2)$

$FL_{dual\ 3\text{-}inch} = (0.2)(35.76)(2)$

$FL_{dual\ 3\text{-}inch} = 14.3$ or 14 psi

$EP = (0.5\ psi/ft)(H)$

$EP = (0.5\ psi/ft)(60)$

$EP = 30\ psi$

$FL_{3\text{-}inch} = CQ^2L$

$FL_{3\text{-}inch} = (0.677)\left(\frac{598}{100}\right)^2\left(\frac{100}{100}\right)$

$FL_{3\text{-}inch} = (0.677)(5.98)^2(1)$

$FL_{3\text{-}inch} = (0.677)(35.76)(1)$

$FL_{3\text{-}inch} = 24.2$, or 24 psi

$TPL = FL_{dual\ 3\text{-}inch} + FL_{3\text{-}inch} + EP$ + appliance loss (siamese) + appliance loss (nozzle)

$TPL = 14 + 24 + 30 + 10 + 25$

$TPL = 103\ psi$

Step 2: Determine the nozzle pressure.

The nozzle pressure for an aerial solid stream nozzle is 80 psi.

Step 3: Determine the pump discharge pressure.

$PDP = NP + TPL$

$PDP = 80 + 103$

PDP = 183 psi

Example 13.9

A large-capacity pumper is supplying an aerial device equipped with two master stream nozzles. The first nozzle is a bed ladder pipe attached to a fixed waterway system and elevated 25 feet. The bed ladder pipe nozzle is equipped with a 1⅝-inch-tip solid stream nozzle, and it is being supplied with 200 feet of 4-inch hose. The detachable fly ladder pipe nozzle is equipped with a 500 gpm fog nozzle and is elevated 50 feet. The fly nozzle is being supplied by 250 feet of 3½-inch hose. Determine the pump discharge pressure.

Step 1: Determine the total pressure loss.

First, determine the pressure loss for the bed ladder pipe:

$$GPM = 29.7\ d^2 \sqrt{NP}$$

$$GPM = (29.7)(1.625)^2 (\sqrt{80})$$

$$GPM = (29.7)(2.64)^2(8.94)$$

$$GPM = 701 \text{ gpm}$$

$$FL_{4\text{-inch}} = CQ^2L$$

$$FL_{4\text{-inch}} = (0.2)\left(\frac{701}{100}\right)^2\left(\frac{200}{100}\right)$$

$$FL_{4\text{-inch}} = (0.2)(7.01)^2(2)$$

$$FL_{4\text{-inch}} = (0.2)(49.14)(2)$$

$$FL_{4\text{-inch}} = 19.65 \text{ or } 20 \text{ psi}$$

$$EP_{\text{bed ladder pipe}} = (0.5 \text{ psi/ft})(H)$$

$$EP_{\text{bed ladder pipe}} = (0.5 \text{ psi/ft})(25)$$

$$EP_{\text{bed ladder pipe}} = 12.5 \text{ or } 13 \text{ psi}$$

$$TPL_{\text{bed ladder pipe}} = FL_{4\text{-inch}} + EP_{\text{Bed Ladder Pipe}} + \text{appliance loss}$$

$$TPL_{\text{bed ladder pipe}} = 20 + 13 + 25$$

$$TPL_{\text{bed ladder pipe}} = 58 \text{ psi}$$

Second, determine the pressure loss for the fly ladder pipe:

$$FL_{3½\text{-inch}} = CQ^2L$$

$$FL_{3½\text{-inch}} = (0.34)\left(\frac{500}{100}\right)^2\left(\frac{250}{100}\right)$$

$$FL_{3½\text{-inch}} = (0.34)(5)^2(2.5)$$

$$FL_{3½\text{-inch}} = (0.34)(25)(2.5)$$

$$FL_{3½\text{-inch}} = 21.25 \text{ or } 21 \text{ psi}$$

$$EP_{\text{fly ladder pipe}} = (0.5 \text{ psi/ft})(H)$$

$$EP_{\text{fly ladder pipe}} = (0.5 \text{ psi/ft})(50)$$

$$EP_{\text{fly ladder pipe}} = 25 \text{ psi}$$

$TPL_{fly\ ladder\ pipe} = FL_{3½\text{-}inch} + EP_{fly\ ladder\ pipe}$ + appliance loss (siamese) + appliance loss (nozzle)

$TPL_{fly\ ladder\ pipe} = 21 + 25 + 10 + 25$

$TPL_{fly\ ladder\ pipe} = 81$ psi

Step 2: Determine the nozzle pressures.

The nozzle pressure for the bed ladder (solid stream) is 80 psi.

The nozzle pressure for the fly ladder (fog stream) is 100 psi.

Step 3: Determine the pump discharge pressure.

$PDP_{bed\ ladder\ pipe} = NP + TPL$

$PDP_{bed\ ladder\ pipe} = 80 + 58$

$PDP_{bed\ ladder\ pipe} = 138$ psi

$PDP_{fly\ ladder\ pipe} = NP + TPL$

$PDP_{fly\ ladder\ pipe} = 100 + 81$

$PDP_{fly\ ladder\ pipe} = 181$ psi

The pump discharge pressure will be based on the higher or these two figures, or in this case, the pressure for the fly ladder pipe, which is 181 psi.

Determining Net Pump Discharge Pressure

Virtually all fire pumps in use today are centrifugal pumps. Centrifugal pumps can take advantage of water pressure coming into the pump. Thus, if a pumper has to discharge 150 psi and the water coming into the pump is under a pressure of 50 psi, the pump only needs to "create" an additional 100 psi to meet the demand. The amount of pressure the pump is actually creating is called *net pump discharge pressure* (NPDP). Net pump discharge pressure takes into account all factors that contribute to the amount of work the pump must do to produce a fire stream. Knowing the net pump discharge pressure is not particularly important from a tactical standpoint, but knowing its concepts may aid the pump operator in critical flow situations.

When the pumper is receiving water from a pressurized source, the following formula is used to determine the net pump discharge pressure:

Equation 13.2

$NPDP_{PPS} = PDP - \text{intake reading}$

Given:

$NPDP_{PPS}$ = Net pump discharge pressure from a positive pressure source

PDP = Pump discharge pressure in psi

Example 13.10

A pumper operating from a hydrant is discharging water at 140 psi. The incoming pressure from the hydrant registers 50 psi on the intake gauge. Determine the net pump discharge pressure.

$NPDP_{PPS}$ = PDP - intake reading

$NPDP_{PPS}$ = 140 psi - 50 psi

$NPDP_{PPS}$ = 90 psi

Example 13.11

A pumper is connected to a fire hydrant in front of a burning house. The pumper is supplying an attack line comprising one 200-foot section of 2½-inch hose and a solid stream nozzle with a 1¼ inch tip. A 200-foot section of 1¾-inch hose equipped with a 150-gpm fog nozzle is stretched as a backup line. Determine the net pump discharge pressure if the intake reading from the hydrant is 60 psi.

$GPM_{2\frac{1}{2}\text{-inch}} = 29.7\ d^2\sqrt{NP}$

$GPM_{2\frac{1}{2}\text{-inch}} = (29.7)\ (1.25)^2(\sqrt{50})$

$GPM_{2\frac{1}{2}\text{-inch}} = (29.7)\ (1.5625)^2(7.07)$

$GPM_{2\frac{1}{2}\text{-inch}} = 328\ gpm$

$FL_{2\frac{1}{2}\text{-inch}} = CQ^2L$

$FL_{2\frac{1}{2}\text{-inch}} = (2)\left(\frac{328}{100}\right)^2\left(\frac{200}{100}\right)$

$FL_{2\frac{1}{2}\text{-inch}} = (2)(3.28)^2(2)$

$FL_{2\frac{1}{2}\text{-inch}} = (2)(10.76)(2)$

$FL_{2\frac{1}{2}\text{-inch}} = 43\ psi$

$PDP_{2\frac{1}{2}\text{-inch}} = FL_{2\frac{1}{2}\text{-inch}} + NP$

$PDP_{2\frac{1}{2}\text{-inch}} = 43 + 50$

$PDP_{2\frac{1}{2}\text{-inch}} = 93\ psi$

$FL_{1\frac{3}{4}\text{-inch}} = CQ^2L$

$FL_{1\frac{3}{4}\text{-inch}} = (15.5)\left(\frac{150}{100}\right)^2\left(\frac{200}{100}\right)$

$FL_{1\frac{3}{4}\text{-inch}} = (15.5)(1.5)^2(2)$

$FL_{1\frac{3}{4}\text{-inch}} = (15.5)(2.25)(2)$

$FL_{1\frac{3}{4}\text{-inch}} = 69.75$, or 70 psi

$PDP_{1\frac{3}{4}\text{-inch}} = FL_{1\frac{3}{4}\text{-inch}} + NP$

$PDP_{1\frac{3}{4}\text{-inch}} = 70 + 100$

$PDP_{1\frac{3}{4}\text{-inch}} = 170\ psi$

The required pump discharge pressure will be based on the 1½-inch hose line. The driver/operator then uses this figure to determine the net pump discharge pressure:

$NPDP_{PPS}$ = PDP - Intake reading

$NPDP_{PPS}$ = 170 psi - 60 psi

$NPDP_{PPS}$ = 110 psi

As we have noted, the net pump discharge pressure takes into account all factors that affect how much work the pump must do to produce a fire stream. When at draft, the net pump discharge pressure is *more* than the pressure shown on the discharge gauge; it is the sum of the pump discharge pressure and the intake pressure correction. The intake pressure correction must take into account friction loss in the intake hose and the height of the lift (**Figure 13.5**). **Table 13.1, p. 336** gives these allowances. These allowances are then used in the following formula to calculate the pressure correction:

Equation 13.3

$$\text{Pressure correction} = \frac{\text{lift} + \text{total intake friction loss}}{2.3}$$

This correction can be used in the following equation to calculate the net pump discharge pressure at draft:

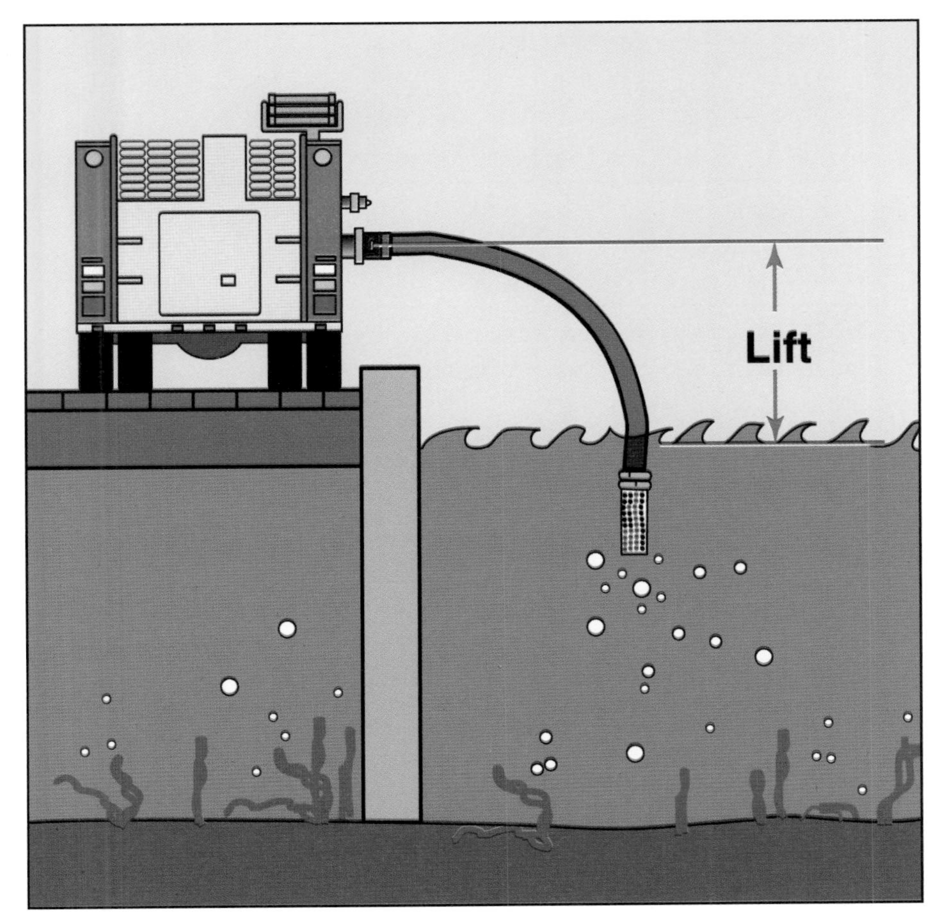

Figure 13.5 Lift is the distance between the pump and the source of water.

Table 13.1
Allowances for Friction Loss in Intake Hose

Rated Capacity of Pumper gpm	Diameter of Intake Hose in Inches	For 10 ft. of Intake Hose	Allowance (feet) for Each Additional 10 ft. of Intake Hose
500	4	6	plus 1
	4½	3½	plus ½
750	4	7	plus 1½
	5	4 ½	plus 1
1000	4½	12	plus 2½
	5	4½	plus 1½
	6	4	plus ½
1250	5	12 ½	plus 2
	6	6 ½	plus ½
1500	6	9	plus 1
1500	4½ (dual)	7	plus 1½
1500	5 "	4½	plus 1
1500	6 "	2	plus ½
1750	6	12½	plus 1½
1750	4½ (dual)	9½	plus 2
1750	5 "	6½	plus 1
1750	6 "	3	plus ½
2000	4 ½ "	12	plus 1½
2000	5 "	8	plus 1½
2000	6 "	4	plus ½

NOTE: The allowance computed above for the capacity test should be reduced by 1 psi for the allowance on the 200-psi test and by 2 psi for the allowance on the 250-psi test.

Equation 13.4

$NPDP_{Draft}$ = PDP + intake pressure correction

Given:

$NPDP_{Draft}$ = Net pump discharge pressure at draft

PDP = Pump discharge pressure in psi

Example 13.12

A 1,000 gpm pumper operating from draft is discharging water at 120 psi. The required lift is 9 feet through 20 feet of 5-inch intake hose. Calculate the net pump discharge pressure.

$$\text{Pressure correction} = \frac{\text{lift} + \text{total intake friction loss}}{2.3}$$

$$\text{Pressure correction} = \frac{9 + (4.5 + 1.5)}{2.3}$$

$$\text{Pressure correction} = \frac{15}{2.3}$$

Intake pressure correction = 6.5, or 7 psi

$NPDP_{Draft}$ = PDP + Intake pressure correction

$NPDP_{Draft}$ = 120 psi + 7 psi

$NPDP_{Draft}$ = 127 psi net pump discharge pressure at draft

SUMMARY

The essence of being a professional fire pump operator is the ability to determine the pumping requirements for a specific incident and then setting the pump to achieve those requirements. This is both a tactical and safety issue. It is a tactical issue in the sense that it allows the driver/operator to provide the amount of water necessary to control the incident; it is a safety issue in that providing either too much or too little water and pressure could hinder firefighter safety. Performing these tasks sets apart a professional from a "lever puller."

REVIEW QUESTIONS

1. Define pump discharge pressure.

2. Determine the pump discharge pressure required to adequately supply a 250-foot section of 2-inch hose equipped with a 175 gpm fog nozzle.

3. Determine the pump discharge pressure required to adequately supply a 200-foot 1¾-inch attack line that is equipped with a ⅞-inch solid stream nozzle tip and is operated in the basement of a store.

4. Determine the pump discharge pressure required to adequately supply 400 feet of 4-inch hose that is supplying a manifold to which two 150 foot 2½-inch attack lines are connected, each equipped with a 250 gpm fog nozzle.

5. Determine the pump discharge pressure required to adequately supply 200 feet of 3½-inch hose that is supplying a wye to which two attack lines are attached. One line is 200 feet of 2½-inch hose equipped with a 1⅛-inch solid stream tip and operated at ground level. The other attack line is 150 feet of 1 3/4-inch hose equipped with a 125 gpm fog nozzle and is operated from a ladder into a third- floor window.

6. Determine the pump discharge pressure required to adequately supply a portable master stream device equipped with a 1,000 gpm fog nozzle. This device is being supplied by three 2½-inch hose lines, each 250 feet long.

7. A car fire starts on the seventh level of a parking garage that is not equipped with a standpipe system. Fire apparatus cannot access the upper levels of the parking garage due to height restrictions. To attack the fire, firefighters stretch 150 feet of 1½-inch hose equipped with a 95-gpm fog nozzle from the standpipe hose outlet on an aerial platform. What is the required pump discharge pressure?

8. A pumper is supplying an aerial ladder equipped with a detachable ladder pipe that has a 1⅝-inch solid stream nozzle and is elevated 60 feet. The ladder pipe is supplied by 100 feet of 3-inch hose with 3-inch couplings. This hose is in turn being supplied by a dual lay of 3-inch hose equipped with 2½-inch couplings each 300 feet long. Determine the required pump discharge pressure.

---

9. A pumper is supplying multiple hose lines at a structure fire. A portable master stream device equipped with a 500 gpm fog nozzle is being supplied by two 2½-inch hose lines, each 250 feet long. A 400–foot five-inch hose is being used to supply a manifold attached to three attack handlines. Line 1 is 150 feet of 2½-inch hose equipped with a 1-inch solid stream nozzle. Line 2 is 200 feet of 2-inch hose equipped with a 150 gpm fog nozzle; it is being operated on the roof of a 40-foot-high building. Line 3 is 200 feet of 1¾-inch hose equipped with a ⅞-inch tip solid stream nozzle being operated on the second floor. Determine the pump discharge pressure for this pumper.

---

10. Suppose a relay pumper was supplying the pumper in review question 9 and the intake pressure was 35 psi. What would be the net pump discharge pressure for the pumper?

---

11. Suppose the 1,250 gpm pumper in review question 5 was operating from a draft at a farm pond and the required lift was 11 feet through 20 feet of 6-inch hard intake hose. What would be the net pump discharge pressure for the pumper.

Answers found on page 436.

FESHE Course Objectives

The information in this chapter is intended to meet some of the objectives outlined for the Fire Protection Hydraulics and Water Supply course by the United States Fire Administration's National Model Core Curriculum as developed by the Fire and Emergency Services Higher Education (FESHE) initiative. The information in this chapter also allows the student to meet the requirements of NFPA® 1002, *Standard for Fire Apparatus Driver/Operator Professional Qualifications* (2009 edition) listed below.

FESHE Course Objectives

1. Describe the conditions that necessitate relay pumping operations.
2. List the types of apparatus and equipment used for relay pumping operations.
3. Describe the operational considerations for establishing and operating a relay pumping operation.
4. Describe the maximum distance relay method.
5. Describe the constant pressure relay method.
6. Describe the operational considerations for shutting down a relay pumping operation.

NFPA® 1002 Requirements

5.2.2 Pump a supply line of 65 mm (2½-inch) or larger, given a relay pumping evolution, the length and size of the line, and the desired flow and intake pressure, so that the correct pressure and flow are provided to the next pumper in the relay.

5.2.2 (A) Requisite Knowledge. Hydraulic calculations for friction loss and flow using both written formulas and estimation methods.

Chapter 14

Relay Pumping

INTRODUCTION

We have thus far assumed that the pumper supplying attack lines is either operating from its water tank or connected directly to a reliable water supply source. In the real world we know this is not always the case. In many instances, whether urban, suburban, or rural, the attack pumper may be remote from a reliable water supply source. In rural and even some suburban situations there may be no water supply system serving the immediate area of the emergency incident. Large incidents in urban settings may require relay pumping when water supply from the hydrants closest to the scene is insufficient to control the incident.

When the attack pumper cannot be connected directly to a reliable water supply source, firefighters have three options:

1. Use the available water in apparatus water tanks to mount an attack or protect exposures and then allow the fire building to burn if the fire cannot be extinguished.
2. Assemble and operate a water shuttle operation.
3. Assemble and operate a relay pumping operation.

Obviously, Option 1 is the least desirable. The citizens we protect expect us to use all means necessary to protect their lives and property. To simply "write off" a structure and make no reasonable effort to obtain the necessary water for mounting an effective fire fighting operation is unprofessional and contrary to the mission of the fire service.

Figure 14.1 Water shuttle operations should be used where relay pumping is not practical.

In jurisdictions or at specific incidents where water supply sources are a considerable distance from the fire scene, the best option may be to assemble and operate a water shuttle operation **(Figure 14.1, p. 342)**. Water shuttle operations involve the use of water tankers or tenders to shuttle water between a water supply source (called the fill site) and an incident scene (called the dump site). Tankers dump their initial loads into portable tanks at the incident scene and then drive to a fill site where they are quickly reloaded. They then repeat the process until the incident is controlled. Water shuttle operations are commonly used in situations where the water supply source is more than one mile from an incident scene or the departments involved in the incident do not have sufficient hose to establish a relay pumping operation between the water supply source and the incident scene. For more detailed information on conducting water shuttle operations, consult Chapter 14 of IFSTA's *Pumping Apparatus Driver/Operator Handbook.*

Most experts prefer the third option, establishing a relay pumping operation, when possible. A *relay operation* uses a pumper at the water supply source to pump water under pressure through one or more hoselines to the next pumper in line. This pumper, in turn, boosts the pressure to supply the next pumper, and so on, until water reaches the fireground apparatus.

These experts prefer relay pumping over shuttle operations for a number of reasons. Relay pumping operations establish a constant flow of water between the water supply source and the incident scene, assuming that none of the pumpers in the relay has a mechanical breakdown. Unless a water shuttle runs at optimum efficiency, interruptions in the attack pumper's water supply are possible as tankers come and go from the fill site. From a safety standpoint, certainly a stationary relay pumping operation presents considerably fewer hazards than tankers traveling back and forth on roadways, often among civilian traffic. Relay pumping is also more fuel-efficient than operating a water shuttle operation. Pumpers in stationary pumping positions will not use as much fuel as tankers operating in a shuttle.

For the purpose of this chapter, we will consider relay pumping operations to include a pumper at the water supply source, a pumper at the fire scene, and at least one pumper connected into the hoseline and relaying water between the two terminal pumpers. Technically, a pumper connected to a water supply source and pumping directly to an attack pumper is performing relay pumping; some jurisdictions and textbooks also call this two-pumper relay a *tandem pumping* operation. Many of the concepts are the same as those we discuss in this chapter, but we will focus on longer relays that require at least three pumping apparatus, as these scenarios are more challenging.

This chapter explores the various components of a relay operation, examines the factors that influence the capability of the relay operation, and discusses various methods for establishing a relay operation.

RELAY APPARATUS, EQUIPMENT, AND TERMINOLOGY

As with any aspect of fire protection, successful relay pumping operations require a variety of types of apparatus, hose, and equipment. Some of these elements have special names in the context of a relay pumping operation. This section examines the equipment needed for a properly designed relay pumping operation and highlights special terms applied to these items.

Most jurisdictions use standard fire department pumpers to lay hose and perform the pumping duties in a relay operation. NFPA® 1901 requires standard fire department pumpers to carry at least 800 feet of 2½-inch or larger supply hose. Jurisdictions that commonly perform relay pumping operations may carry more supply hose than this minimum. Pumpers with 2,000 or more feet of large diameter hose (LDH) are common.

Operating in a relay pumping operation does not have to be limited to standard fire department pumpers. Any type of apparatus equipped with a sufficient fire pump can be placed into a relay pumping operation if sufficient pumpers are not available. This includes rescue-pumpers, pumper-tankers (pumper-tenders), and quint aerial apparatus equipped with fire pumps of comparable capacity to the other pumpers in the relay **(Figure 14.2)**. In industrial fire protection situations, relay pumping operations commonly use trailer-mounted fire pumps **(Figure 14.3)**. The trailer pumps are towed to their needed location and then connected into the hose lay to boost the water pressure.

Any type of apparatus equipped with a sufficient fire pump can be placed into a relay pumping operation if sufficient pumpers are not available.

Figure 14.2 If they are not being used to shuttle water, pumper-tankers may be inserted into a relay pumping operation.
Courtesy of Ron Jeffers

Figure 14.3 Trailer pumps are commonly used in industrial facilities.

This chapter uses a number of terms specific to relay pumping. They describe specific functions for pumping apparatus placed into the relay pumping operation **(Figure 14.4, p. 344)**:

- **Source, or supply, pumper**. This is the pumper connected to the water supply at the beginning of the relay operation. The water supply may be a fire hydrant or a static water supply source. The source pumper then pumps water to the next apparatus in line.
- **Relay pumper**. This pumper is sometimes also referred to as the *in-line pumper*. Relay pumpers are connected within the relay and receive water from the source pumper or another relay pumper. They boost the pressure and then supply water to the next relay pumper or the attack pumper.
- **Attack pumper**. This is the pumper at the fire scene; it receives water from the relay and supplies attack lines and fire stream appliances as needed for fire suppression.

Fire departments in jurisdictions that frequently perform relay pumping operations may have special apparatus designed primarily for this purpose. Most commonly referred to as *hose tenders,* these apparatus may look like a standard pumper or may be quite different **(Figure 14.5, p. 345)**. Hose tenders may or may not be equipped with a fire pump that allows them to participate in the pumping operation once the hose is laid. They usually carry a mile or more of large diameter (4-inch or larger) hose **(Figure 14.6, p. 345)**. This hose may be carried in a traditional style hose bed or on a large, mechanically oper-

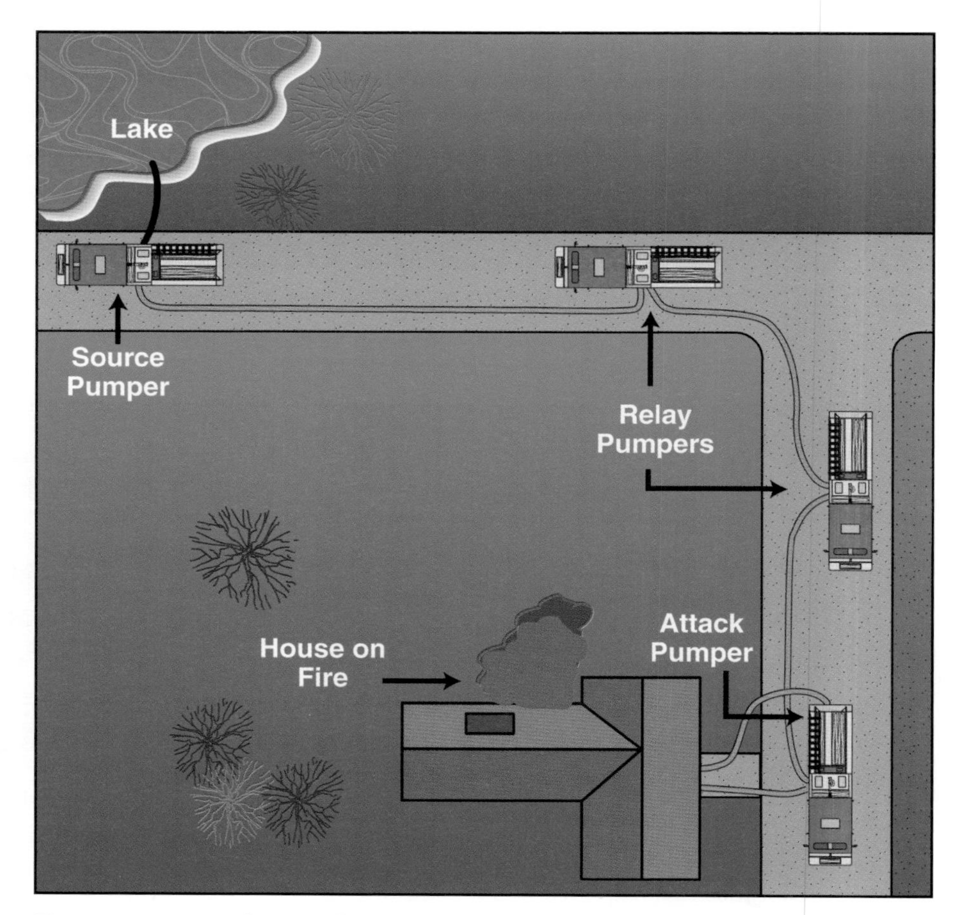

Figure 14.4 Hose tenders usually carry large amounts of LDH. *Courtesy of Ron Jeffers*

ated reel. Hose tenders also generally carry a wide assortment of relay valves, discharge manifolds, and other special water supply equipment for relay pumping operations.

The use of hose tenders is not limited to rural and suburban jurisdictions without water supply systems. Many urban fire departments utilize hose tenders to lay hose lines at large-scale incidents where first-arriving apparatus have already utilized all the available fire hydrants close to the incident. Longer hose lays may then be required to supply the additional water needed to control the incident. In other cases, a portion of a jurisdiction may be protected by an insufficient water supply system. If a large fire occurs within this area, long hose lays may be required to reach more reliable parts of the water supply system. Lastly, some urban jurisdictions maintain hose tenders as a contingency plan in the event that a water main break or a natural disaster such as an earthquake disrupts the water supply system.

Both medium and large diameter supply hose may be used for relay pumping operations. Medium diameter hose (MDH) includes 2½- and 3-inch hoselines. When MDH is used for relay operations, generally two or three of these hoselines are laid. LDH ranges in size from 3½ to 12 inches, with 3½-, 4-, and 5-inch being the most common sizes used by municipal fire departments.

Figure 14.5 Departments that frequently have long hose lays use hose tenders. *Courtesy of Ron Jeffers*

Figure 14.6 Hose tenders usually carry large amounts of LDH. *Courtesy of Ron Jeffers*

The use of 6-inch and larger supply hose is typically limited to industrial fire brigade operations, in particular those protecting petrochemical facilities. As we will see later in this chapter, using a single LDH hose for relay operations has numerous advantages over using MDH hoselines. These include less hose to pick up after the incident, reduced friction loss, and the capacity to move increased volumes of water.

Chapter 8 of this text discussed the various components that make up the fire pump on a fire apparatus, including pressure control devices. Pressure control devices become extremely important during relay pumping operations because of the force and pressure exerted by the large volumes of water being moved. Perhaps the most crucial safety device for relay pumping operations is the *intake pressure relief valve*, sometimes referred to as the *relay relief valve*. A properly adjusted intake pressure relief valve reduces the possibility of damage to the pump and discharge hoselines caused by water hammer when valves are closed too quickly or by other significant rises in intake pressure.

Figure 14.7 Relief valves may be added to the large pump intake.

There are two basic kinds of intake pressure relief valves. One type is supplied by the pump manufacturer and is an integral part of the pump intake manifold. The second type is an add-on device that is screwed onto the pump intake connection **(Figure 14.7)**. The add-on relief valves have become extremely popular in recent years with the increasing use of large diameter supply hose. The truth of the matter is that these valves are unnecessary if the pump is equipped with an integral intake pressure relief valve, as almost all modern fire pumps are. While add-on valves may add some measure of safety, they are rather expensive, and jurisdictions that have budget constraints may choose to forego them.

Regardless of which type of intake pressure relief valve is used, their operation is essentially the same. The valves are preset to allow a predetermined amount of pressure into the fire pump. If the incoming pressure exceeds the pre-set level, the valve activates and dumps the excess pressure/water until the amount of pressure/water entering the pump lowers to the pre-set level. Different jurisdictions have different policies for setting intake pressure relief valves. At a minimum, these valves should be set within 10 psi of the static pressure of the water system supplying the pumper or 10 psi above the discharge pressure of the previous pumper in the relay.

Most add-on intake pressure relief valves also have a manual shut-off valve that allows the pump operator to stop the water supply if desired. These may be either a quarter-turn valve or a screw-down valve operated by a hand wheel. Additionally, the valve will be equipped with a bleeder valve that allows air to bleed off as the incoming supply hose is charged **(Figure 14.8)**. This is particularly important when using these devices with large diameter supply hose. A large amount of air is pushed through the hose until a solid column of water reaches the valve. Bleeder valves may be located directly in the intake piping to the pump itself.

Relays dependent upon later-arriving mutual aid companies can set up an initial relay of lesser volume and greater spacing with in-line relay valves placed in the relay line for the incoming pumpers **(Figure 14.9)**. These valves allow the late-arriving pumpers to hook up after the relay is operating and boost the pressure (and corresponding volume) without interrupting operations **(Figure 14.10)**. Their operational concept is the same as that of four-way hydrant valves, such as the Humat® valve, which allow a hydrant supply line to be initially charged and then have the pressure boosted at a later time without having to shut down the hose.

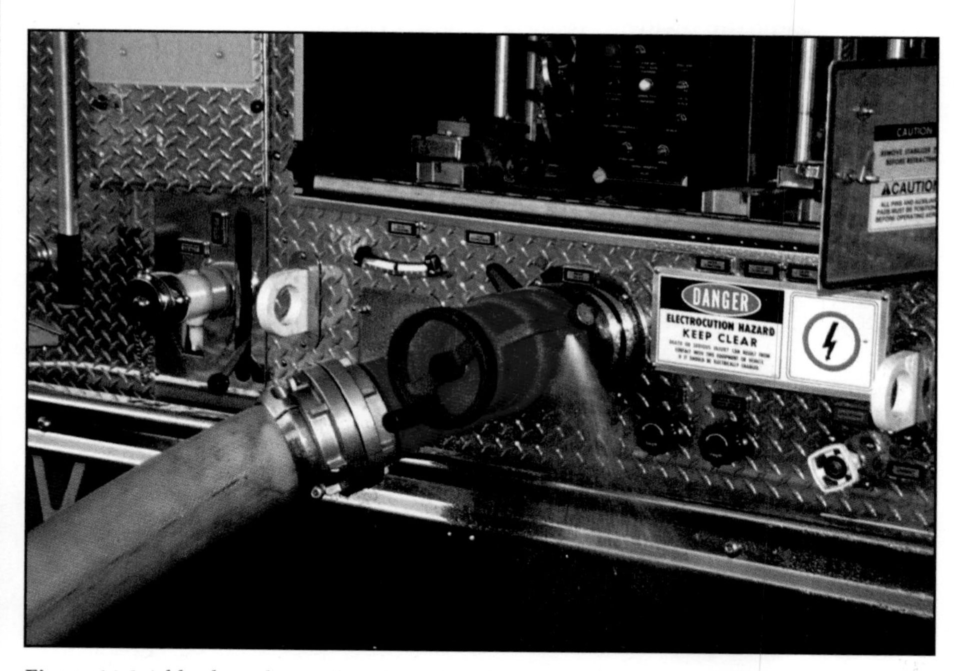

Figure 14.8 A bleeder valve on the relief valve allows air to be released before it enters the pump.

Figure 14.9 A typical in-line relay valve.

If an LDH relay pumping operation is intended to support more than one attack pumper at the fire scene, a discharge manifold may be used to break down the LDH into two or more hoselines that may then be connected to the attack pumpers **(Figure 14.11)**. These discharge manifolds are available in a number of different designs. They typically have one LDH inlet and any combination

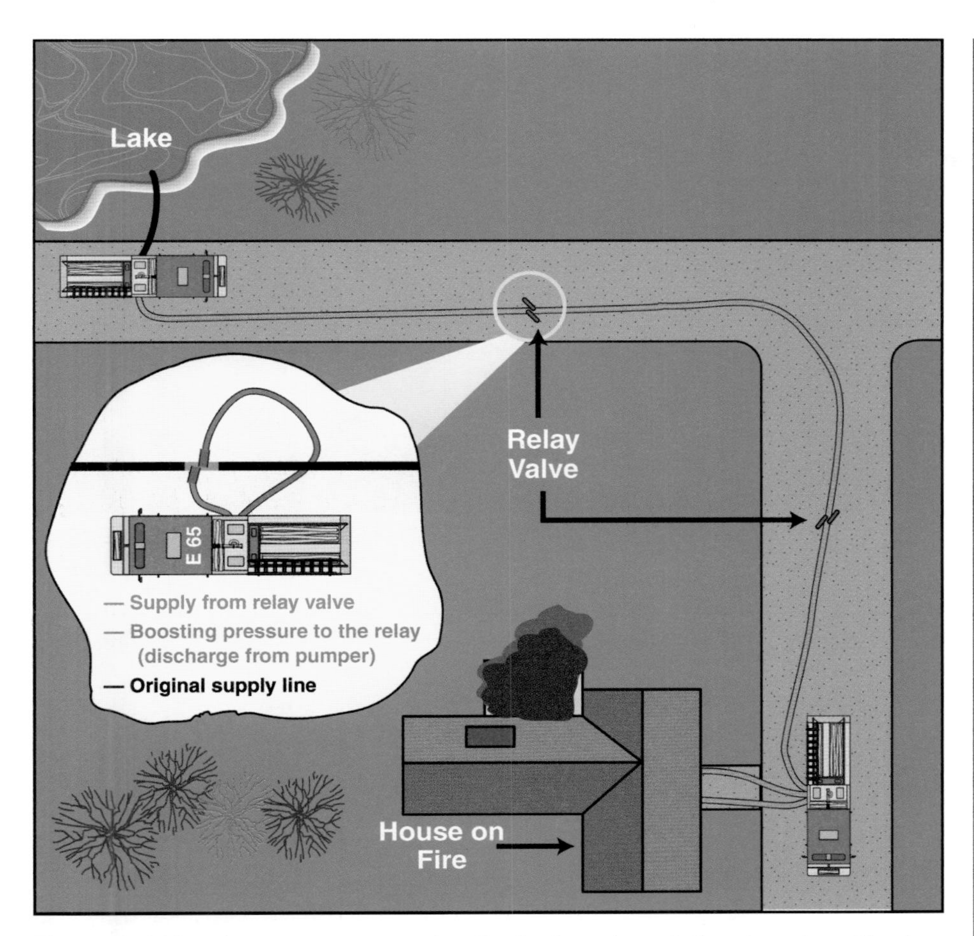

Figure 14.10 The relay pumper connects into flowing hose through the relay valve without interrupting the flow of water.

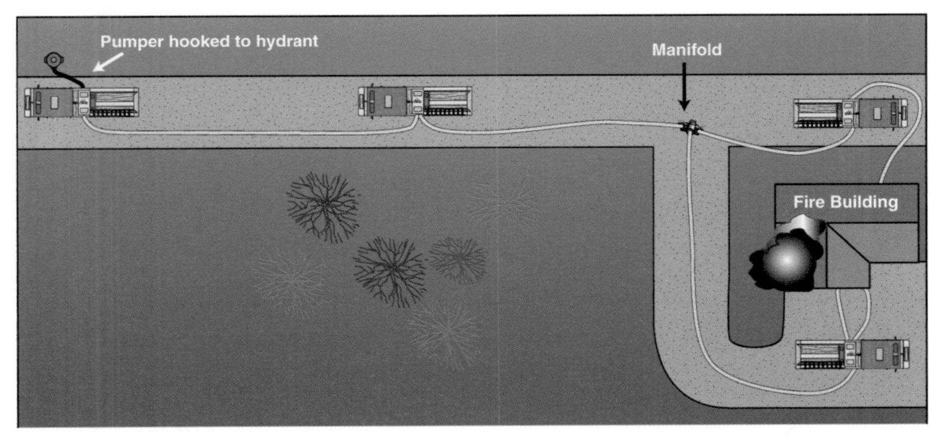

Figure 14.11 The manifold may be used to distribute water to more than one vehicle at the scene.

of MDH and LDH discharges **(Figure 14.12 a and b, p. 348)**. (Chapters 12 and 13 cover calculating required pump discharge pressures for operating with manifolds.) If the relay is using multiple MDH lines, each line may support a different attack pumper at the scene **(Figure 14.13, p. 348)**.

Figure 14.12a A typical LDH manifold.

Figure 14.12b An example of a large-scale LDH operation manifold. *Courtesy of Ron Jeffers*

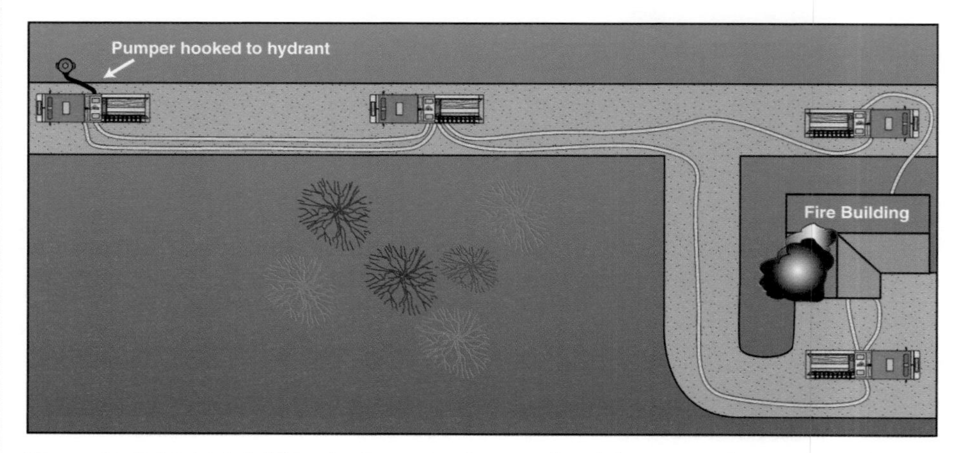

Figure 14.13 Multiple MDH relay hoses may be routed to different attack pumpers at the scene.

RELAY PUMPING OPERATIONAL CONCEPTS

Some basic operational concepts apply to all types of relay pumping operations. Regardless of which method of relay pumping will be used at an emergency scene, its design must always be predicated on two basic requirements:

- The amount of water needed at the emergency scene
- The distance between the emergency scene and the water source

The amount of water needed at the emergency scene has a major impact on the design of the relay. (Chapter 6 explains in detail how to determine the amount of water required for fire fighting operations.) Once the required fire flow has been determined, relay pumping operations may be established to supply either all or part of the water required for the incident. No minimum flow rate is required for the establishment of relay pumping operations. However, going to the trouble to establish relay pumping operations for flows of less than 500 gpm are probably not worth the effort. Incidents that require flows that small can typically be handled by tank water from the attack apparatus or tankers.

The second basic requirement that must be considered is the distance the water must be relayed. Friction loss in a hose lay is primarily affected by two factors: the amount of water being flowed and the size of the hose being used (see Chapters 12 and 13). As the length of the hose lay increases, so does the friction loss. This increased friction loss will reduce the volume of water flowing through the hose. This could pose a serious operational problem if the length of the hose lay was increased to the point that the relay could not supply the required flow.

Increasing the Flow Through the Relay

Realistically, the Incident Commander cannot shorten the distance between the water supply and the incident scene (unless a closer, suitable water flow is located). Thus, if the flow through the hose lay drops below the required level, one of four options to increase the existing relay's flow will be necessary:

1. Increase the size of the hose in the relay.
2. Add one or more additional hoselines in the relay.
3. Increase the pump discharge pressure of the pumpers operating in the relay.
4. Increase the number of pumpers in the relay.

Option 1, increasing the size of the hose in the relay, is more theoretical than practical. It generally would be impractical to shut down an existing relay to replace the hose being used with a larger hose. The time and effort to achieve this option usually would not be worth the effort, unless the relay was anticipated to be in operation for an extended time. For this reason, it is important that driver/operators always lay the largest hose available at the beginning of the operation. For example, if the apparatus is equipped with both 3- and 5-inch supply lines, lay the 5-inch. You can flow as little as 200 or 300 gpm through the 5-inch if that is all that is required. On the other hand, you cannot flow 1,000 gpm through the 3-inch (for any reasonable distance) if that flow is needed.

More practical than replacing a hoseline that is too small would be to add one or more parallel hoselines to the existing relay. This would be accomplished by having engine companies or hose tenders not already committed to pumping lay an additional hoseline between the relay pumpers. Each pumper may have this additional hoseline attached to its pump and may begin flowing the hose when all pumpers are ready. If the position of apparatus and hose make laying the second hoseline along the route of the original relay hose impractical, simply establishing a second relay pumping operation to another suitable water supply source may be necessary.

Like Option 1, Option 3 does not have a significant amount of practical application. In a properly designed relay the pumpers already will be operating near their maximum discharge pressure. This minimizes the number of pumpers needed in the relay. Even if it were possible to increase the pumpers' pump discharge pressure, doing so would not necessarily increase the volume of water through the relay. All pumpers are rated to pump their maximum volume at a net pump discharge pressure of 150 psi. If the pump operates at a pressure higher than 150 psi, the pumper's volume capability decreases proportionally. Depending on the length of the hose lay and the volume of water being flowed, the pumps eventually will reach a point where increasing the pressure will not increase the volume.

At no time should a relay pumping operation result in discharge pressures that exceed the maximum operating pressure for the hose being used.

Another factor to keep in mind when considering increasing the pressure is that you are also limited by the pressure to which the fire hose in use is rated and annually tested. **Tables 14.1** and **14.2** show the annual service test and recommended maximum operating pressures for various types of fire hose. These figures are from NFPA® 1961, *Standard on Fire Hose,* and NFPA® 1962, *Standard for the Care, Use, and Service Testing of Fire Hose Including Couplings and Nozzles.* At no time should a relay pumping operation result in discharge pressures that exceed the maximum operating pressure for the hose being used. Fire departments may specify and purchase hose designed for higher pressures than the NFPA® minimums. This is particularly true of fire departments that frequently use LDH for long relays or to supply fire department connections on fixed suppression systems and/or portable master stream devices. If that is the case, the hose should be pumped at pressures that do not exceed 90 percent of the annual service test pressure.

Table 14.1
Annual Service Test and Maximum Operating Pressures for Fire Hose Manufactured Prior to 1987

Type of Hose	Annual Service Test Pressure in psi	Maximum Operating Pressure in psi
Attack Hose	250	225
Relay Hose – 3½- to 5-inch	200	185
Relay Hose – 6-inch and larger	150	135

Table 14.2
Annual Service Test and Maximum Operating Pressures for Fire Hose Manufactured After 1987

Type of Hose	Annual Service Test Pressure in psi	Maximum Operating Pressure in psi
Attack Hose	300	275
Relay Hose – 3½- to 5-inch	200	185
Relay Hose – 6-inch and larger	200	185

The fourth option for increasing the flow in an existing relay is to place additional pumpers in the relay. By shortening the length of hose each pumper has to supply, maximum pressures (and accordingly, maximum flows) may be maintained throughout the hose assembly. The down side to this possibility is that if in-line relay valves were not placed in the hose lay from the outset, it would be necessary to shut down the relay when the additional pumper(s) was (were) added.

If it becomes necessary to add a pumper to the relay the following procedure should be used:

Step 1: Position the pumper at the appropriate location in the relay, as close to a hose connection as possible.

Step 2: Remove one section of hose from the pumper for each intake and discharge that will be connected.

Step 3: Connect the sections of hose to the intake(s) and discharge(s) that will be used and place the other ends by the relay hose connection that will be broken (**Figure 14.14, p. 352**).

Step 4: Notify the other pumpers in the relay (or the water supply officer) to shut down pumping operations.

Step 5: Put the pump in gear and ready for pumping.

Step 6: Once the relay hose is shut down, break the hose connection and connect the relay hoses to the hose attached to the pumper.

Step 7: Notify the other pumpers in the relay (or the water supply officer) to resume pumping operations.

In the rare situations where low flow rates are required and LDH is available, the required spacing between pumpers may exceed the amount of hose carried on each pumper. In these cases, it may be necessary to call other pumpers solely to lay their hose but not actually to participate in the pumping process. In these cases it is advisable to keep one or more of these extra pumpers available

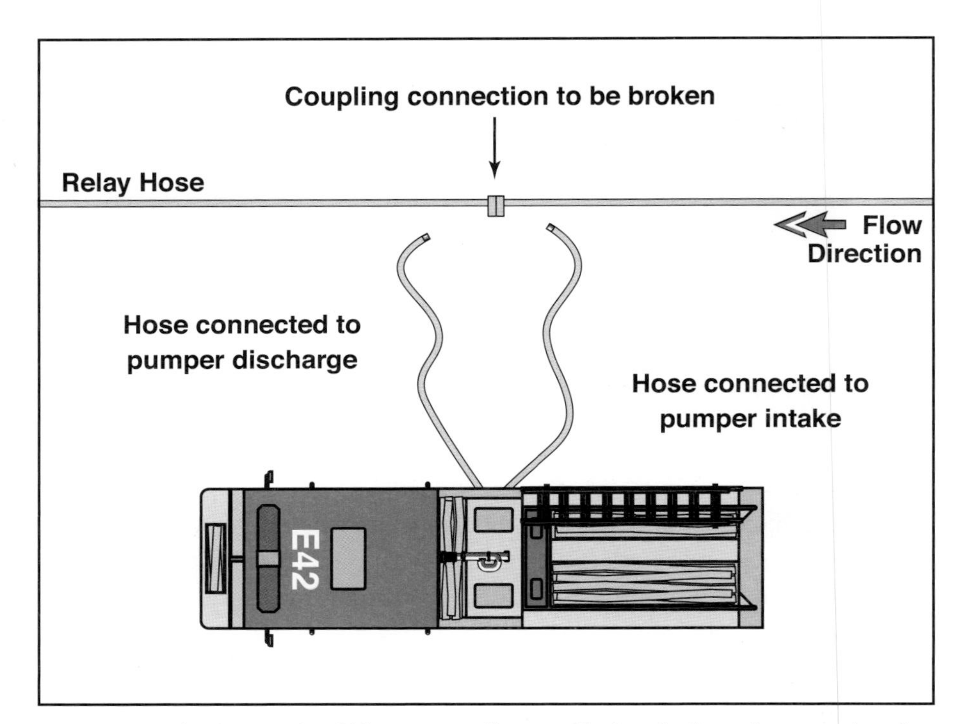

Figure 14.14 The pumper should be set up as illustrated before the flow of water in the relay is interrupted.

in the event that the flow rate needs to be increased. If in-line relay valves are not available, one or more of the extra pumpers may be inserted into the relay without engaging their pumps unless the extra pressure is needed. This connection will increase the friction loss in the hose lay, but in low-flow situations the flow may not be critically decreased.

Types of Relay Pumping Operations

The two basic designs for relay pumping operations are the maximum distance relay method and the constant pressure relay method. Of the two, the maximum distance relay method is the more commonly used. Each individual fire department, or group of fire departments that regularly perform relay pumping operations together, should predetermine the type of relay pumping operation it will use should the need arise and train on this operation regularly.

Where relay pumping operations are commonly required, preincident dispatch procedures may be established for providing relay pumping capabilities at an emergency scene. When a relay is necessary to provide an adequate amount of water to the scene, the Incident Commander notifies the dispatch center to respond a relay task force or strike team. The dispatch center then dispatches from three to five pumpers, each with large capacity fire pumps and usually large diameter hose, as well as a chief officer to oversee the relay to the scene. These companies establish the water supply independent of those already operating on the scene. Depending on local procedures, either the IC or the responding water supply officer will determine the water supply source and the relay route.

Maximum Distance Relay Method

The *maximum distance relay method* involves predetermining the volume of water to be flowed by the relay and then spacing the pumpers appropriately for the maximum distance that they can pump that amount of water through the planned hose lay. Jurisdictions that always lay the same hose for relay pumping operations should make a chart that lists various intervals of water flow correlated to maximum distances between pumpers in the relay. If the department might use different numbers or sizes of hose, the chart should reflect all of these possibilities. **Table 14.3** contains sample information on predetermined amounts of water that may be pumped through various types of hose lays.

Table 14.3
Maximum Distance Relay Lengths in Feet

	Hose size in inches						
Flow in gpm	**One 2½**	**One 3**	**One 4**	**One 5**	**Two 2½**	**One 2½ & One 3**	**Two 3s**
250	1,440	3,600	13,200	33,000	5,760	9,600	14,400
500	360	900	3,300	8,250	1,440	2,400	36,000
750	160	400	1,450	3,670	640	1,050	1,600
1000	90	225	825	2,050	360	600	900
1250	50	140	525	1,320	200	375	500

Table 14.3 is based on calculations using the formulas and methods discussed in Chapters 12 and 13 of this text. Fire departments choosing to develop their own chart may use either the friction loss coefficients in Chapter 12 to determine these distances or other coefficients that they have developed on their own or that have been provided by hose or equipment manufacturers.

One factor built into the figures in Table 14.3 is the consideration for assuring that a minimum 20 psi residual pressure would be available at the next pumper in the relay. Departments that develop their own charts also must factor in a residual pressure for the next pumper. IFSTA recommends a minimum residual pressure of 20 psi, but each jurisdiction can certainly adjust that figure to its preference. Another factor that has to be considered when developing this type of chart is the rated test pressure of the hose being used. Table 14.3 is based on the NFPA® rated test pressures of 200 psi for 2½- and 3-inch hose and 185 psi for 4- and 5-inch hose. Again this may have to be adjusted for local needs.

Also to be considered is the relationship between the pump discharge pressure and volume. Most standard fire department pumpers are rated by Underwriter's Laboratories (UL) to flow their maximum volume at 150 psi, 70 percent of their maximum volume at 200 psi, and 50 percent of their maximum volume at 250 psi.

Figure 14.15 FDNY utilizes a number of high-pressure pumpers in front-line service. *Courtesy of Ron Jeffers*

The exception would be pumpers specially designed to supply high pressures of water. These are common in cities that have high-rise buildings requiring fire suppression system support.

One example of this can be found in New York City. The Fire Department of New York (FDNY) protects more high-rise structures than any other fire department in the United States. Historically, all standard FDNY pumpers were equipped with 1,000 gpm fire pumps. In addition to the standard UL ratings, regular FDNY pumpers are required to be able to pump 250 gpm at 600 psi from a draft. Special high-rise pumpers are located strategically throughout those sections of the city that have large numbers of high-rises (**Figure 14.15**). These pumpers have special three-stage 1,000 gpm centrifugal pumps that can supply 500 gpm at 700 psi and 350 gpm at 800 psi. The high-pressure discharges are color-coded to avoid confusion on the fireground. In recent years FDNY has switched to ordering all new pumpers to be equipped with 2,000 gpm pumps. At the time this edition was written about one-fourth of all front line engines were equipped with the higher volume pumps. Eventually all front-line pumpers will be equipped with the larger pumps.

If you are using standard fire department pumpers and are developing or using a chart that involves discharge pressures in excess of 150 psi, such as Table 14.3, you will have to assume that the pumpers will not be able to flow their highest rated volumes. For example, if you were to use Table 14.3 based on discharge pressures of 185 and 200 psi, you would have to use minimum pump capacities listed in **Table 14.4** to achieve the needed flows/distances. If the local procedure uses distances less than those on the chart, it will be possible to use smaller capacity pumps than those listed or to remove one or more pumpers from the relay.

Table 14.4
Required Pump Sizes to Achieve the Flows in Table 14.3

Desired Flow	Minimum Pump Size
250 and 500 gpm	750 gpm rated pumper
750 gpm	1,250 gpm rated pumper
1,000 gpm	1,500 gpm rated pumper
1,250 gpm	1,750 gpm rated pumper

Regardless of whether a department uses Table 14.3 or a locally-developed maximum distance relay chart, the number of pumpers needed to relay a given amount of water can be determined by using the following formula:

Equation 14.1

$$\frac{\text{Relay Distance}}{\text{Distance from Table 14.3}} + 1 = \text{Total Number of Pumpers Needed}$$

The "1" added to relay distance in Equation 14.1 represents the attack pumper that is being supplied at the end of the relay. The equation assumes that the attack pumper will be receiving an adequate residual pressure to supply the attack lines needed to control the incident. If Equation 14.1 results in any answer other than a whole number you *must round up* to the nearest whole number. For example, if the result is 6.3, you need 7 pumpers to achieve the flow. This is because any amount above a whole number means that the flow cannot quite be achieved with that whole number of pumpers.

The following examples demonstrate the application of Equation 14.1, using Table 14.3.

Example 14.1

If a single line of 3-inch hose is used, how many pumpers will be needed to supply 1,000 gpm to a fire scene that is 1,200 feet from the water source **(Figure 14.16)**?

$$\frac{\text{Relay Distance}}{\text{Distance from Table 14.3}} + 1 = \text{Total Number of Pumpers Needed}$$

$$\frac{1{,}200}{225} + 1 = 5.3 + 1 = \mathbf{6.3, or\ 7\ pumpers\ needed}$$

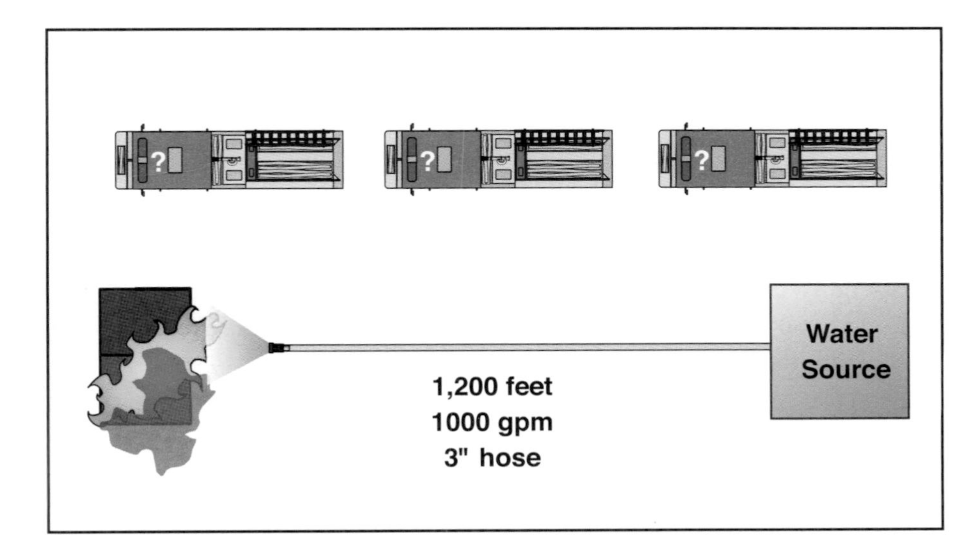

Figure 14.16 Example 14.1

Realistically, tying up seven pumpers to provide 1,000 gpm to a fire that is only 1,200 feet away is impractical and rarely, if ever, would be put into place. A better solution would be to lay a parallel hoseline (dual 3-inch lines, if that is all you have) or to use LDH. To show the difference, Examples 14.2 and 14.3 will calculate the same requirements using dual 3-inch lines or one 5-inch line respectively. Keep in mind that laying dual 3-inch lines and a single 4-inch line has roughly the same results hydraulically.

Example 14.2

If dual lines of 3-inch hose are used, how many pumpers will be needed to supply 1,000 gpm to a fire scene that is 1,200 feet from the water source?

$$\frac{\text{Relay Distance}}{\text{Distance from Table 14.3}} + 1 = \text{Total Number of Pumpers Needed}$$

$$\frac{1{,}200}{900} + 1 = 1.33 + 1 = \mathbf{2.33, \text{ or } 3 \text{ pumpers needed}}$$

In a real-life setting for Example 14.2, the relay pumper would not be spaced 900 feet from one pumper and 300 feet from the other. The relay pumper would assume a position closer to the middle of the hose lay and would be able to supply 1,000 gpm to the attack pumper without having to pump at maximum discharge pressure. The formulas contained in Chapters 12 and 13 could be used to determine the necessary pump discharge pressure.

Example 14.3

If a 5-inch hose is used, how many pumpers will be needed to supply 1,000 gpm to a fire scene that is 1,200 feet from the water source?

$$\frac{\text{Relay Distance}}{\text{Distance from Table 14.3}} + 1 = \text{Total Number of Pumpers Needed}$$

$$\frac{1{,}200}{2{,}050} + 1 = 0.59 + 1 = \mathbf{1.59, \text{ or } 2 \text{ pumpers needed}}$$

This example clearly shows the benefit of using LDH for relay water supply operations. In this situation the supply pumper at the water source can provide the needed 1,000 gpm to the attack pumper without even having a relay pumper inserted into the hose lay.

It is possible to establish a maximum distance relay even if all the pumpers are not equipped with the same size of hose. This may simply require adjusting the distance between the pumpers to compensate for the different sized hose. For example, if several engine companies are setting up a long 1,000 gpm relay using 5-inch hose, as we saw in example 14.3, the pumpers can be 2,050 feet apart. If another engine shows up with a dual lay of 3-inch hose, it can still participate as long as its dual 3-inch lay is limited to 900 feet.

Constant Pressure Relay Method

The constant pressure relay method establishes the maximum flow available from a particular relay setup by using a constant pressure in the system. Each pumper in the relay lays out the same type and length of hose. Once the connections are established, all of the pumpers then pump the lines at the same pump discharge pressure. The constant pressure relay depends on a consistent flow being provided on the fireground. The attack pumper can maintain this flow by using an open discharge or a secured waste line to handle any flow in excess of that being used in the attack lines.

When performed properly, the constant pressure relay has several advantages over the maximum distance relay:

- It speeds relay activation. Each driver/operator knows exactly how much hose to lay out and how to pump it without awaiting orders.
- It requires no calculations on the emergency scene.
- It minimizes radio traffic and confusion among pump operators.
- The attack pumper driver/operator can govern fire lines with greater ease.
- Operators in the relay only have to guide and adjust pressure to one constant figure.

Forming the Constant Pressure Relay

The following step-by-step method may be used to form a constant pressure relay operation:

Step 1: The Incident Commander sizes up the incident and determines the volume of water flow needed. At this time one or more attack pumpers may be positioned at the incident and begin an initial attack with the water they carry.

Step 2: Position a pumper at the water supply source and make the necessary connection. Most jurisdictions place the pumper with the largest pumping capacity at the water source if possible.

Step 3: Lay out the hose between the supply pumper and the attack pumper according to department procedures and make the connections to the relay pumpers. Considering that the NFPA® only requires pumpers to carry a minimum of 800 feet of supply hose, **Table 14.5** on page 358, shows the flows that may be achieved through various 750-foot hose lays. This length may be increased or decreased according to local policy. Always leave at least two sections of hose in reserve in the hose bed in the event that a hose failure occurs during the operation.

Step 4: The driver/operator for each pumper, except the supply pumper: Open an unused discharge gate if the pump does not have a relay relief valve. This allows the air from the hoselines to escape as the water advances up the hoseline.

Table 14.5
Constant Pressure Relay Available Flows for Various Hose Lays

Hose Lay (750 Feet)	Flow Available (GPM)
One 2½ -inch	321
One 3-inch	508
Two 2½ -inch	643
One 2½ -inch and one 3-inch	830
Two 3-inch	1,017
Two 3-inch and one 2½ -inch	1,312
One 4-inch	1,017
One 5-inch	1,607

Step 5: The pumper operator at the water supply source: throttles up the engine until the pumper discharge pressure reaches 175 psi.

Step 6: The driver/operator at the first relay pumper: Close the unused discharge gate once a steady stream of water flows from it; then advance the throttle until that pumper is discharging 175 psi. Each successive driver/operator follows the same procedure.

Step 7: Each driver/operator should set the pressure regulating device on his or her pumper according to the policies established by that department.

Step 8: The pump operator at the attack pumper must then adjust the discharge pressure(s) for the attack line(s) so that appropriate nozzle pressures are achieved.

Step 9: The pump operator on the attack pumper should see that a constant flow is maintained during temporary shutdowns of hoselines by using one or more discharge gates as waste, or dump, lines. If a hoseline bursts, open a discharge gate on the relay pumper before the rupture to dump water until the length is replaced.

Step 10: Lay additional hoselines between the apparatus in the relay if additional water supplies are needed on the fireground.

When a constant pressure relay is in operation, the pump operators should keep correcting their pump discharge pressure to 175 psi until one of the following conditions occurs:

- Intake pressure from a pressurized water supply source may drop to 20 psi or lower. If the intake pressure drops below 20 psi, there is a danger that the pump will go into cavitation. Cavitation can be recognized by the fact that increasing the engine rpm does not result in an increase in discharge pressure. This is a signal that the relay's maximum capacity has been reached. The results of cavitation can be pump damage and/or disruption of the flow to the fireground.

- Operating the throttle does not result in an increase in rpm. This indicates that the engine driving the pump has reached its maximum governed speed.

The above procedure used 175 psi as the constant discharge pressure because that is a common figure among departments that use this style of relay. However, the constant pressure figure of 175 psi can be modified in some cases that may require a different pressure. These situations include:

- Variations in relay pumper spacing (increase pressure for longer spacing, decrease for shorter spacing)
- Severe elevation differences between the source and the fire (decrease discharge pressure when pumping downhill, increase when pumping uphill)
- Increases in needed fire flow
- Large diameter hose (requires lower discharge pressure to supply the same volume of water)

When increasing the relay pressure, first the pressure at the supply pumper is adjusted until the desired level is reached. Then each successive pumper is similarly adjusted. When a decrease in the flow is required, the attack pumper throttles down. One way to do this is by opening the dump line to relieve excessive water. The supply pumper should discharge its dump line back into the water supply source. The relay pumpers then successively throttle down to the desired pressure, beginning with the pumper closest to the water source. The water supply officer or incident commander must be aware of the flow and pressure limitations of a given relay setup and should not attempt to exceed the capabilities of the apparatus and hose.

GENERAL GUIDELINES FOR RELAY OPERATIONS

Regardless of what type of relay pumping operation is being used, driver/operators should follow some basic guidelines for establishing the relay, maintaining the pumping operations, and shutting it down in a safe and organized manner. All driver/operators who are expected to operate in relay pumping operations must be thoroughly trained in these guidelines so that the relay can be expediently constructed and effectively operated.

Establishing a Relay Operation

One basic tenet of relay pumping operations that driver/operators must always remember is that the relay's maximum capacity will be determined by the capacity of its smallest pump and smallest hoseline. It is not uncommon for a relay operation to be comprised of pumpers with different pump capacities, and this is perfectly acceptable as long as the pumper(s) with the smallest pump capacity can flow the amount desired in the relay. For example, if it is determined that

A relay's maximum capacity will be determined by the capacity of its smallest pump and smallest hoseline.

Figure 14.17 The pumper should connect to all available hydrant discharges.

five pumpers will be required to establish a 1,200 gpm water supply and one of the five pumpers assigned to the relay task force is a 1,000 gpm pumper equipped with 3-inch supply hose, the relay will have problems meeting its goal.

If all the pumpers to be placed in a relay have sufficient pumping capacity to meet the demands of the desired relay, the order in which those pumpers are placed into the relay makes little difference. The one exception to this rule is that the largest capacity pumper should be used at the water supply source. This is particularly true if the relay is being supplied by drafting from a static water supply source. Having the larger pump at the water source is desirable because the source pumper will have to develop a higher net pump discharge pressure than the other pumpers in the relay. This higher net pump discharge pressure is needed because the relay pumpers will have a residual pressure at their intake to reduce the amount of pressure needed from the pump.

The source pumper begins the operation by connecting to the water supply source. If the source is a static water supply, this pumper will have to use hard intake hose. If the source is a fire hydrant, the pumper should make as many connections to the hydrant as possible. Often the pumper is only connected by a single large-diameter intake hose to the large (steamer) connection on the hydrant. This alone may not allow the operation to take advantage of the hydrant's full capacity. Other hoses should be connected between the hydrant's remaining (usually 2½-inch) connections and the auxiliary pump intake connections to ensure that maximum water is available for the pump **(Figure 14.17)**.

Figure 14.18 The dump line discharges excess water.

Once the water supply has been established, the driver/operator on the source pumper should open an uncapped discharge or allow water to waste through a dump line until the first relay pumper is ready for water **(Figure 14.18)**. This is particularly important if operating from a static water supply source so that the pump will not lose its prime while waiting for the relay pumping to begin. The driver/operator may increase the throttle at this time to build the pump discharge pressure to the necessary level. If this occurs before sending the water to the next pumper in the relay, the driver/operator should slowly close the dump line discharge valve to avoid wasting all the water from the water supply or damaging the surrounding area.

The relay pumper awaiting water from the source pumper, and all successive pumpers, should have their dump line or extra, uncapped discharge valve open.

If it is anticipated that the relay pumper(s) will be receiving water in just a few minutes, the other pumps may be engaged at a low rpm until the water arrives. If a longer delay is anticipated, they should not be engaged until they begin to receive water. Otherwise they may overheat during the delay.

When both the source pumper and the first relay pumper are ready, the driver/operator opens the discharge supplying the relay hose and closes the dump-line valve in a coordinated action. The driver/operator must open the discharge to the relay hose slowly to prevent a sudden discharge into the empty hose. Such sudden discharges can damage the hose. As the water fills the relay hose, the air will be forced through the relay pumper's pump and out its open dump line. When clear water comes out of the dump line, the pump on the relay pumper can be engaged.

Some jurisdictions' standard operating procedures allow the relay pumpers' driver/operators to start a relay by pumping water from their booster tanks through the relay hose while awaiting water from the source pumper. At the very least this fills the hoselines with water and reduces the time needed to remove air from the lines when the water supply is established. In short relays, it also may provide some additional water to attack apparatus until the main water supply is established. This procedure works best on short relays using MDH. On long lays of LDH, it may not even be possible to completely fill the hose between the two pumpers. One hundred feet of 5-inch hose holds about 102 gallons of water. Thus, if the pumper has a 500-gallon water tank and it is connected to more than 500 feet of 5-inch hose, it could not even fill the entire length of hose.

The relay pump operator(s) should try to maintain the pump intake pressure at between 20 and 30 psi. If the intake pressure rises above 50 psi, the relay pump operator should open the dump-line valve until the residual pressure drops below 50 psi. As the pump operator increases the throttle and pump discharge pressure to send the water to the next pumper in line, the dump-line valve will have to be gated down to maintain the 50 psi residual pressure. Otherwise, the friction loss would increase on the intake-side relay hose to the point that the pump might go into cavitation.

Once the water being discharged by the relay pumper has reached the desired pressure, this portion of the relay has been established and no further adjustments should be necessary. When the next relay pumper is ready for water, its driver/operator will follow the same procedure. The first relay pumper opens the discharge valve supplying the second pumper while closing the dump line on a coordinated basis. The operator does this while carefully observing the intake gauge to maintain the 50 psi maximum residual pressure. The second relay pumper allows water to discharge through the dump line and follows the same procedure used by the first relay pumper in receiving water from the source pumper. This process continues until all pumpers between the water supply source and the attack pumper are in operation.

When the water reaches the attack pumper, the operator should bleed out the air from the supply line by opening the bleeder valve on the intake being used. The operator can then open the intake valve on the attack pumper and establish a water supply through the relay. The attack pumper also needs a dump line as well; when one of the attack lines is shut down, an alert operator can open the dump line to allow water to flow, thus preventing a dangerous pressure buildup in the relay.

Operating the Relay

Once the relay is in operation and water is moving, the source and relay pump operators should have minimal work other than monitoring conditions and making adjustments as required. Initially, once the water is moving and the relay pumping operation seems stable, the pump operators should set their automatic pressure control devices to an appropriate level. This will negate the damaging effects of overpressurization should there be an interruption of water flow somewhere in the system. As well, they should adjust their apparatus' auxiliary cooling systems to maintain the proper engine operating temperature over the extended periods that are often necessary during a relay.

If the source and relay pumpers are equipped with intake pressure relief valves, they should be put in service according to manufacturer's instructions. Adjustable intake pressure relief valves should be set to discharge at 10 psi above the static pressure of the water system to which they are attached (source pumper) or 10 psi above the discharge pressure of the previous pumper in the relay (relay pumpers). At this setting, the valve will not open and cause excessive fluctuations when minor changes in flow occur. Keep in mind that the set relief pressure should never be higher than the safe working pressure of the hose. Otherwise the hose could burst before the relief valve is activated.

Maintaining a good system of communications during relay pumping operations is essential.

If the attack pumper is equipped with an adjustable intake pressure relief valve, set it between 50 and 75 psi to establish a stable operating condition for the attack pumper. If an attack line is shut down or the amount of discharge otherwise decreases, the friction loss in the relay hose decreases and the residual pressure increases. This causes the intake pressure relief valve to open allowing excess water to dump out. When this happens, the flow through the relay hose increases and pressures return to their original settings. When the attack pumper's flow demand again increases (such as when an attack line is reopened) the residual pressure reduces and the relief valve will close.

Less-experienced pump operators involved in relay pumping operations often tend to overreact. Small variations in pressure occur frequently but are not significant, and operators should not attempt to maintain exact pressures. As long as the intake pressure does not drop below 10 psi or increase above 100 psi, no action should be required. Changing the pressure at any of the pumpers in a relay operation always has some effect on the other pumpers. Excessive changes can result in constantly varying pressures throughout the relay. In some cases, a pressure change will take a long time to actually occur in a long relay. This delay often causes overcorrection errors that can impair the entire relay operation.

Maintaining a good system of communications during relay pumping operations is essential. Because the actions of one unit may affect all the others in the relay, the pump operators must stay in constant contact with each other. Radios are the best means of communication. However, caution must be exercised to ensure that relay radio operations are not so excessive that they hamper fireground communications and endanger firefighters in hazardous positions. Where multiple radio frequencies are available, a channel separate from the fireground channel should be dedicated to coordinating the water supply operation. When units involved in the relay are equipped with incompatible radios, portable radios may be useful. Radio-equipped ambulances or utility units that are not otherwise occupied can be used to establish communications throughout the relay. These units can be positioned next to the relay pumper and their radios used as needed.

The length of a relay pumping operation is limited only by the number of available pumpers and the amount of hose they carry.

Shutting Down the Relay

Relay operations should be shut down in order from the fire scene to the source pumper. If the source pumper is shut down while the rest of the relay is still operating, the relay pumpers will run out of water and cavitation can result. Starting with the attack pumper, each operator should slowly decrease the throttle, open the dump line, and take the pump out of gear. Once all of the pumpers are shut down, the hose may be drained and readied for reloading.

SUMMARY

Relay pumping operations are the most desirable method for providing water to a fire or emergency scene that is remote from the water supply source. Assuming that equipment stays operational, relay pumping operations provide a continuous flow of water between the water supply source and the attack pumper(s). They are more fuel efficient, less apparatus-intensive, and safer than operating a water shuttle operation with fire department tankers (tenders).

The length of a relay pumping operation is limited only by the number of available pumpers and the amount of hose they carry. Relay pumping operations of more than one mile are common in some jurisdictions. As with virtually every facet of fire fighting, proper training and regular drilling is essential to successfully setting up and operating relay pumping operations during emergency conditions. All new pump operators must be trained in the principles of relay pumping operations, and companies that are expected to operate from them must participate regularly in live pumping training exercises.

REVIEW QUESTIONS

1. List the three choices available to firefighters when the attack pumper cannot be connected to a reliable water supply source.

   1. ______________________________
   2. ______________________________
   3. ______________________________

2. List at least two reasons that relay pumping operations are more desirable than water shuttle operations.

   1. ______________________________
   2. ______________________________

3. What is another term commonly used to describe a two-pumper relay pumping operation?

   ____

   ____

4. What are the three names for the functional positions of pumpers in a relay pumping operation?

   1. ______________________________
   2. ______________________________
   3. ______________________________

5. List at least two advantages of using LDH rather than MDH for relay pumping operations.

   1. ______________________________
   2. ______________________________

6. The design of relay pumping operations is always predicated on which two factors?

   1. ______________________________
   2. ______________________________

7. List the four options for increasing the flow through an existing relay pumping operation.

   1. ______________________________
   2. ______________________________
   3. ______________________________
   4. ______________________________

8. What are the two basic methods for operating a relay pumping operation?

   1. ______________________________
   2. ______________________________

9. If a single line of 4-inch hose is used, how many pumpers will be needed to supply 1,000 gpm to a fire scene that is 4,000 feet from the water source?

   ____

   ____

10. List at least three advantages of the constant pressure relay over the maximum distance relay.

    1. ______________________________

    2. ______________________________

    3. ______________________________

11. List at least two reasons why the 175 psi figure may be adjusted during a constant pressure relay pumping operation.

    1. ______________________________

    2. ______________________________

12. Given pumpers of differing capacities, which pumper should be placed at a static water supply source that will be used to supply a relay pumping operation? Why?

    ____

    ____

    ____

13. How far above the water system operating pressure or the previous relay pumper's discharge pressure should the intake pressure relief valve be set on a relay pumper?

    ____

    ____

    ____

14. At what pressure range should the intake pressure relief valve on an attack pumper be set?

    ____

    ____

15. From which end of the relay is a relay pumping operation first shut down?

    ____

    ____

Answers found on page 442.

FESHE Course Objectives

The information in this chapter is intended to meet some of the objectives outlined for the Fire Protection Hydraulics and Water Supply course by the United States Fire Administration's National Model Core Curriculum as developed by the Fire and Emergency Services Higher Education (FESHE) initiative. The information in this chapter also allows the student to meet the requirements of NFPA® 1002, *Standard for Fire Apparatus Driver/Operator Professional Qualifications* (2009 edition) listed below.

FESHE Course Objectives

1. Explain the designs and operational principles of wet, dry, preaction, and deluge sprinkler systems.
2. List and describe the major components of an automatic sprinkler system.
3. Calculate the pump discharge pressure necessary to supply an automatic sprinkler system.
4. Explain the designs and operational principles of wet and dry standpipe systems.
5. List and describe the major components of a standpipe system.
6. Calculate the pump discharge pressure necessary to supply a standpipe system.

NFPA® 1002 Requirements

5.2.4 Supply water to fire sprinkler and standpipe systems, given specific system information and a fire department pumper, so that water is supplied to the system at the correct volume and pressure.

(A) Requisite Knowledge. Calculation of pump discharge pressure; hose layouts; location of fire department connection; operating principles of sprinkler systems as defined in NFPA® 13, NFPA® 13D, and NFPA® 13R; fire department operations at sprinklered properties as defined in NFPA® 13E; and operating principles of standpipe systems as defined in NFPA® 14.

Chapter 15

Supplying Fixed Fire Suppression Systems

INTRODUCTION

Most fire departments today protect at least one occupancy that has a fixed fire suppression system, such as an automatic sprinkler system or a standpipe hose system. Some jurisdictions have hundreds or thousands of these occupancies. A properly maintained and operating sprinkler system remains the most effective way of controlling and/or extinguishing fires in their incipient or early stages. Standpipe systems allow firefighters to rapidly deploy handlines for fire attack in remote portions of a structure that would otherwise take considerable time and energy to access.

This chapter surveys the various types of automatic sprinkler and standpipe systems that firefighters are likely to encounter. Firefighters must understand the basic operational principles and components of these systems in order to utilize them effectively. Firefighters should conduct preincident surveys of all occupancies containing these systems so that they will be readily familiar with them in the event of a fire emergency.

The information gained from these preincident surveys can be incorporated into standard operating procedures utilized at any response to a facility with an automatic sprinkler or standpipe system. This information is particularly crucial for pumping apparatus driver/operators who will be responsible for supplying water to the fixed suppression system and possibly to other attack lines being used at the incident. The success of the incident and the safety of the firefighters operating in forward positions will depend on the pump operator's proficiency in supplying these systems correctly.

AUTOMATIC SPRINKLER SYSTEM OPERATIONS

Sprinkler systems are designed to provide automatic fire suppression using a properly designed system of underground and overhead piping designed to meet appropriate codes and standards, most commonly NFPA® 13, *Standard for the Installation of Sprinkler Systems*. These systems include one or more automatic water supplies. The aboveground portion of the sprinkler system includes a network of specially sized piping installed in a building, structure, or area, to which the actual sprinklers are attached. The valve controlling each sprinkler system riser is located in the system riser or its supply piping. Each system includes a device for actuating an alarm when the system is operating. Sprinkler systems are typically activated by heat from a fire and discharge water over the involved area. To correctly utilize sprinkler systems firefighters and pump operators must have a working understanding of every facet of these systems.

While the fire service and fire protection systems are subject to new advances and new technologies every day, building protection by sprinkler systems is not a new concept. The earliest sprinkler systems were developed to protect industrial buildings more than 125 years ago. In fact, the very first of what we now know as the NFPA® standards was titled *Rules and Regulations of the National Board of Fire Underwriters for Sprinkler Equipments, Automatic and Open Systems* and dates back to 1896. This document ultimately became the current NFPA® 13 standard.

Automatic sprinklers are generally considered the most useful and reliable method of providing fire protection.

This early standard was developed because there was no prior way to ensure that sprinkler systems were being designed and installed properly. Many early sprinkler systems were crude and unreliable. The influence of this standard has resulted in evolving sprinkler technology to the point that today's improved systems are very effective and reliable when properly installed and maintained. As well, sprinkler systems are no longer limited to industrial occupancies. Modern sprinkler systems are installed in schools, health care facilities, high-rise buildings, and commercial and residential occupancies. Automatic sprinklers are generally considered the most useful and reliable method of providing fire protection. Statistics compiled from several sources have demonstrated the effectiveness of automatic sprinklers. One major insurance carrier reported that during a three-year period, automatic sprinklers failed to control the fire in only 24 of 1,225 losses involving sprinklers—a failure rate of approximately 2 percent. In 29 percent of the reported fires, one sprinkler controlled the fire. In 75 percent of the reported fires, ten or fewer sprinklers controlled the fire.

Generally, three principle reasons motivate building owners to invest in and install automatic sprinklers:

- Code requirements
- Insurance purposes
- General fire protection

Recognizing the measure of protection afforded by sprinkler systems, many jurisdictions have enacted building and fire codes that require their installation. The reasons for mandating sprinklers in buildings arise from a need to protect the community as a whole or to protect the occupants in individual buildings. Depending on the jurisdiction's preference, the laws requiring sprinklers may be based on a combination of building occupancy, construction type, and/or size. Most commonly these laws require sprinklers when a particular occupancy exceeds a given area. Communities that aggressively promote fire prevention and protection require sprinkler systems in all types of occupancies, including residential.

Because of the proven reliability and effectiveness of automatic sprinklers, most insurance companies offer significantly reduced fire insurance rates on sprinklered properties. This rate reduction usually pays for the cost of the sprinkler system within just a few years. From that point on, the rate reduction becomes pure savings for the building owner.

Realistically the least common reason for installing a sprinkler system falls into the third category: an interest in general fire safety even where codes do not require automatic sprinklers and insurance is not a factor. While this is certainly a good business and ethical practice, most property owners are more influenced by the first two reasons than simply by their own good will.

In view of the prominent role sprinkler systems play in fire protection, an understanding of these systems and their components is essential to the firefighter and pump operator. It is especially important to understand sprinkler system operation so that fire fighting procedures can be carried out more efficiently in buildings protected with sprinklers. The following sections will examine:

- Common types of sprinkler systems that fire personnel will encounter
- Basic components of most sprinkler systems
- Proper preincident inspection and planning procedures for sprinkler systems
- Fire department operations at sprinklered occupancies
- Hydraulic calculations for pump operators supplying sprinkler systems

Common Types of Sprinkler Systems and Their Designs

All sprinkler systems are not the same. Building designers and fire protection engineers responsible for the selection and design of a sprinkler system have a number of options. The choice of a particular type of sprinkler system depends on a number of variable factors that will be discussed later in this section. In general, fire professionals will encounter four basic types of sprinkler systems:

- Wet-pipe systems
- Dry-pipe systems
- Deluge systems
- Preaction systems

Before describing each of these systems individually, it is important to explain their basic design process. For any sprinkler system to function properly it must be properly designed, installed, and maintained. Although the fundamental concept of the automatic sprinkler system is simple, ensuring proper design can become very complex because of the variety of buildings and hazards encountered in actual practice. An inadequately designed system cannot be expected to control fires.

The sprinkler system designer must establish two basic factors when determining the appropriate type of system to design and install: the type of building and the occupancy to be protected. This information will determine the most appropriate type of system for that building (wet, dry, deluge, or preaction) and the required water supply needed to support the system.

The designer must also determine a myriad of other variables that go into the system design. These include such considerations as:

- The type of sprinklers to be used
- The spacing between the sprinklers

- The type and sizes of pipe to be used
- The type of pipe hangers that will support the system
- The type of valves, alarms, drains, and other system components to be used

NFPA® 13 provides the basis for the design of sprinkler systems for most types of occupancies. Other standards that designers may need to consult concerning exceptions to NFPA® 13 include:

- NFPA® 13D, *Standard for the Installation of Sprinkler Systems in One- and Two-Family Dwellings and Mobile Homes.* This standard lists the requirements for small, fast-response sprinkler systems that increase the life safety factor in private homes.
- NFPA® 13R, *Standard for the Installation of Sprinkler Systems in Residential Occupancies up to Four Stories in Height.* This standard provides requirements for residential-type sprinkler systems in low-rise, multifamily dwellings.
- NFPA® 409, *Standard on Aircraft Hangars.* This standard provides special information on protecting the extreme fuel hazards in these occupancies.
- NFPA® 231, *Standard for General Storage.* This standard covers general warehousing of merchandise and its protection **(Figure 15.1)**.
- NFPA® 231C, *Standard for Rack Storage of Materials.* High rack storage of materials is addressed in this standard **(Figure 15.2)**.
- NFPA® 231D, *Standard for Storage of Rubber Tires.* Bulk storage of rubber tires requires special considerations because of their high fuel load.
- NFPA® 231E, *Recommended Practice for the Storage of Baled Cotton.*
- NFPA® 231F, *Standard for the Storage of Roll Paper.*

When using NFPA® 13 to design a sprinkler system, the designer begins by determining the occupancy hazard classification to be protected. NFPA® 13 divides occupancies among three very broad classifications:

- Light hazard
- Ordinary hazard
- Extra hazard

Light hazard occupancies are those that probably will have a low rate of heat release if they catch fire. These include occupancies such as churches, hospitals, offices, and apartments, where the quantity of combustible contents is low **(Figure 15.3)**.

Ordinary hazard occupancies are subdivided further into two subclasses depending on the amount and configuration of combustibles they contain. Ordinary hazard group 1 includes occupancies such as bakeries, canneries, and electronic plants, where combustibles do not exceed 8 feet in height **(Figure 15.4)**. Ordinary hazard group 2 includes occupancies such as cereal mills, confectionery products manufacturing, printing processes, and tire manufacturing, where combustibles do not exceed 12 feet in height.

Figure 15.1 NFPA® 231 addresses general warehouse storage practices.

Figure 15.2 NFPA® 231C addresses high-rack storage of materials.

Figure 15.3 Churches are light hazard occupancies.

Figure 15.4 This car stereo manufacturing plant would be an ordinary hazard group 1 occupancy.

Extra hazard occupancies are those where severe fires can be anticipated. Extra hazard occupancies are further subdivided into group 1 and group 2. Extra hazard group 1 occupancies include particle board manufacturing, saw mills, and textile packing. Extra hazard group 2 includes occupancies that contain flammable liquid spraying, varnish dipping, and flow coating operations.

The delineations among these three classifications are not precise or, clear-cut. On occasion occupancies that fall primarily within one class also have characteristics of another. For example, offices are generally classified as light hazards, but storerooms within an office building may have a higher fuel load than found within an office. Nonetheless, these classifications serve as the starting point for sprinkler system design.

Much of the work in designing a sprinkler system involves the proper placement and sizing of the piping. Two basic methods of system design are used to ensure that the piping is properly designed and sized: the pipe schedule method and the hydraulic calculation method.

Designing a system by the pipe schedule method involves using pipe schedule information contained in NFPA® 13. These pipe schedules specify the pipe size that must be used to supply a given number of sprinklers, based on occupancy classification. The standard includes different pipe schedules for light, ordinary, and extra hazard occupancies. The design process using the pipe schedule method is typically less expensive than the custom design process using hydraulic calculations. However, in almost every case this initial savings is erased by the expense of the larger sizes of pipe that the pipe schedule method generally requires. Few of today's systems are designed using the pipe schedule method.

The hydraulic calculation method uses scientific formulas to calculate the pressure losses of water flowing through sprinkler pipe. These calculations ensure that the system will meet minimum design discharge requirements. Most sprinkler systems being installed today use the hydraulic design method. In addition to providing better fire protection, hydraulic design usually permits the use of smaller pipe. Complex computer programs that factor in both the occupancy and water supply information do most hydraulic calculations of sprinkler systems.

Wet-Pipe Sprinkler Systems

The very first sprinkler systems developed in the nineteenth century were wet-pipe systems. Wet-pipe systems remain the most common and reliable type of sprinkler system in service today. These systems usually succeed in controlling fires with a minimum of sprinklers actually opening.

In a wet-pipe system, the entire piping system is filled with water at all times **(Figure 15.5)**. This allows water to immediately discharge from individual sprinklers when they are fused. Wet-pipe systems are typically installed in locations where they will not be subjected to freezing temperatures. Otherwise the water in the system could freeze and damage the piping. Wet-pipe systems of limited size may be installed in areas subject to freezing if they are freeze-protected using an antifreeze solution.

Wet-pipe systems contain an alarm check valve or a waterflow indicator in their risers. These devices sense the movement of water in the system and trigger an alarm to sound. Depending on the design, this alarm may sound only locally or it may be broadcast to a monitoring center so that the fire department can be dispatched to check the situation.

The basic operational sequence for a wet-pipe sprinkler system is:

- Heat from a fire causes the sprinkler's heat-actuating plug to drop from the frame.
- Water contained in the piping immediately discharges from the open sprinkler.
- As water begins to flow through the system, the alarm check valve on the water supply riser opens and activates the water motor gong and/or electronic signaling equipment.
- The alarm is transmitted to a supervising agency or fire department.
- If needed, the volume and pressure of water in the system is boosted via a fire department pumper through the fire department connection (FDC).

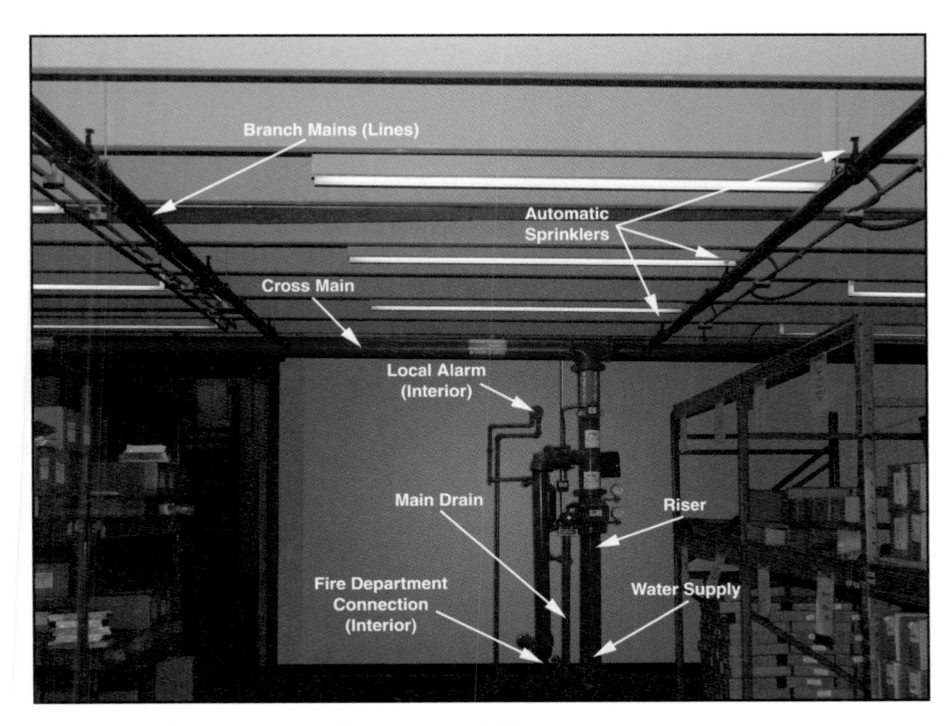

Figure 15.5 The components of a wet pipe sprinkler system.

Dry-Pipe Sprinkler Systems

In a dry-pipe sprinkler system air under pressure replaces the water in the pipes above the sprinkler valve in the riser **(Figure 15.6, p. 374)**. Dry-pipe sprinkler systems are most commonly used in buildings that are not sufficiently heated to keep the water in the sprinkler pipes from freezing. The riser and dry-pipe valve must be located in a heated portion of the structure. Otherwise, if the valve is subjected to freezing conditions, it may not activate properly when required.

The force created by the air pressure in the system holds the clapper in the dry pipe valve closed and keeps water from entering the system until a sprinkler operates. The reduction in pressure when air escapes from the open sprinkler permits the dry-pipe valve to open and admit water to the system. The system's compressed air may be obtained from one of two sources. One alternative is an air compressor/tank arrangement that is dedicated exclusively for use on the system. The second alternative is to hook the sprinkler system into a facility-wide compressed-air system. Dry-pipe systems can be equipped with either electric or hydraulic (water) alarm signaling equipment.

The basic operational sequence for a dry-pipe sprinkler system is:

1. Heat from a fire causes the heat actuating plug in the sprinkler to drop from the frame.
2. Pressurized air in the piping begins to flow through the open sprinkler.
3. After a slight drop in air pressure, the quick-opening device (if present) activates to accelerate the removal of air from the piping.
4. Once the air pressure is reduced sufficiently, the dry-pipe valve trips open. The interior clapper is held in the open position by a latch.

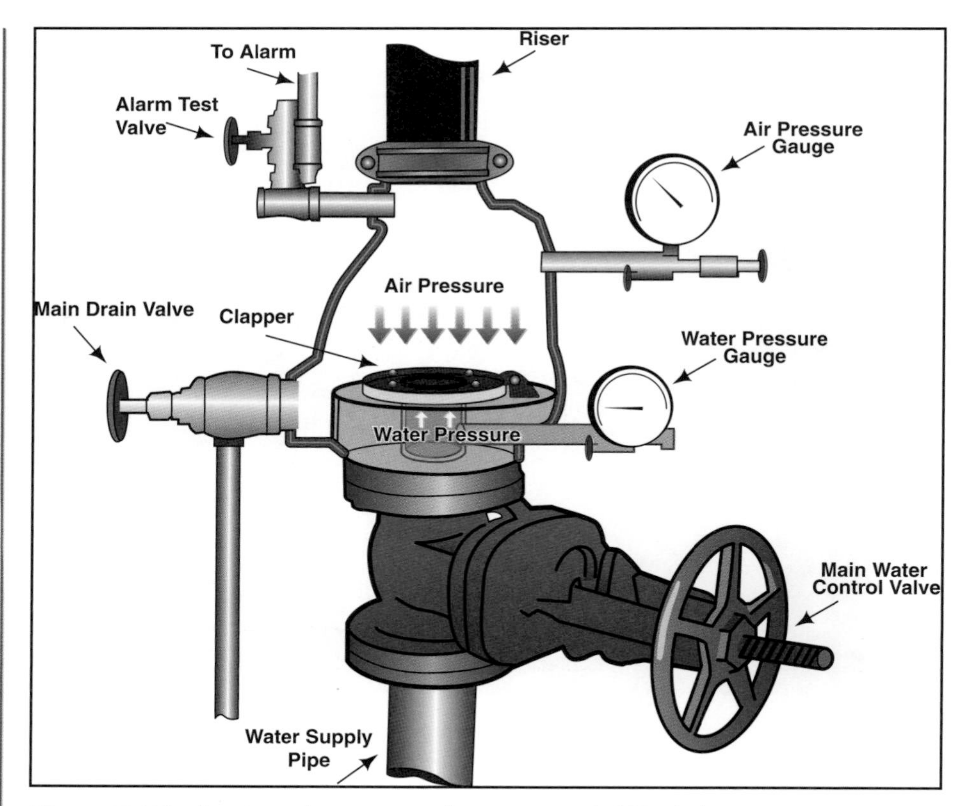

Figure 15.6 The dry pipe valve uses air under pressure to hold back the water.

5. Water enters the intermediate chamber of the dry-pipe valve. This automatically forces the automatic drip valve closed and begins the flow of water through alarm-signaling equipment.
6. Water flows through the entire piping system and is discharged through the open sprinkler.

Deluge Sprinkler Systems

A deluge system is similar to a dry-pipe system in that the piping contains no water before the activation of the deluge valve. In a deluge system, however, all of the sprinklers have open heads with no fusible links. This means that when the valve is tripped and water enters the system, the water will discharge through all sprinklers simultaneously **(Figure 15.7)**. An electrically, pneumatically, or hydraulically operated deluge valve controls the flow of water to the system. Unlike wet- or dry-pipe systems, the sprinklers do not function as the detection device in a deluge system. The deluge valve is activated by heat, smoke, or flame detectors that are installed in the same area as the sprinklers.

Deluge systems are installed in situations that require quickly applying a large volume of water to the protected area. Deluge systems are normally used to protect extra hazard occupancies, such as aircraft hangars, woodworking shops, cooling towers, ammunition storage, or manufacturing facilities. They are also frequently used to protect large electrical transformers and other oil-containing equipment at large electrical substations. Because deluge systems require a large volume of water, they are almost always supplied by fire pumps.

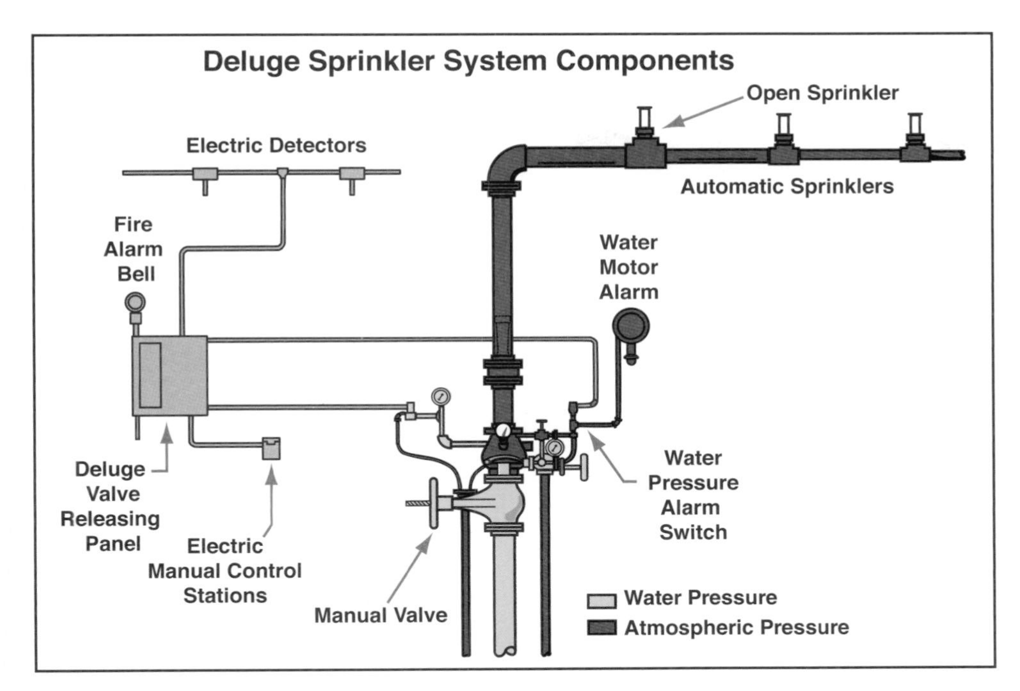

Figure 15.7 A typical deluge sprinkler valve assembly.

The sequence of operation for a deluge sprinkler system is:

1. A product-of-combustion (heat, smoke, or flame) detector senses the presence of a fire condition, *or* an individual working in the area discovers a fire in progress.
2. The fire detection system sends a signal to the deluge valve, causing the valve to open, *or* the individual who discovers the fire manually trips the deluge valve.
3. As water enters the deluge valve and the piping, waterflow indicator alarms are transmitted to signaling stations, and the water motor gong (if present) is activated.
4. Water flows through all open sprinklers simultaneously.

Preaction Sprinkler Systems

Preaction systems are used in occupancies where preventing water damage is especially important. Preaction systems are frequently used to protect computer rooms and document-storage areas. Preaction systems combine the features of both deluge and dry-pipe systems. They employ a deluge-type valve and fire detection devices along with a dry-pipe system's closed sprinklers. As with deluge systems, the preaction system valve will not discharge water into the sprinkler piping until it receives an indication from fire detection devices (other than the sprinklers) that a fire may exist. Once water is in the system, it may then be discharged through any sprinkler that is fused. This provides an added measure of protection from accidental discharges of water.

The piping in a preaction system is normally dry; therefore, it can be used in freezing environments such as cold storage warehouses. Usually air under a low pressure is maintained in the piping as a supervisory function. In the event of a leak or break in the piping, the supervisory air pressure drops and transmits an alarm without admitting water to the system.

The sequence of operation for a preaction sprinkler system is:

1. A product-of-combustion (heat, smoke, or flame) detector senses the presence of a fire condition.
2. The fire detection system sends a signal to the preaction valve, causing the valve to open.
3. Sensors in the piping system detect the flow of water into the system and trigger the waterflow fire alarm.
4. When the level of heat at a sprinkler reaches the appropriate temperature, the sprinkler fuses and water flows through the open orifice.

Automatic Sprinkler System Components

In general, all sprinkler systems are constructed of relatively the same components. These basic components include:

- Sprinklers
- Piping
- Valves
- A water supply
- A fire department connection (FDC)

The following section summarizes each of these components.

Sprinklers

The sprinkler, or sprinkler head as it is sometimes called, is that portion of the sprinkler system that delivers the water to the fire area. Closed sprinklers, which are used in wet-pipe systems, dry-pipe systems, and preaction systems, also sense the fire and then react to it. Three main components of a closed sprinkler are of interest to firefighters: the heat-sensing device, the deflector, and the discharge orifice **(Figure 15.8)**. Open sprinklers, which are used only in deluge sprinkler systems, do not contain the heat-sensing device.

The sprinkler's heat-sensing device is located between the nozzle orifice and the deflector. It holds a plug over the sprinkler nozzle orifice. When the heat-sensing device is subjected to sufficient heat, it allows the plug to drop off, permitting air and/or water to discharge from the sprinkler (Figure 15.9). A variety of heat-sensing devices are used in sprinklers. They include fusible links, fusible alloy pellets, chemical pellets, and frangible glass bulbs. Each of these may be designed to activate at a wide variety of temperatures depending on the occupancy's need. Operating temperatures vary from 135° to 575°F, and are determined by the maximum temperature expected at the level of the sprinkler under normal conditions in the particular application. Under conditions that normally fall within the range of ordinary room temperatures, a sprinkler with a temperature rating of 165°F is most frequently used. If the ambient temperature exceeds 100°F (38°C), as in an attic or near a heater, a sprinkler with a higher temperature rating is used.

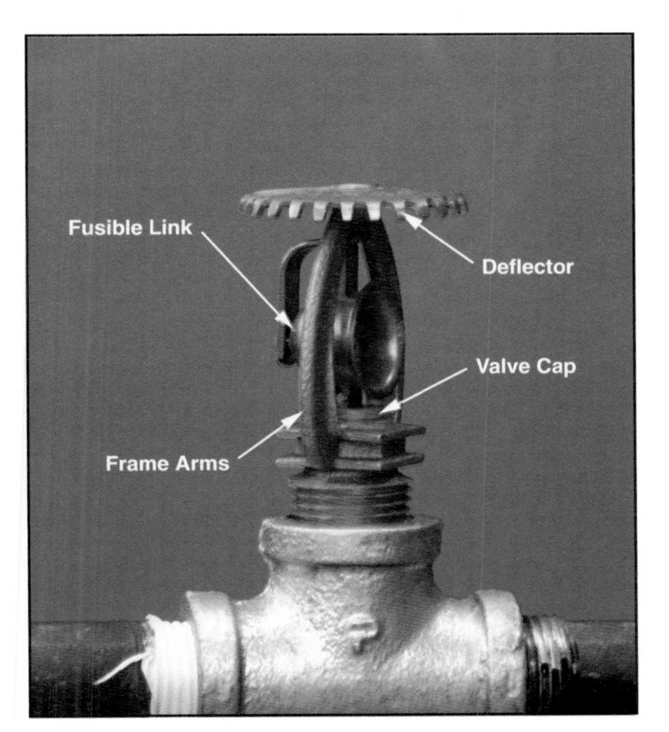

Figure 15.8 The components of a closed sprinkler.

Figure 15.9 When the heat sensitive device falls out, air and/or water are allowed to escape from the discharge orifice.

TABLE 15.1
Sprinkler Temperature Ratings, Classifications, and Color Codings

Max. Ceiling Temp. °F	Temperature Rating °F	Temperature Classification	Color Code	Glass Bulb Colors
100	135 to 170	Ordinary	Uncolored or black	Orange or red
150	175 to 225	Intermediate	White	Yellow or green
225	250 to 300	High	Blue	Blue
300	325 to 375	Extra high	Red	Purple
375	400 to 475	Very extra high	Green	Black
475	500 to 575	Ultra high	Orange	Black
625	650	Ultra high	Orange	Black

When inspecting sprinklers, fire professionals can quickly identify the temperature ratings of sprinklers by the color-coding on the sprinkler frame or the frangible glass bulb **(Table 15.1)**. The temperature rating also is stamped on the link in solder sprinklers. On other types of sprinklers, the temperature rating will be stamped on some other part of the sprinkler.

Because all sprinklers, with the exception of deluge sprinklers, are thermal devices, there is an inherent delay between the fire's ignition and the sprinkler's operation. The delay occurs as the thermal element's temperature rises from room temperature to its melting or operating temperature. It is a function of several variables including sprinkler design. The relative speed of operation can be evaluated by a number known as the response time index (RTI). The lower a sprinkler's RTI, the faster it responds.

To speed the operation of sprinklers, engineers have designed a type of sprinkler known as a quick-response sprinkler. While ordinary sprinklers have RTIs that vary from 225 to 700 $feet^{½}second^{½}$, quick-response sprinklers may have RTIs as low as about 50 $feet^{½}second^{½}$ (28 $m^{½}sec^{½}$), which results in significantly faster operation.

Another primary component of a sprinkler is the deflector, which is attached to the sprinkler frame. The deflector reshapes the solid stream of water that leaves the sprinkler orifice into a spray pattern. There are three basic types of sprinkler design: upright, pendant, and sidewall **(Figure 15.10)**. Each of these designs has a specific use, and they cannot be used interchangeably. Upright sprinklers are designed to deflect the spray of water downward in a hemispherical pattern. Pendant sprinklers are used in locations where upright sprinklers would be impractical or unsightly. The deflector on a pendant sprinkler breaks the stream of water into a circular pattern of small water droplets. Sidewall sprinklers discharge most of their water to one side and are useful in such applications as corridors, offices, hotel rooms, and residential occupancies.

Figure 15.10 (a) An upright sprinkler, **(b)** a pendant sprinkler, and **(c)** a sidewall sprinkler.

The actual amount of water discharged by a sprinkler onto a fire is determined by the size of the discharge orifice and the pressure available at the sprinkler head. The most common size of orifice in automatic sprinklers is ½-inch. As with other parts of the sprinkler, the size of the orifice can be varied for different applications. Sprinklers are available with orifice sizes of ⅜, $^{7}/_{16}$, ½, $^{17}/_{32}$, and ¾. The amount of water being discharged from a given sprinkler head can be easily calculated using a simple equation:

Equation 15.1

$Q = (K)(\sqrt{P})$

Given:

Q = The quantity of water discharged in gpm

P = The discharge pressure at the sprinkler in psi

K = A discharge coefficient

The actual amount of water discharged by a sprinkler onto a fire is determined by the size of the discharge orifice and the pressure available at the sprinkler head.

The discharge coefficient (K) varies with different sprinklers and takes into account such factors as the orifice diameter and smoothness. For example, a ½-inch sprinkler will typically have a K factor of about 5.5, while a sprinkler with an orifice of $^{17}/_{32}$-inch would typically have a K factor of around 8.

Example 15.1

Determine the flow in gpm from a ½-inch sprinkler with a K of 5.5 when the discharge pressure at the sprinkler orifice is 16 psi.

$Q = (K)(\sqrt{P})$

$Q = (5.5)(\sqrt{16})$

$Q = (5.5)(4)$

Q = 22 gpm

Example 15.2

Determine the flow in gpm from a $^{17}/_{32}$-inch sprinkler with a K of 8 when the discharge pressure at the sprinkler orifice is 16 psi.

$Q = (K)(\sqrt{P})$

$Q = (8)(\sqrt{16})$

$Q = (8)(4)$

Q = 32 gpm

While these two examples ignore the effects of velocity, they do show the orifice diameter's importance relative to sprinkler discharge. In this example, an increase of only $^{1}/_{32}$-inch in orifice diameter results in an increase in discharge of 10 gpm (45 percent) at the same pressure.

An increase in pressure also results in an increase in discharge for a given orifice diameter. In the previous example of a ½-inch diameter sprinkler, using the equation $Q = K\sqrt{P}$, a pressure increase from 16 psi to 25 psi results in an increase in discharge from 22 gpm to 27.5 gpm. Because the sprinkler discharge varies with the *square root* of the pressure, an increase in pressure causes a smaller increase in flow than does an increase in sprinkler orifice diameter.

Sprinkler System Piping

The various NFPA® sprinkler system standards specify which types of piping are permissible in sprinkler systems. Three basic types of pipe are used: ferrous metal, copper tubing, and plastic. Of these three, ferrous metal (black steel) piping is the most commonly used. Four types of ferrous metal piping are in use:

Schedule 5, Schedule 10, Schedule 40, and special lightweight (XL). Schedules 5 and 10 may be joined by welding or with rolled-grooved fittings. Schedule 40 and XL may be joined by welding, with rolled-grooved fittings, or with threaded fittings. Under normal conditions, steel piping has a life expectancy of several hundred years. Its primary disadvantages are that it will corrode and that it is very heavy.

Copper piping is highly resistant to corrosion, has low friction loss, weighs less than steel pipe, and is neat in appearance where exposed. It is joined by soldering or brazing. Since its introduction in the 1960s, copper piping has been used primarily where there is a concern about corrosion damage to ferrous metal pipes. Of the three pipe materials, copper is the most expensive. Its use has declined in recent years with the increased use of less expensive plastic pipe, which will also resist corrosion.

Two types of plastic pipe, polybutylene and chlorinated polyvinyl chloride (CPVC), have been approved for wet-pipe sprinkler systems in light hazard and residential occupancies since the early 1980s. In dry-pipe systems, plastic pipe could be damaged before it had the opportunity to discharge water as designed. This type of pipe is the least expensive, the lightest in weight, and the easiest to install. Plastic pipe is joined together with plastic cement. Its main drawback is that it may not be installed in areas where the ambient temperature exceeds 120°F for polybutylene pipes or 150°F for CPVC pipes.

Regardless of the pipe material, the various types and sizes of pipes in a sprinkler system are given specific functional names based on their role (refer back to Figure 15.5):

- Water supply main
- Riser
- Feed main
- Cross main
- Branch line

Typically, the water supply main is considered to comprise all piping below the sprinkler valve. The *riser* is the vertical piping that extends up from the sprinkler valve to the first horizontal feed main. *Feed mains* are large pipes that supply water to each of the cross mains. In turn, the *cross mains* feed the branch lines. The *branch lines* are those pipes that contain the sprinklers.

Valves

In addition to the main sprinkler valves in the riser (the wet-pipe, dry-pipe, deluge, or preaction valves), every sprinkler system is equipped with a variety of other water control valves and various operating (test and drain) valves. Each of these valves has an important, specific function.

Control valves are used to shut off the water supply to the system when replacing sprinklers or performing maintenance is necessary. They are located between the water supply source and the sprinkler system. After maintenance has been performed, the control valves must always be returned to the open position. To help ensure that they are not inadvertently left closed, most sprinkler system control valves are indicating valves, which allow personnel to visually determine whether they are open or closed. Four types of indicating valves commonly are used as sprinkler control valves:

- **Outside screw and yoke (OS&Y) valve.** The OS&Y valve has a yoke on the outside with a threaded stem that opens and closes the gate **(Figure 15.11)**. The threaded portion of the stem is outside the yoke when the valve is open and inside the yoke when the valve is closed.
- **Post indicator valve (PIV).** The PIV is used to control underground sprinkler valves **(Figure 15.12).** It consists of a hollow metal post attached to the valve housing. The valve stem is inside this post. On the stem is a movable target bearing the words *OPEN* and *SHUT*. The operating handle is fastened and normally locked to the post. When the valve is closed, the word *SHUT* appears in the opening.
- **Wall post indicator valve (WPIV).** The wall post indicator valve is similar to a PIV. However, the WPIV extends through the building wall with the target and valve operating wheel on the outside of the building **(Figure 15.13)**.
- **Indicating butterfly valve.** The indicating butterfly valve has a paddle indicator or a pointer arrow that shows the valve's position **(Figure 15.14, p. 382)**.

To further ensure that sprinkler control valves are kept open, they either are chained or locked in the open position or are electrically supervised. An electrically supervised valve is attached to a switch. Movement of the valve causes an electrical circuit in the switch to open and transmit a signal to a watch service or manned guard post.

In addition to control valves, sprinkler systems employ various operating valves such as check valves, automatic drain valves, globe valves, and stop (or cock) valves. Check valves limit the flow of water to one direction. They are placed in water sources to prevent recirculation or backflow of water from the sprinkler system into the municipal water supply system. Automatic drain valves drain piping when pressure in the pipe is relieved. This prevents the part of the pipe that extends through the wall from freezing in cold weather. Automatic drain valves' most common application is to drain water from fire department connections after use. Globe valves are small handwheel-type valves that typically are designed to be turned clockwise to open and counterclockwise to close. They are used primarily on drains and test valves; the inspector's test valve is commonly controlled by a globe valve **(Figure 15.15, p. 382)**. Stop, or cock, valves are also used for drains and alarm testing. They are most commonly ball-type valves that open or close with a quarter-turn.

Water Supply

Every sprinkler system must have an automatic water supply of adequate volume, pressure, and reliability. Sources of supply include municipal water systems, standby pumps connected to large static sources, elevated tanks (gravity tanks), and pressure tanks.

Figure 15.11 An OS&Y valve.

Figure 15.12 A post indicator valve (PIV).

Figure 15.13 A wall post indicator valve (WPIV).

Figure 15.14 An indicating butterfly valve.

Figure 15.15 A globe-type inspector's test valve.

The volume of water required to adequately supply a sprinkler system depends upon the occupancy classification. Although it gives no hard requirements, NFPA® 13 does recommend guidelines for needed water supply volume based on occupancy classification. The only firm code requirements are for water pressure at the most remote sprinkler in the system. A water supply must be able to deliver the required volume of water at a residual pressure of 15 psi at the highest sprinkler in pipe-schedule-designed systems. In hydraulically calculated systems, the minimum residual pressure is 7 psi. If the water supply system cannot deliver these pressures, a fire pump must be added to the system to boost pressures.

Of the supply sources, municipal water mains are used most frequently. They tend to be the most reliable source of supply for sprinkler systems, but they must be of adequate pressure and volume. To ensure that a municipal water distribution system has adequate capacity, flow tests are performed before the system is installed.

Elevated tanks, also known as gravity tanks, are one of the oldest methods used to supply on-site water for sprinkler systems. While elevated tanks provide a reliable source of water, they must be placed on a tower or platform many feet above the highest sprinklers in order to supply adequate pressure. You will remember from Equation 2.3 that Pressure = 0.433 x height (feet) (Chapter 2). Thus, if a sprinkler system required an operating pressure of 40 psi, the tank would have to be elevated 92 feet **(Figure 15.16)**. In some cases, tanks are placed on the roof of the structure, which results in higher construction costs because of the tanks' added weight.

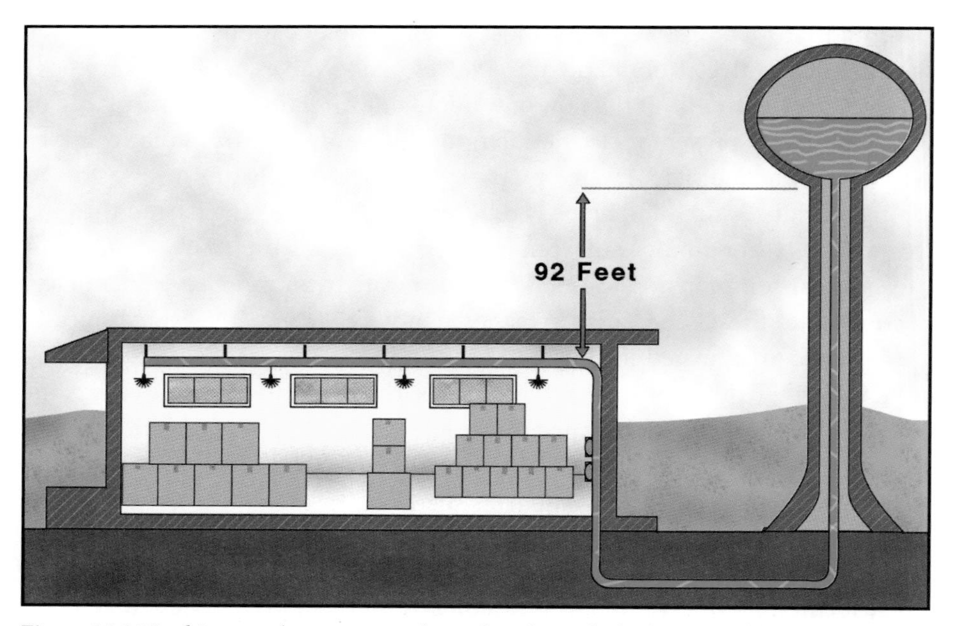

Figure 15.16 In this case, the water must be 92 feet above the highest sprinkler in order to supply the necessary pressure to the system.

In other cases a fire pump at an aboveground reservoir supplies the sprinkler system. This method has the advantage of providing a large amount of water without having to install a costly tower. Aboveground reservoirs are frequently used to protect industrial complexes where the amount of water needed may reach 250,000 gallons or more. Fire pumps can also draft from such natural water sources as rivers, ponds, or lakes. The downside of these systems is that if for any reason the pump fails to start, no water will be supplied to the sprinkler system. Also, if the reservoirs are supplied from a natural source, they may not hold sufficient quantities of water during droughts.

Pressure tanks consist of large tanks of water and air under pressure. When a sprinkler opens, the air pressure in the tank forces the water out of the tank and into the sprinkler piping. Pressure tanks normally contain two-thirds water and one-third air. The tanks are usually pressurized to 75 psi. Their primary disadvantages are that they store a limited volume of water and have a limited discharge time.

Figure 15.17a A typical sprinkler system fire department connection (FDC).

Fire Department Connections

In addition to the primary sources of water supply, sprinkler systems are fitted with one or more fire department connections (FDC), sometimes known as siamese connections. The FDC enables a fire department pumper to pump additional water into the system, thereby boosting both the pressure and volume of water in the system **(Figures 15.17 a and b)**. Fire department connections are especially important in protecting buildings where the primary water supply is weak or where a large number of sprinklers have opened.

Figure 15.17b Some jurisdictions require large diameter hose connections to sprinkler systems.

The most common type of fire department connection consists of a 4-inch pipe equipped with a siamese fitting on the outside of the building. The siamese fitting has two, 2½-inch inlets to which fire hose are connected. Large systems may have three or more inlets. Some jurisdictions require new systems to be equipped with inlets for large diameter (4- or 5-inch) hose.

A check valve inside the building on the 4-inch pipe keeps system water out of the portion of pipe that extends outside the building. This prevents damage from freezing temperatures and keeps the system pressure off the clappers in the hose inlets. An automatic drain valve (ball drip valve) is also installed in the fire department connection to prevent problems associated with freezing temperatures.

In a single-riser sprinkler system, the fire department connection is attached on the sprinkler system side of the main supply valve. In multiple-riser systems, the fire department connection enters the supply piping between the main supply valve and the individual riser valves. This arrangement allows water to be supplied to the system even when one riser valve has been closed either deliberately or inadvertently. Water still will be pumped into the remaining risers.

Preincident Inspection and Planning Procedures for Sprinkler Systems

All fire department personnel, particularly pump operators, must be familiar with the basics of sprinkler system operation and support, as well as with the particulars of the specific sprinkler systems in their response district. This knowledge ensures that systems will be used and supported according to their design, and any problems that arise can be quickly identified and corrected before they cause any harm. Excellent sources of information on what to look for and

All fire department personnel must be familiar with the basics of sprinkler system operation and support, as well as with the particulars of the specific sprinkler systems in their response district.

how to react to certain situations are NFPA® 13E, *Recommended Practice for Fire Department Operations in Properties Protected by Sprinkler and Standpipe Systems,* and NFPA® 25, *Standard for the Inspection, Testing, and Maintenance of Water-Based Fire Protection Systems.*

The three principle causes of unsatisfactory performance in sprinkler systems are:

- A closed valve in the water supply or some other part of the system
- Inadequate water supply to the sprinkler system
- Changes to the occupancy that negate the sprinkler system's effects

Each of these conditions can be identified relatively easily during fire inspection and/or preincident planning visits to an occupancy. The best time to identify and correct these situations is before an incident occurs. In many emergencies they cannot be identified and/or corrected fast enough to prevent catastrophic losses.

Closed valves may occur for a variety of reasons. The most common cause is failure to turn the valves back on after maintenance work on the system. In other cases the system develops a leak, and the property owner turns off the valve to avoid water damage. In rare instances, the valves are turned off as a prank or by arsonists attempting to ensure maximum damage to the structure. When performing inspections or preincident planning, fire personnel should check all valves to verify that they are fully opened.

The adequacy of the water supply source should be established prior to construction of the building or the installation of the sprinkler system. The certificate of occupancy should not be signed unless the water supply system tests to be sufficient. However, in some cases an otherwise adequate water supply source is adversely affected during fire fighting operations. If pumping apparatus supplying attack lines and/or master stream devices are connected to the same water supply source that is supplying the sprinkler system, they can reduce the flow through the water system to the point that inadequate volume or pressure reaches the sprinkler system. Preincident planning operations should identify the potential for these problems and locate separate water supply sources for pumpers supplying attack lines and/or master streams.

Inspections and preincident planning should identify potential problems with automatic sprinkler systems.

A common failure in sprinkler performance occurs when an occupancy changes significantly but the sprinkler system is not updated to meet the needs of an additional fire load. For example, suppose a building and sprinkler system were originally designed for a light hazard office space. This is a relatively low fire load. A few years down the line, the office space is converted into a paint store. The paint store will have a significantly higher fire load than the office, and the sprinkler system very possibly will not provide sufficient water to control a fire in the higher fire load. Again, inspections and preincident planning should identify these situations and initiate corrections.

When performing preincident inspections, fire personnel should note the following details concerning sprinkler systems:

- The construction, contents, and layout of the structure and the presence of sprinklers in all areas
- The type of sprinkler system(s) in the building
- The water supply to the sprinkler system and the location and position of all water supply valves
- The consequences of shutting off various valves in the system
- The location and condition of fire department connections and sources of water to supplement the system that will not hinder system operation
- Sources of water supply for other fire fighting operations that will not jeopardize sprinkler system operation
- Alternative means for supplying the sprinkler system in the event of damage to the fire department connection
- The location of spare sprinklers and wrenches
- The location of waterflow indicators and annunciator panels associated with the sprinkler system

When checking fire department connections, fire personnel should ensure that the swivels operate properly and that gaskets are in place in the intakes. Any debris that has been maliciously inserted into the connections should be removed. On new occupancies it is a good idea to verify that the threads on the inlets are the same as those on fire-department hose. The time of a fire is not the time to find out that hose-thread incompatibility prevents you from hooking up to the fire department connection.

All of the information noted during preincident planning visits should be used to develop preincident tactical plans for the structure and/or to strengthen the department's standard operating procedures (SOPs). Serious deficiencies noted during these visits should be reported immediately to code enforcement personnel so that they can be rectified.

Serious deficiencies noted during preincident inspections should be reported immediately to code enforcement personnel so that they can be rectified.

Fire Department Operations at Sprinklered Occupancies

It is important that fire departments develop effective preincident plans and standard operating procedures (SOPs) for sprinklered occupancies long before they are ever called to respond to those occupancies for a report of an emergency. The Incident Commander and various other officers at the scene should ensure that all companies and personnel on the scene follow the SOPs for the duration of the incident unless a planned deviation is necessary.

Fire departments should develop effective preincident plans and standard operating procedures for sprinklered occupancies in their jurisdiction.

Most jurisdictions have an SOP that requires either the first or second arriving pumper at the scene of a sprinklered occupancy to be responsible for supporting the water supply to the sprinkler system **(Figure 15.18, p. 386)**. While en route, the company that will be supporting the standpipe system should review preincident plan information to determine the location of the fire department connection and the water supply source they will use to boost pressure in the sprinkler system.

Figure 15.18 The first- or second-arriving pumper should connect to the water supply source and the FDC.

Upon arriving at the scene, the sprinkler-support engine company should proceed immediately to the fire department connection and make hose connections to both the water supply source and the FDC. In the vast majority of cases the water supply source will be a fire hydrant in relatively close proximity to the FDC. The pump operator should ensure that the intake hose connected to the hydrant will be large enough to pump the rated capacity of the fire pump if necessary. If the water supply source is 200 feet or more from the FDC or if it is a static water supply source, it may be necessary to reverse lay from the FDC to the water supply source in order to adequately supply the FDC.

Whenever possible the water supply used for the sprinkler system should be separate from the water supply used for attack lines and master stream devices.

Whenever possible the water supply used for the sprinkler system should be separate from the water supply used for attack lines and master stream devices. Otherwise the water supply to the sprinkler system could be reduced to the point where it is ineffective. If the fire mains in the proximity of the fire building are large enough to supply both needs adequately, no special actions need to be taken. Otherwise, pumpers that will supply attack lines and master streams should secure a different water supply source. This could be hydrants on different supply lines or a static water supply source.

Do not use a static water supply source to boost pressure in a sprinkler system if it can at all be avoided.

Do not use a static water supply source to boost pressure in a sprinkler system if it can at all be avoided. Dirt and rocks that may be picked up in the drafting process could clog important parts of the sprinkler system. Another concern is that if the sprinkler system is connected to a municipal water supply system, untreated water might accidentally backflow into the municipal system and contaminate it. All sprinkler systems are equipped with backflow preventers and check valves intended to stop contamination from occurring. However, the possibility of these devices failing during an incident is always present, and they should not be entirely relied upon.

The pump operator should connect at least one 2½-inch or larger hose between an apparatus discharge and the FDC. Most departments require the pump operator to connect a hose between the pump and every FDC inlet of the FDC. Most sprinkler FDCs have two inlets, though some have three or more. The hose lines between the water source and the pump and between the pump and the FDC should not be charged with water unless the Incident Commander directs. Also, the apparatus fire pump should remain out of gear until the need to pump is confirmed. Otherwise the pump could overheat while fire conditions are being investigated and verified.

If the Incident Commander orders the pump operator to charge the sprinkler system through the FDC, the pump operator should place the pump in gear, open the water supply source, and begin discharging water into the FDC. Most fire departments use a rule-of-thumb of 150 psi at the FDC inlets as the desired FDC intake pressure. In most cases this will adequately supply the sprinklers unless the building is a very tall high-rise. In some instances the designer of the sprinkler system will have specified an intake pressure for FDC. If this is the case, that required intake pressure should be provided to the fire department during preincident planning, and it also should be posted on a plate at the FDC.

Supporting a sprinkler system in a high-rise building may require greater discharge pressure than a single standard fire department pumper can develop. As mentioned in Chapter 14, some cities that protect numerous high-rises have special high-pressure pumpers. These pumpers can supply water to the highest levels of a high-rise in sufficient pressure and volume to properly support the sprinkler system.

If the fire department does not have high-pressure pumpers, two pumpers connected together at the FDC may generate sufficient pressure. This is accomplished by having one pumper connect to the water supply source and pump water into the intake of a second pumper that is connected to the FDC. The second pumper boosts the pressure from the first pumper to a much greater level than either pumper could create alone. For example, if the first pumper is discharging water into the second pumper at 200 psi, and the second pumper is set to provide a net pump discharge pressure of 200 psi, it will actually discharge 400 psi into the FDC (**Figure 15.19**). Of course, before performing this maneuver or using a single high-pressure pumper, fire personnel should ensure that the hoses used and the sprinkler system piping are rated for these very high pressures.

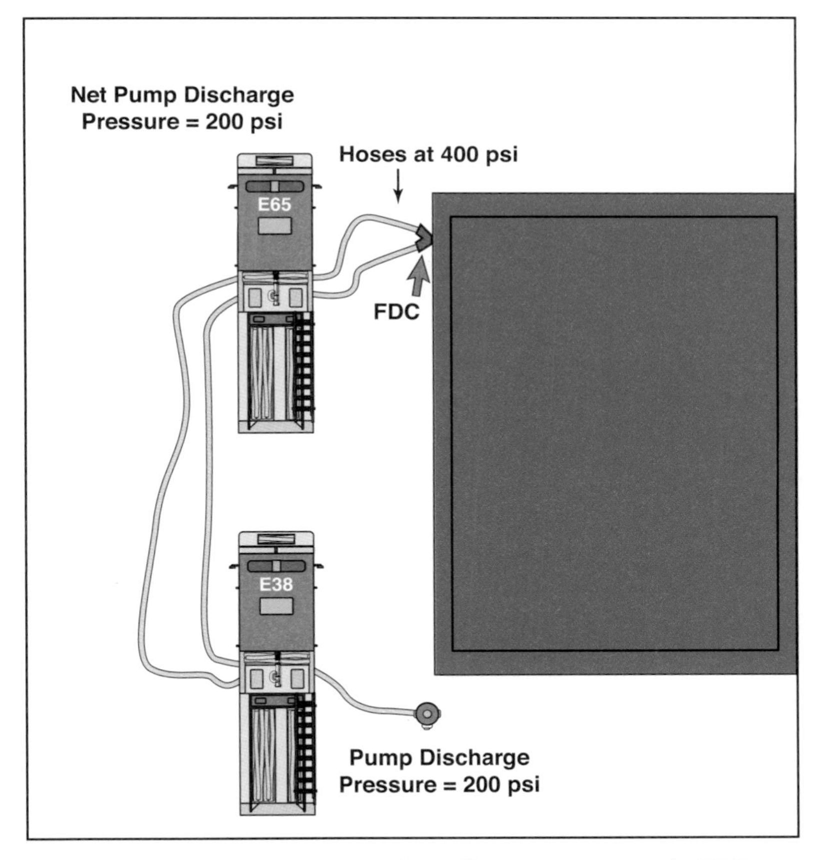

Figure 15.19 These two pumpers together will pump 300 psi into the FDC.

When a sprinkler system is adequately supported by a pumper, firefighters operating in the proximity of the flowing sprinkler should see a noticeable increase in the pressure and volume of water flowing through the open sprinklers. This is one method of ensuring that water is flowing into the FDC and the sprinkler system. Another way is to observe the flowmeter if the pumper is equipped with one; the flow being discharged through the sprinklers will show on the readout. Keep in mind that if only a few sprinklers have been activated, the flow may be very small, possibly less than 100 gpm. Low flows do not necessarily indicate a problem.

Good preincident plans should show the location of important valves, lessening the time required to search for closed valves during an incident.

If no noticeable increase in the sprinkler discharge occurs or the flowmeter is showing no flow, the most likely culprit is a closed valve somewhere in the sprinkler system. Personnel should be assigned to search for and open any valves that are closed. Good preincident plans will show the location of important valves, lessening the time this search should take.

Sprinkler operation should not be discontinued until personnel in the vicinity of the fire have ensured that the fire has been extinguished.

Sprinkler operation should not be discontinued until personnel in the vicinity of the fire have ensured that the fire has been extinguished. In existing case histories, fire personnel have requested that sprinkler systems be shut down before the total extinguishment of the fire was verified. Most often they did this because of poor visibility due to steam and smoke pushed down to the floor level by operating sprinklers. In some of these cases, the fire quickly grew to unmanageable proportions before the sprinklers could be turned back on. Rather than turning the sprinklers off, firefighters should use standard ventilation practices in these circumstances. Once ventilation is established, visibility will improve.

When firefighters have ascertained that the fire is out, the Incident Commander can order shutting down the sprinkler system. This process should begin by having the pumper that is supplying the FDC stop pumping water into the system. Once the outside water supply has been stopped, personnel can shut down the flowing sprinklers. They should shut down as few sprinklers as possible so that the rest of the system will remain in service. If just a few sprinklers have been activated, their orifices can be plugged with sprinkler plugs or wooden chocks/wedges. If a valve that controls the flow of water only to that affected section or floor can be located, that valve should be used to isolate the open sprinklers. Only as a last resort should the entire sprinkler system be shut down at the main control valve.

The pumpers that were assigned to supply the FDC should remain in place with their hoses attached until salvage and overhaul operations are complete. This will reduce the time needed to recharge the sprinkler system should the fire flare up and the sprinklers be needed again.

Whether or not fire personnel restore an automatic sprinkler system, it is always good practice to have a trained sprinkler system technician inspect and service the system after it has been activated.

Different jurisdictions have different policies regarding the fire department's role in restoring sprinkler systems to readiness. Some jurisdictions allow fire personnel to do such tasks as resetting valves and replacing sprinklers. Because of liability concerns, however, most jurisdictions prohibit their personnel from performing these functions. If the system failed to operate properly in the future, the fire department could be held liable for improperly restoring the system. Whether or not fire personnel restore the system, it is always good practice to have a trained sprinkler system technician inspect and service the system after is has been activated.

NFPA® 13E, *Recommended Practice for Fire Department Operations in Properties Protected by Sprinkler and Standpipe Systems,* recommends any postincident report involving a sprinkler system activation should include the following specific data:

- The location of the operating sprinklers
- The number of sprinklers that operated
- The overall result of sprinkler operation
- The perceived reason for any unsatisfactory operation
- The employee assigned to check the control valve
- The engine company, if any, that supplied the FDC
- The size and number of hoselines connected to the FDC
- The duration and pressure of water pumped into the FDC
- Any valves that were closed after the fire and who closed them
- The restoration functions, if any, that were performed by the fire department
- The status of the system when the fire department left the scene
- The operation of any private water supplies used to supply the sprinkler system
- Who was notified of the system activation (fire marshal, sprinkler company, etc.)
- Building representatives who were notified or contacted during the incident.

Hydraulic Calculations for Pump Operators Supplying Sprinkler Systems

In general, pump operators supplying sprinkler systems will have to perform few hydraulic calculations. The vast majority of the pump operator's sprinkler support duties will involve connecting the pumper to a fire hydrant very close to the fire department connection for the sprinkler system and then simply pumping 150 psi into the system. If the fire department wishes to calculate a more accurate required pressure for sprinkler system support, the process is essentially the same as described for determining pressure loss and pump discharge pressure in Chapters 12 and 13.

Although determining pressure loss in piping is an important part of designing sprinkler systems, it is generally considered inconsequential in fire department pumping operations. The friction loss coefficients for various types of piping are extremely low (see Table 12.2). Pump operators therefore do not generally need to be concerned with determining friction loss in sprinkler piping. Most departments use a standard rule-of-thumb of 25 psi to account for all pressure losses in sprinkler piping, including the friction loss in piping, and pressure losses created by the FDC, valves, bends, and fittings in the piping system. Thus, when calculating pressure loss in sprinkler systems, other than the rule-of-thumb system loss, the pump operator's primary concerns will be elevation loss in the system and friction loss in the hoselines between the pump and the FDC. The following examples show various calculations that can be made in support of sprinkler systems.

Example 15.3

Determine the total pressure loss when a fire department pumper is supplying a FDC with two 3-inch hoselines with 2½-inch couplings, each 50 feet long. Seven sprinklers are discharging 15 gpm each on the ninth floor of the structure.

$Q = \text{(No. of sprinklers flowing)(gpm/sprinkler)}$

$Q = (7)(15)$

$Q = 105 \text{ gpm}$

$FL_{\text{3-inch}} = CQ^2L$

$FL_{\text{3-inch}} = (0.2)\left(\frac{105}{100}\right)^2\left(\frac{50}{100}\right)$

Note: The 3-inch lines are essentially siamese lines, so C is taken from Table 12.3.

$FL_{\text{3-inch}} = (0.2)(1.05)^2(0.50)$

$FL_{\text{3-inch}} = (0.2)(1.1025)(0.50)$

$FL_{\text{3-inch}} = 0.1 \text{ psi}$

$EP = \text{(5 psi/floor)(Number of floors} - 1)$

$EP = \text{(5 psi/floor)}(9 - 1)$

$EP = \text{(5 psi/floor)}(8)$

$EP = 40 \text{ psi}$

$FL_{\text{Total}} = FL_{\text{3-inch}} + EP + \text{System Loss}$

$FL_{\text{Total}} = 0.1 + 40 + 25$

$\mathbf{FL_{Total} = 65.1, \text{ or } 65 \text{ psi}}$

Example 15.4

A fire department is preincident planning a new high-rise structure in their city. The fire department needs to determine the total pressure loss that will occur when 20 sprinklers with a ½-inch orifice and a K of 5.5 are all flowing water at 15 psi on the 18th floor of the structure. To supply the FDC, a pumper will have two reverse lines of 2½-inch hose 300 feet to a fire hydrant and pump the lines.

$Q_{\text{per sprinkler}} = (K)(\sqrt{P})$

$Q_{\text{per sprinkler}} = (5.5)(\sqrt{15})$

$Q_{\text{per sprinkler}} = (5.5)(3.87)$

$Q_{\text{per sprinkler}} = 21.3, \text{ or } 21 \text{ gpm}$

Q_{total} = (No. of sprinklers flowing)(gpm/sprinkler)

$Q_{total} = (20)(21)$

$Q_{total} = 420$ gpm

$FL_{2\ 1/2\text{-inch}} = CQ^2L$

$FL_{2\ 1/2\text{-inch}} = (0.5)(420/100)^2(300/100)$

Note: The 2½-inch lines are essentially siamese lines, so C is taken from Table 12.3.

$FL_{2½\text{-inch}} = (0.5)(4.2)^2(3)$

$FL_{2½\text{-inch}} = (0.5)(17.64)(3)$

$FL_{2½\text{-inch}} = 26.5$, or 27 psi

EP = (5 psi/floor)(Number of floors – 1)

EP = (5 psi/floor)(18 – 1)

EP = (5 psi/floor)(17)

EP = 85 psi

$FL_{Total} = FL_{2½\text{-inch}}$ + EP + System Loss

$FL_{Total} = 27 + 85 + 25$

$\mathbf{FL_{Total} = 137}$ **psi**

Example 15.5

Determine the required pump discharge pressure for the pumper in Example 15.4.

PDP = TPL + NP

PDP = 132 + 15

PDP = 147 psi

Example 15.6

A fire department is preincident planning a new high-rise structure in their city. The fire department needs to determine the required pump discharge pressure to supply 12 sprinklers with a 17/32-inch orifice and a K of 8, all flowing water at 18 psi on the 38th floor of the structure. To supply the FDC, a pumper will supply two lines of 3½-inch hose, each 100 feet long.

$Q_{per\ sprinkler} = (K)(\sqrt{P})$

$Q_{per\ sprinkler} = (8)(\sqrt{18})$

$Q_{per\ sprinkler} = (8)(4.24)$

$Q_{per\ sprinkler} = 33.9$, or 34 gpm

Q_{total} = (No. of sprinklers flowing)(gpm/sprinkler)

$Q_{total} = (12)(34)$

$Q_{total} = 408$ gpm

$FL_{3\frac{1}{2}\text{-inch}} = CQ^2L$

$$FL_{3\frac{1}{2}\text{-inch}} = (0.34)\left(\frac{204}{100}\right)^2\left(\frac{50}{100}\right)$$

Note: Since there is no coefficient for dual 3½-inch lines, assume one-half the total is flowing through each.

$FL_{3\frac{1}{2}\text{-inch}} = (0.34)(2.04)^2(1)$

$FL_{3\frac{1}{2}\text{-inch}} = (0.34)(4.16)(1)$

$FL_{3\frac{1}{2}\text{-inch}} = 1.4$, or 1 psi

EP = (5 psi/floor)(Number of floors – 1)

EP = (5 psi/floor)(38 – 1)

EP = (5 psi/floor)(37)

EP = 185 psi

$FL_{Total} = FL_{3\frac{1}{2}\text{-inch}}$ + EP + System Loss

$FL_{Total} = 1 + 185 + 25$

$FL_{Total} = 211$ psi

PDP = TPL + NP

PDP = 211 + 18

PDP = 229 psi

STANDPIPE SYSTEM OPERATIONS

Standpipe systems are an arrangement of piping, valves, hose connections, and related equipment installed in a building with hose discharge connections located to facilitate connecting attack lines for manual fire attack in remote parts of the building. Standpipe systems are most common in high-rise structures, but may also be found in large-area one-story structures such as large warehouses, factories, and shopping malls. Vertical standpipes are provided in most buildings over four stories tall. Standpipes are required by code in high-rise buildings beyond the reach of fire department aerial ladders.

Standpipe systems are not an acceptable substitute for sprinkler systems.

The value of standpipes in large-area one-story buildings is primarily one of expediency and convenience. Horizontal standpipes reduce the effort required to advance hoselines long distances, often several hundred feet, into a structure. This facilitates easier fire attack or mop-up and overhaul of fires that were controlled by sprinklers.

In most high-rise buildings, a standpipe is the *only* means of manually controlling a fire on the upper floors and is an essential aspect of the building's design. In some situations, a standpipe system can be very simple, consisting only of a single vertical pipe (called a riser) with hose connections and a fire department connection. Other situations call for a very complex building system consisting of multiple pumps and risers, valves, and reservoirs on the upper floors.

Fire protection and code enforcement professionals alike should always keep in mind that standpipe systems are not an acceptable substitute for sprinkler systems. Sprinkler systems still provide the highest degree of protection in a structure by sensing and discharging water onto a fire in its earliest stages. Standpipes should be considered only a back up to a properly designed and installed sprinkler system.

Standpipe System Design

There are a variety of different designs for standpipe systems. The type of system chosen for a building is dependent on many variables, including the size of the building, the atmospheric conditions to which the building will be subject, and the reliability of the building's water supply system. The five types of standpipe systems recognized by NFPA® 14 are:

- Dry automatic standpipes
- Wet automatic standpipes
- Semiautomatic standpipes
- Dry manual standpipes
- Wet manual standpipes

Dry Automatic Standpipes

Dry automatic standpipes operate much like the dry-pipe sprinkler systems discussed earlier in this chapter. Dry automatic standpipes contain air under pressure that holds the water beneath a dry-pipe valve. When a discharge is opened, air first escapes from the pipe. The drop in air pressure in the system allows the dry-pipe valve to open, which in turn allows water to flow into the piping. Dry automatic standpipe systems have a water supply that provides sufficient volume and pressure for operating hoselines without the need to boost pressure from an outside source, such as a fire department pumper connected to the FDC. These systems may be used in locations where the piping above the dry-pipe valve could be subjected to freezing temperatures. Their primary disadvantage is that firefighters will have to bleed off an extraordinary amount of air through their hoses before they can achieve a good fire stream.

Wet Automatic Standpipes

Wet automatic systems are the most commonly encountered design of standpipe system. They are very similar in design and concept to wet-pipe sprinkler systems. Wet automatic systems are filled with water at all times and can supply water in sufficient volume and pressure to establish effective attack lines anywhere in the protected occupancy. The opening of a valve to charge a hoseline will trigger a water-flow alarm to signal the building occupants and, in most cases, a central alarm company, which will notify the fire department.

Semiautomatic Dry Standpipes

Semiautomatic dry standpipes are similar in design and operation to automatic dry standpipe systems. The primary difference is that a remote control activation device must be operated in order to allow water into the piping system. In this way the semiautomatic dry standpipe system is somewhat similar to the preaction sprinkler system described earlier in this chapter. A remote-control activation device must be provided adjacent to each hose connection, and each device must be supervised in accordance with NFPA® 72, *National Fire Alarm Code*. Once the system is charged with water, it is capable of supplying water in sufficient volume and pressure to establish effective attack lines anywhere in the protected occupancy.

Manual Dry Standpipes

Manual dry standpipe systems do not have an attached permanent water supply. They comprise a series of pipes, valves, and discharges that are empty when not in use. To be used for fire fighting operations, the manual dry standpipe must be charged and supplied by a fire department pumper attached to the FDC. Once the standpipe is charged, firefighters will have to bleed a considerable amount of air from their hoses before an effective fire stream will flow from their nozzle(s).

Manual Wet Pipe Standpipes

Manual wet standpipe systems are attached to a small water supply source. The source is only capable of keeping the piping system filled with water at all times. A fire department pumper will still have to charge the FDC in order for the system to be operated. These systems' primary advantage over manual dry standpipe systems is that firefighters will not have to bleed as much air through hoselines when they charge the system to fight a fire. Some manual wet standpipe systems are connected directly to the building's sprinkler system. However, the water supply to the dual system could not supply the demand for both systems; thus, the need for a pumper to supply the FDC.

Classification of Standpipe Systems

NFPA® 14, *Standard For the Installation of Standpipe and Hose Systems,* regulates the design and installation of standpipe systems. NFPA® 14 recognizes three classes of standpipe systems—Class I, Class II, and Class III—based on the system's intended use.

Class I Standpipe Systems

Class I standpipe systems are primarily for use by fire fighting personnel trained in handling large handlines, such as 2½-inch attack lines. Class I systems must be capable of supplying effective fire streams during the more advanced stages of fire within a building or for fighting a fire in an adjacent building. Class I systems have 2½-inch hose connections attached to the standpipe riser **(Figure 15.20)**. Some jurisdictions require the 2½-inch connection to have caps that include a reducer with a 1½-inch hose connection as well. NFPA® 14 requires Class I standpipes in high-rise buildings to be of the automatic or semiautomatic type. Buildings that are not high-rises may have manual, semiautomatic, or automatic Class I systems.

In buildings equipped with a single Class I standpipe system, the system must be capable of flowing 500 gpm at 175 psi through the most remote, hydraulically challenging discharge. If the building is equipped with more than one Class I standpipe, each additional standpipe must be capable of flowing 250 gpm. The maximum flow required for any level in the building is 1,000 gpm.

Figure 15.20 A Class I standpipe connection.

Class II Standpipe Systems

Class II standpipe systems are designed for use primarily by building occupants who have little or no training in fire fighting. Class II systems may also be used by firefighters during an initial fire attack or mop-up operations. They typically have hose stations located at the hose connections. These hose stations most frequently contain 1½-inch lightweight hose **(Figure 15.21)**. This hose is usually single-jacket, linen hose and is equipped with a lightweight, twist-type shutoff nozzle. Fire department personnel should not use the supplied hose and nozzle for fire attacks because it may be old and in poor condition. Fire personnel should carry their own hose packs into the structure, disconnect the house line, and connect their own equipment.

Figure 15.21 A Class II standpipe hose cabinet.

Some jurisdictions allow 1-inch hose on Class II standpipe in buildings considered light hazard occupancies. This 1-inch hose may be single-jacket linen hose or noncollapsible booster-type hose on a reel **(Figure 15.22)**.

Class II standpipe systems must be capable of supplying 100 gpm at 100 psi to the most remote, hydraulically challenging discharge on the system. These requirements do not change if multiple systems are located in the building.

Experts disagree on the value of Class II standpipe and hose systems. The presence of the small hose may give a false sense of security to building occupants and give them the impression that they should attempt to fight a fire when their safer course would be to flee. In any case, Class II hose on a standpipe cannot be depended on for fire control operations.

Class II hose on a standpipe cannot be depended on for fire control operations.

NFPA® 14 states that Class II standpipes shall not be required in buildings that are fully protected by a sprinkler system. However, in those cases the building should have a Class I standpipe system and the 2½-inch caps on the standpipe discharges must be equipped with 1½-inch reducers and chained caps.

Figure 15.22 A Class II standpipe hose reel.

Figure 15.23 A Class III hose cabinet and connections.

Figure 15.24 The standpipe hose connection must be no less than 3 and no more than 5 feet from the floor.

NFPA® 14 also requires both Class II and Class III systems to be wet systems unless they are in facilities that are subject to freezing conditions and have fire brigade personnel trained to operate the system until the fire department arrives and intervenes.

Class III Standpipe Systems

Class III standpipes combine the features of Class I and Class II systems. Class III systems have both 2½-inch connections for fire department personnel and 1½-inch hose stations and connections for use by the building occupants (**Figure 15.23**). The system must allow the Class I and Class II services to be used simultaneously. The minimum required flow rates for Class III standpipes are the same as those for Class I standpipes.

Standpipe System Components

As with sprinkler systems, all standpipe systems regardless of type or class have common components with which firefighters and driver/operators should be familiar. This knowledge enables personnel to utilize the systems optimally and to troubleshoot problems should they occur.

Piping

All standpipes are required to use ferrous metal, steel, or copper piping. This piping must be capable of withstanding operating pressures of up to 300 psi. The piping must be protected from mechanical damage. The piping either must be located in exit stairways or must be protected to the same degree as exit stairways.

If they are stand-alone systems, Class I and III standpipe risers must be at least 4 inches in diameter. If they are combined with sprinkler systems, they must be at least 6 inches in diameter. Horizontal branch lines that extend to discharge on individual floors must be at least 2½-inches in diameter.

NFPA® 14 does not specify minimum piping sizes for Class II systems. These systems simply must be hydraulically designed to meet the flow requirements described above.

Hose Connections

Regardless of the class of system, hose connections on standpipes for attack lines must be no less than 3 feet and no more than 5 feet from the floor on which they are located (**Figure 15.24**). All discharges should be equipped with the same hose threads that the local fire department uses. Each discharge should be equipped with a cap that is chained to the piping. The locations and sizes of the hose connections vary, depending on the class of the system.

Class I standpipes must be equipped with 2½-inch hose connections at the following locations:

- Each intermediate landing between floor levels in every required exit stairway
- On each side of the wall adjacent to the exit openings of horizontal exits
- In each exit passageway at the entrance from the building areas into the passageway (except for covered malls)

- At the entrance to each exit passageway or corridor in covered malls and at the interior side of public entrance from the exterior to the mall
- At the highest landing of stairways with stairway access to the roof, and on the roof where stairways do not access the roof

The travel distance to any point on a floor protected by a Class I standpipe system shall not exceed 150 feet in unsprinklered occupancies and 200 feet in sprinklered occupancies. If the building is sprinklered, each hose connection should be equipped with a cap that has a 1½-inch reducer and cap on top of the 2½-inch connection **(Figure 15.25)**.

Figure 15.25 If the building is sprinklered, the 2½-inch hose connection should be outfitted with a 1½-inch reducer and cap.

The requirements for Class II hose connection placement are not as stringent as those for Class I systems. Class II discharges and hose cabinets should be located so that the travel distance on any part of the floor is no more than 130 feet for cabinets equipped with 1½-inch hose or 120 feet for cabinets equipped with hose smaller than 1½-inch. These distances are based on the assumption that each discharge will be equipped with 100 feet of hose and a 1½-inch fire stream will shoot at least 30 feet while a smaller stream will shoot 20 feet.

Class III systems must have hose connections that meet the requirements for both Class I and II systems.

Pressure-Regulating Devices

Where the discharge pressure at a hose outlet exceeds 100 psi for 1½-inch discharges and 175 psi for 2½-inch discharges, NFPA® 14 requires a pressure-regulating device to limit the discharge to those pressures unless otherwise approved by the fire department **(Figure 15.26)**. A pressure-regulating device prevents pressures that make hose difficult or dangerous to handle. It also enhances system reliability because it extends individual zones to greater heights. In some instances, it may improve system economy by eliminating the need for booster pumps on the system. However, pressure-regulating devices make the system design more complex.

Figure 15.26 A standpipe pressure reducing valve.

There are several different types of pressure-regulating devices. One type consists of a simple restricting orifice inserted in the waterway. The pressure drop through the orifice plate depends on the orifice diameter and the flow. The individual restricting orifices must be sized for their specific application and will not be the same for each floor of a given building.

Another type of pressure-regulating device may consist of vanes in the waterway that can be rotated to change the cross-sectional area through which the water flows. A pressure-regulating device may also take the form of a pressure-reducing valve. Several different pressure-reducing valves are available from various manufacturers. Some of the valves are field adjustable, and others are set at the factory.

A pressure-regulating device must be specified and/or adjusted to meet the pressure and flow requirements of the individual installation. Factory-set devices must be installed on the correct hose outlet to ensure proper installation. When field-adjustable devices are used, the installer must follow carefully the manufacturer's instructions for making adjustments.

Pressure regulating devices in standpipe systems must be specified and/or adjusted to meet the pressure and flow requirements of the individual installation.

Preincident planning and inspection procedures for standpipe systems must focus particularly on the presence and proper operation of pressure regulating devices.

Preincident planning and inspection procedures must focus particularly on the presence of pressure regulating devices and their proper operation. If a pressure-regulating device is not correctly installed or properly adjusted for the required inlet pressure, outlet pressure, and flow, it may greatly reduce the available flow and seriously impair fire fighting capabilities. These problems were a major factor in the 1991 Philadelphia, Pennsylvania, One Meridian Plaza fire, which resulted in the deaths of three firefighters. In that fire, improperly installed and adjusted pressure regulators severely hampered fire fighting efforts. Firefighters could not overcome those problems during the fire fighting operation, and they ultimately stretched large diameter hoselines up the stairwells to form improvised standpipes. The fire's progress was not halted until it reached a floor that had an operating sprinkler system.

Figure 15.27 This high-rise building in Montreal has one standpipe that serves the first 47 floors and another that serves floors 48 and above.

Fire Department Connections

Each Class I or Class III standpipe system requires one or more fire department connections through which a fire department pumper can supply water into the system. High-rise buildings having two or more zones require a fire department connection for each zone **(Figure 15.27)**.

In high-rise buildings with multiple zones, the upper zones may be beyond the height to which a fire department pumper can effectively supply water. This height will be around 450 feet, depending on available hydrant pressure and other factors. For standpipe system zones beyond that height, a fire department connection is not of any value unless the fire department is equipped with special high-pressure pumpers.

NFPA® 14 specifies that there shall be no shutoff valve between the fire department connection and the standpipe riser. In multiple-riser systems, however, gate valves are provided at the base of the individual risers.

Figure 15.28 The standpipe FDC name plate should have raised letters.

The hose connections to the fire department connection must be female and equipped with standard caps. It is important that the hose coupling threads conform to those used by the local fire department. The exception to this rule would be for fire departments that require standpipes to be equipped with large diameter Storz-type connections. The fire department connection must be designated by a raised-letter sign on a plate or fitting reading "STANDPIPE" **(Figure 15.28)**. If the fire department connection does not service all of the building, the sign must indicate which floors are serviced.

Preincident Inspection and Planning Procedures for Standpipe Systems

Good policies and procedures for preincident planning and inspection of standpipe systems is no less important than for sprinkler systems. In fact, the procedures for inspecting and developing preincident plans for standpipes and sprinkler systems are generally the same. Fire personnel should take these activities especially seriously, as they or their comrades will be the ones whose safety depends on the standpipe system's proper operation.

The safety of fire personnel depends on a standpipe system's proper operation.

Preincident planning and inspection of standpipe systems should start before building construction begins. Detailed design plans should be submitted to the local fire department or building department in accordance with local codes. As construction proceeds, the installation should be checked for conformity with the plans. In a high-rise building, the standpipe will have to be in partial operation as construction proceeds. This will protect the structure should fire occur on the upper levels during construction.

When the installation is complete, the following tests and inspections should be performed:

- The system should be hydrostatically tested at a pressure of at least 200 psi for two hours to ensure the tightness and integrity of fittings. If the normal operating pressure is greater than 150 psi, the system should be tested at 50 psi greater than its normal pressure.
- The system should be flow-tested to remove any construction debris and to ensure that there are no obstructions.
- On systems equipped with an automatic fire pump, a flow test should be performed at the highest outlet to ensure that the fire pump will start when the hose valve is opened.
- The fire pump should be tested to ensure that it will deliver its rated flow and pressure.
- All devices used should be inspected to ensure that they are listed by a nationally recognized testing laboratory.
- Hose stations and connections should be checked to ensure that they are in cases within 6 feet of the floor and are positioned so that the hose can be attached to the valve without kinking.
- Each hose cabinet or closet should be inspected for a conspicuous sign that reads "FIRE HOSE" and/or "FIRE HOSE FOR USE BY OCCUPANTS OF BUILDING" (**Figure 15.29**).
- Fire department connections should be checked for the proper fire department thread and for a sign indicating "STANDPIPE" with a list of the floors served by that connection.
- When a dry standpipe is used, check for a sign indicating "DRY STANDPIPE FOR FIRE DEPARTMENT USE ONLY."

Figure 15.29 Standpipe hose connections should be within 6 feet of the floor.

To avoid potential liability, fire department personnel should not perform actual tests. A fire protection contractor or, if they are sufficiently knowledgeable, the building's operating staff should carry out the actual testing of standpipes. However, fire department personnel should witness the tests.

After the initial inspection, testing, and approval, standpipe systems need to be inspected and tested at regular intervals. The building management should make a visual inspection at least monthly. Because interior fire fighting depends on the standpipe system, the fire department must also inspect standpipes at regular intervals. All companies that may be expected to operate in the occupancy should make preincident planning trips to examine the system and include it in their SOPs. They should note the following items:

All companies that may be expected to operate in a standpipe equipped occupancy should make preincident planning trips to examine the system and include it in their SOPs.

- All water supply valves are sealed in the open position.
- Power is available to the fire pump, if the system is so equipped.
- Individual hose valves are free of paint, corrosion, and other impediments.
- Hose discharge threads are not damaged and caps are present.
- Hose discharge valve wheels are present and not damaged.
- Hose cabinets are accessible.
- Hose in Class II and Class III hose cabinets is in good condition, has proper dryness, and is properly positioned on the rack.
- Discharge outlets in dry systems are closed.
- Dry standpipe is drained of moisture.
- Access to the fire department connection is not blocked.
- The fire department connection is free of obstruction, the swivels rotate freely, and caps are in place.
- Water supply tanks are at the proper level.
- If the system is equipped with pressure-regulating devices, those devices are tested as required by the manufacturer. NFPA® 14 recommends they be tested yearly.
- Hose valves on dry systems are closed.
- Pump discharge pressures that will be required to supply the system are determined.
- Dry systems are hydrostatically tested every five years.

Fire Department Operations at Occupancies Equipped with Standpipes

Fire departments must develop and enforce stringent SOPs for utilizing standpipes in fire fighting operations.

Fire departments must develop stringent SOPs for utilizing standpipes in fire fighting operations, and they must ensure that all personnel adhere to these standards. SOPs for operating at occupancies equipped with standpipes should address at least the following considerations:

- The equipment to be carried into the structure by companies that will hook attack lines to standpipes
- The pump discharge pressure that pumpers assigned to supply the FDC will need to use
- How to identify which standpipe (if the building has more than one) will be used for fire fighting operations so that the pumper connects to the correct FDC for that standpipe
- Methods for flaking out the hose between the standpipe hose connection and the entrance to the fire floor
- Procedures for adjusting pressure reducing devices if that becomes necessary
- Methods for bypassing the FDC to supply the system in the event that the FDC is blocked or otherwise inoperable

Fire personnel should always carry their own hose and equipment into the structure for use with the standpipe system. They should never rely on the hose in the structure's hose cabinets. This hose, even when in perfect condition, is not designed for the rigors of heavy-duty fire fighting. In many cases it is old and subject to dry rot or other degradation that makes it likely to fail when charged with water. Most fire departments develop high-rise packs, bundles, or carts that contain all of the items necessary to place a standpipe attack line in service **(Figure 15.30)**. When entering a structure with a standpipe, firefighters simply grab the pack or bundle and enter the structure. These packs typically include 100–150 feet of 1½-inch or larger hose, an appropriate nozzle, a 2½-inch to 1 ½-inch reducer, a spanner wrench, door chocks, and other items specified by the department.

Fire personnel should always carry their own hose and equipment into the structure for use with the standpipe system.

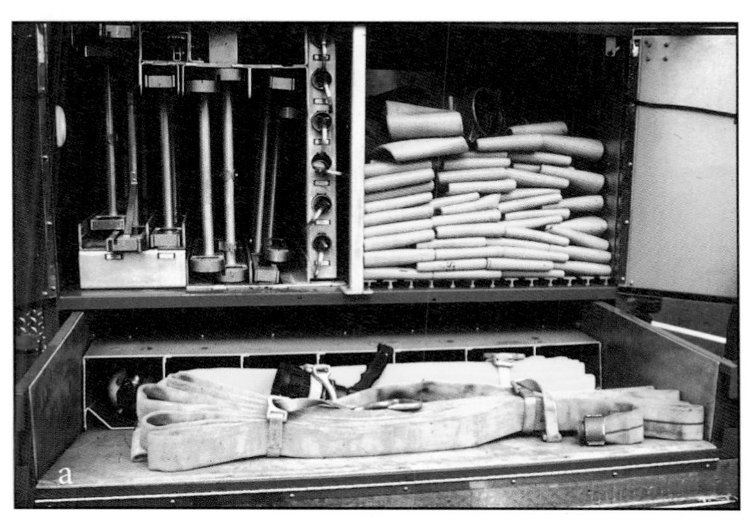

Figure 15.30 (a) A typical high-rise hose bundle *(courtesy of Mesa, AZ Fire Department)*; **(b)** some departments use wheeled high-rise hose carts to make transporting hose less physically demanding on firefighters.

Before a company leaves the lobby, it should review the building floor plan, if available, and the type of standpipe. Extra hose and air cylinders should be brought into the building and kept in the lobby until needed. Additional personnel may be needed to carry these items to the fire or staging floor. Personnel should be assigned to check riser valves and fire pump operation.

Firefighters should proceed to the floor below the fire. They must wear full protective clothing and self-contained breathing apparatus. If elevators are necessary to reach the upper floors of tall buildings, they should be on manual or fire service operation. When possible, walking up the stairs is safest. Firefighters should connect their hose to the standpipe on the floor below the fire and advance the hose up to the fire floor **(Figure 15.31)**. This provides a safe working area in the event of heavy involvement on the fire floor. (**NOTE:** One firefighter must remain at the standpipe valve to charge the line after it has been moved into operating position.) Excess hose can be laid up the stairway past the fire floor. This will make it feed easier onto the fire floor as it is pulled. Firefighters must exercise care to create as little obstruction as possible for occupants exiting the building via the stairwell.

Figure 15.31 Connect the hose to the standpipe and flake it up the stairs. *Courtesy of Mesa, AZ Fire Department*

When heavy fire has developed on the fire floor, the line should be charged and the stairwell door carefully opened to begin the attack. If needed, additional hoselines can be attached to the hose valves on lower floors and advanced to the fire floor. The firefighter operating the standpipe valve should watch for fire developing behind the attack team and warn them before their position becomes untenable.

Firefighters must decide in advance what type of nozzle to use. Fog nozzles produce fine water droplets that afford better protection to advancing firefighters and provide more efficient heat absorption. However, if scale is present in the standpipe, it may obstruct the fog nozzle. Although a solid stream does not produce a spray, it provides good reach and penetration into burning debris.

When a building has multiple standpipes, firefighters can use more than one riser for a multiple line attack. Firefighters advancing onto a floor from different directions must keep in contact by radio to coordinate their efforts and avoid endangering each other with their streams.

The fire department can use a number of simple operations to overcome various standpipe impairments. For example, if a fire department connection has a frozen swivel, a double male can be used with a double female. If a fire department connection is totally unusable because of vandalism, the standpipe riser can be charged at the first-floor level by attaching a double female to a hose valve at the first-floor level. If an individual hose valve on an upper floor is inoperative, a valve on the next floor down can be used. To overcome a problem in a single-riser building where the standpipe is totally unserviceable, firefighters may have to hoist a line up the outside of the building. Often this can be accomplished by unrolling hose down the side of the building from every second or third floor rather than attempting to hoist the line several floors. Standpipes in adjacent buildings can also be used to protect exposures. As a last resort, supply hose can be laid up the interior stairwell to take the place of the standpipe.

Figure 15.32 Some aerial devices have standpipe pipe hose connections.

If aerial apparatus can reach the fire floor, standpipe connections at the tip of the aerial apparatus can supply hoselines on upper levels **(Figure 15.32)**. If the fire floor is higher than the reach of the aerial device, the aerial device can still be used in place of hoselines for the first 6 to 10 floors. The aerial device will have much less friction loss than would hose used for the same distance. However, once the aerial device is used to supply a hoseline it is essentially out of service. If a window rescue or some other operation becomes necessary, the aerial device connected to the hoseline cannot be used because doing so would endanger firefighters on that hoseline. For this reason many fire departments choose not to use aerial-device standpipe connections to supply attack lines in structures.

On occasion, vandals or curious individuals may open the hose valves in dry standpipes and leave them in that position. When the standpipe is charged, water will discharge on levels below the fire floor and rob the attack line of the volume and pressure required to extinguish the fire. A firefighter then will have to go down the stairwell to close the valves. Having fire personnel check all standpipe valves as they first proceed up the stairwells can prevent this.

Additional information and requirements for fire department operations at occupancies equipped with standpipe systems can also be found in NFPA® 13E, *Recommended Practice for Fire Department Operations in Properties Protected by Sprinkler and Standpipe Systems.*

Hydraulic Calculations for Pump Operators Supplying Standpipe Systems

Supplying water at the correct pressure is perhaps even more important for standpipe FDCs than for sprinkler FDCs. If a standpipe FDC is not properly supported, firefighters in dangerous, forward fire fighting positions without doubt will be in peril. All pump operators must be thoroughly knowledgeable of SOPs related to standpipe support and follow them explicitly.

All pump operators must be thoroughly knowledgeable of standpipe support SOPs and follow them explicitly.

Unlike sprinklers, there is no agreed upon rule-of-thumb for the proper pressure to which a standpipe FDC should be charged. To ensure that each system is supported properly, its required pump discharge pressure must be calculated individually.

Generally, the friction loss in the standpipe is small unless the flows are large, such as when two 2½-inch lines are being supplied from the same riser. Allowance must also be made for pressure losses in the fire department connection and pipe bends. As we did with sprinkler systems, we will simply calculate a straight 25 psi loss to account for friction loss in the standpipe system. It is most desirable if the system designer can provide the fire department with an accurate friction loss figure. Otherwise calculating friction loss is difficult, as all piping bends, fittings, and other factors often cannot be seen or determined.

When a standpipe system is known to be equipped with pressure-reducing valves, the elevation pressure used must be based on the total height of the standpipe or the zone being used. This is because the pressure-reducing valve generally acts to reduce whatever pressure is presented to it. If the pressure at a particular pressure-reducing valve is less than that for which the valve was adjusted, the pressure for the hoselines will be inadequate.

At no time should a standpipe system be pressurized in excess of 200 psi unless it has been designed for the higher pressures.

According to NFPA® 14, at no time should the standpipe system be pressurized in excess of 200 psi unless it has been designed for the higher pressures. Otherwise the system could fail and endanger the firefighters in the fire area. When determining the correct pressure at which to charge a standpipe system, the following factors must be considered:

- Friction loss in the hose between the pumper and the FDC
- Friction loss in the standpipe system (in this text, a flat 25 psi)
- Pressure loss due to elevation
- Number, size, and length of attack lines connected to the standpipe
- Required nozzle pressure

We discussed appropriate nozzle pressures in Chapters 11 and 12. Briefly, handline fog nozzles typically operate at 100 psi nozzle pressure. The exceptions to this are special low-pressure nozzles, which are commonly used in high-rise fire fighting in conjunction with standpipes. They will have a manufacturer-specified nozzle pressure, usually around 75 psi. Solid stream handline nozzles should be operated at a discharge pressure of 50 psi.

The following examples show how to calculate pressure loss and pump discharge pressures for standpipe attack lines.

Example 15.7

A fire is discovered on the eighth floor of a structure. Firefighters from the first-arriving engine company proceed to the seventh floor of the occupancy and connect three 50-foot sections of 2½-inch hose to the standpipe outlet. The hose is equipped with a solid stream nozzle that has a 1¼-inch tip. The pump operator connects the pumper to a fire hydrant and stretches two lines of 3-inch hose, each 100 feet long with 2½-inch couplings, to the FDC. What is the total pressure loss between the pumper connected to the FDC and the nozzle on the attack line?

$$GPM = 29.7\,(d^2)\,(\sqrt{NP})$$

$$GPM = (29.7)(1.25)^2(\sqrt{50})$$

$$GPM = (29.7)(1.5625)(7.07)$$

$$GPM = 328 \text{ gpm}$$

$$FL_{2\frac{1}{2}\text{-inch attack line}} = CQ^2L$$

$$FL_{2\frac{1}{2}\text{-inch attack line}} = (2)\left(\frac{328}{100}\right)^2\left(\frac{150}{100}\right)$$

$$FL_{2\frac{1}{2}\text{-inch attack line}} = (2)(3.28)^2(1.5)$$

$$FL_{2\frac{1}{2}\text{-inch attack line}} = (2)(10.76)(1.5)$$

$$FL_{2\frac{1}{2}\text{-inch attack line}} = 32.3, \text{ or } 32 \text{ psi}$$

$$EP = (5 \text{ psi/floor})(\text{No. of floors} - 1)$$

$$EP = (5)(8 - 1)$$

$$EP = (5)(7)$$

$$EP = 35 \text{ psi}$$

$$FL_{3\text{-inch lines}} = CQ^2L$$

$$FL_{3\text{-inch lines}} = (0.2)\left(\frac{328}{100}\right)^2\left(\frac{100}{100}\right)$$

$$FL_{3\text{-inch lines}} = (0.2)(3.28)^2(1)$$

$$FL_{3\text{-inch lines}} = (0.2)(10.76)(1)$$

$$FL_{3\text{-inch lines}} = 2.1, \text{ or } 2 \text{ psi}$$

$$TPL = FL_{2\frac{1}{2}\text{-inch attack line}} + EP + FL_{3\text{-inch lines}} + \text{System Loss}$$

$$TPL = 32 + 35 + 2 + 25$$

$$\mathbf{TPL = 94\ psi}$$

Example 15.8

A fire is discovered on the 16th floor of a structure. Firefighters from the first two arriving engine companies proceed to the 15th floor of the occupancy and connect one 150-foot 2½-inch attack line equipped with a 250 gpm fog nozzle and one 150-foot 2-inch attack line equipped with a 175 gpm fog nozzle to the same standpipe riser. The pump operator on the first-arriving pumper connects to a fire hydrant and stretches two lines of 2½-inch hose, each 50 feet long, to the FDC. What is the total pressure loss between the pumper connected to the FDC and the nozzles on the attack lines?

$$FL_{2\frac{1}{2}\text{-inch attack line}} = CQ^2L$$

$$FL_{2\frac{1}{2}\text{-inch attack line}} = (2)\left(\frac{250}{100}\right)^2\left(\frac{150}{100}\right) FL$$

$$FL_{2\frac{1}{2}\text{-inch attack line}} = (2)(2.5)^2(1.5)$$

$$FL_{2\frac{1}{2}\text{-inch attack line}} = (2)(6.25)(1.5)$$

$$FL_{2\frac{1}{2}\text{-inch attack line}} = 18.75\text{, or } 19\ psi$$

$$FL_{2\text{-inch attack line}} = CQ^2L$$

$$FL_{2\text{-inch attack line}} = (8)\left(\frac{175}{100}\right)^2\left(\frac{150}{100}\right)$$

$$FL_{2\text{-inch attack line}} = (8)(1.75)^2(1.5)$$

$$FL_{2\text{-inch attack line}} = (8)(3.0625)(1.5)$$

$$FL_{2\text{-inch attack line}} = 36.75\text{, or } 37\ psi$$

$$EP = (5\ psi/floor)(\text{No. of floors} - 1)$$

$$EP = (5)(16 - 1)$$

$$EP = (5)(15)$$

$$EP = 75\ psi$$

$$FL_{2\frac{1}{2}\text{-inch supply lines}} = CQ^2L$$

$$FL_{2\frac{1}{2}\text{-inch supply lines}} = (0.5)\left(\frac{250+175}{100}\right)^2\left(\frac{50}{100}\right)$$

$$FL_{2\frac{1}{2}\text{-inch supply lines}} = (0.5)\left(\frac{425}{100}\right)^2(0.5)$$

$$FL_{2\frac{1}{2}\text{-inch supply lines}} = (0.5)(4.25)^2(0.5)$$

$$FL_{2\frac{1}{2}\text{-inch supply lines}} = (0.5)(18.06)(0.5)$$

$FL_{2½\text{-inch supply lines}} = 4.5$, or 5 psi

$TPL = FL_{2\text{-inch attack line}} + EP + FL_{2\ 1/2\text{-inch supply lines}} + \text{System Loss}$

$TPL = 37 + 75 + 5 + 25$

TPL = 142 psi

Example 15.9

Determine the pump discharge pressure required to supply the attack line in Example 15.7.

$PDP = TPL + NP$

$PDP = 94 + 50$

PDP = 144 psi

Example 15.10

Given the conditions in Example 15.8, assume that the 2½-inch attack line is equipped with a standard fog nozzle designed to operate at 100 psi and the 2-inch attack line is equipped with a low-pressure fog nozzle designed to operate at 75 psi. Determine the appropriate pump discharge pressure for this evolution.

Pressure requirement for 2½-inch attack line:

$PDP_{2½} = TPL + NP$

$PDP_{2½} = 19 + 100$

$PDP_{2½} = 119$ psi

Pressure requirement for 2-inch attack line:

$PDP_2 = TPL + NP$

$PDP_2 = 37 + 75$

$PDP_2 = 112$ psi

Thus, in this case use the 2½-inch line to calculate the PDP.

$PDP_{total} = PDP_{2½} + EP + FL_{2½\text{-inch supply lines}} + \text{System Loss}$

$PDP_{total} = 119 + 75 + 5 + 25$

$\mathbf{PDP_{total} = 224}$ **psi**

SUMMARY

Sprinkler and standpipe systems are an important weapon in the fire fighting arsenal. Firefighters and pump operators must be familiar with these systems' design and operational principles in order to use them to the their fullest advantage during fire fighting operations. Pump operators and personnel making preincident plans for occupancies that have these systems must understand their hydraulic implications and know how to factor them into pressure loss and pump discharge pressure calculations.

For more detailed information on inspecting sprinkler and standpipe systems, see the IFSTA *Fire Inspection and Code Enforcement* manual. For more information on the design and use of sprinkler and standpipe systems, see the IFSTA *Fire Detection and Suppression Systems* manual.

REVIEW QUESTIONS

1. Which NFPA® standard regulates the design and installation of fire sprinkler systems?

2. What are the three principle reasons that motivate building owners to invest in and install automatic sprinklers in their building?
   1.
   2.
   3.

3. What are four basic types of sprinkler systems that fire personnel will encounter?
   1.
   2.
   3.
   4.

4. Determine the flow in gpm from a 1/2-inch sprinkler with a K of 6 when the discharge pressure at the sprinkler orifice is 19 psi.

5. What are the three basic types of pipe that are used for sprinkler systems?
   1.
   2.
   3.

6. What is the minimum residual pressure required at the most hydraulically challenging sprinkler in a pipe schedule system? In a hydraulically calculated system?

7. What is the rule-of-thumb pump discharge pressure that most fire department use for supplying sprinkler system FDCs?

8. What are the three principle causes of unsatisfactory performance in sprinkler systems?

   1. ______________________________

   2. ______________________________

   3. ______________________________

9. A fire department is preincident planning a new high-rise structure in their city. The fire department needs to determine the total pressure loss that will occur when 10 sprinklers with a ½-inch orifice and a K of 5.5 are all flowing water at 17 psi on the 14th floor of the structure. In order to supply the FDC, a pumper will have to reverse two lines of 2½-inch hose 500 feet to a fire hydrant and pump the lines.

   ____

   ____

   ____

10. Determine the required pump discharge pressure for the pumper in Review Question 9.

    ____

    ____

    ____

11. What are the five basic types of standpipe systems?

    1. ______________________________

    2. ______________________________

    3. ______________________________

    4. ______________________________

    5. ______________________________

12. Which class of standpipe system is designed primarily for occupant use?

    ____

13. A fire is discovered on the 12th floor of a structure. Firefighters from the first-arriving engine company proceed to the 11th floor of the occupancy and connect three 50-foot sections of 2-inch hose to the standpipe outlet. The hose is equipped with a solid stream nozzle that has a 1-inch tip. The pump operator connects the pumper to a fire hydrant and stretches two lines of 3-inch hose with 2½-inch couplings, each 100 feet long, to the FDC. What is the total pressure loss between the pumper connected to the FDC and nozzle on the attack line?

    ____

    ____

14. What is the pump discharge pressure for Review Question 13? ______________________________

Answers found on page 444.

Appendix A
List of Equations

Location	Reference	Equation
Eq.2.1, p. 15	Principle 3, pressure formula	$\text{Pressure} = \frac{\text{Force}}{\text{Area}}$
Eq.2.2, p. 16	Principles 3 & 4 combined	Pressure = weight x height or $P = (w)(h)$
Eq.2.3, p. 16	Applying Equation 2.2 to water	$P = \frac{62.4 \text{ lbs/ft}^3}{144 \text{ in}^2/\text{ft}^2} (h) \rightarrow P = (0.433 \text{ psi/ft})(h)$
Eq.2.4, p. 18	Head formula	$h = \frac{P}{W}$
Eq.2.5, p. 18	Inverse Relationship	$(1/x)$
Eq.2.6, p. 19	Potential Energy	$PE = (W)(h)$ Total potential energy within a water supply system can be expressed as: $PE_t = PE_h + PE_p$
Eq.3.1, p. 23	Kinetic Energy	$KE = G = \frac{(m)(v)^2}{2}$ Alternate version: $m = \frac{W}{g}$
Eq.3.2, p. 24	Kinetic Energy of Water	$KE = G = \frac{(w)(v)^2}{(2)(g)}$
Eq.3.3, p. 24	Conservation of Energy or Total Energy	$TE = PE + KE$
Eq.3.4a, p. 25	Bernoulli's Equation	$h_1 + \frac{P_2}{W} h_2 + \frac{P_2}{W} + \frac{(v_1)^2}{(2)(g)}$
Eq.3.4b, p. 25	Bernoulli's Equation, form 2	$z_2 + \frac{P_2}{W} + \frac{(v_1)^2}{(2)(g)} + \frac{P_2}{W} + \frac{(v_2)^2}{(2)(g)} z_2 + h_{1,2}$
Eq.3.5, p.27-28	Conservation of Matter/flow of a liquid through a conduit	$Q = (A)(v)$
Eq.3.6, p. 28	Mathematical Eq. for Equation 3.5	$(A_1)(V_1) = (A_2)(V_2)$
Eq.3.7, p. 36	The Darcy-Weisbach Formula	$h_f = \frac{(f)(v)^2(L)}{(2)(g)(D)}$
Eq.3.8, p. 37	Reynolds number	$R_E = \frac{(v)(D)}{\mu}$

Appendix A

List of Equations (continued)

Location	Reference	Equation
Eq. 3.9, p. 38	Original Hazen-Williams Formula	$v = (1.318)(C)(R)^{0.63}(S)^{0.54}$
Eq.3.10, p. 38	Revised or Fire Protection Version of Hazen-Williams Formula	$P_f = \frac{(4.52)(Q)^{1.85}}{(C)^{1.85}(D)^{4.87}}$
Eq.5.1, p. 76 & 79	Flow in gpm	Flow in gpm = $(29.83)(C_d)(d^2)(\sqrt{P})$
Eq.5.2, p. 85	Determining available water by mathematical computation/uses a variation of Hazen-Williams formula	$Q_r = \frac{(Q_f)(h_r^{0.54})}{h_f^{0.54}}$
Eq.6.1, p. 95	Gallons of water needed to control a fire in a given area	gallons = $\frac{\text{volume of the room (in ft}^3\text{)}}{200 \text{ ft}^3\text{/gallon}}$
Eq.6.2, p. 96	Iowa Rate-of-Flow Formula, of Iowa State Formula	RFF = $\frac{\text{volume of the room (in ft}^3\text{)}}{100}$
Eq.6.3, p. 97	The NFA Fire Flow Formula "Quick-Calculation" Formula	$\frac{L \times W}{3}$ = fire flow in gpm for one floor at 100% involvement
Eq.6.4, p. 100	The basic ISO Required Fire Flow Formula	$F = 18\ C\ (A)^{0.5}$
Eq.9.1, p. 196	Pressure Correction	Pressure correction = $\frac{\text{lift (9 ft.) + intake hose friction loss}}{2.3}$
Eq.11.1, p. 237	Flow from a solid stream nozzle	Flow from a solid stream nozzle in GPM = $29.7 \times d^2 \times \sqrt{NP}$
Eq.11.2, p. 250	Nozzle reaction for a solid stream nozzles	$NR = 1.57 \times d^2 \times NP$
Eq.11.3, p. 251	Simple rule-of-thumb formula for Equation 11.2	$NR = Q/3$
Eq.11.4, p. 252	Nozzle reaction for a Fog Stream nozzles	$NR = 0.050 \times Q \times \sqrt{NP}$
Eq.11.5, p. 252-253	Simple rule-of-thumb for Equation 11.4	$NR = Q/2$
Eq.12.1, p. 262	Basic Friction Loss Formula	$FL = 2Q^2 + Q$
Eq.12.2, p. 264	Friction Loss Formula adjusted for more accurate results for small flows	$FL = 2Q^2 + \frac{1}{2}Q$
Eq.12.3, p. 270	Modern Friction Loss Formula flows	$FL = CQ^2L$

Appendix A
List of Equations (concluded)

Location	Reference	Equation
Eq.12.4, p. 282	Elevation Pressure	$EP = 0.5H$
Eq.12.5, p. 282	Elevation Pressure, form 2	$EP = H/2$
Eq.12.6, p. 282	To estimate pressure loss of gain in multistory building	$EP = (5psi)(\text{number of stories} — 1)$
Eq.12.7, p. 312	Condensed Q Formula	$FL = Q^2$
Eq.12.8, p. 312	Modified Condensed Q Formula for friction loss in 4-inch hose	$FL \text{ per 100 feet} = \frac{Q^2}{5}$
Eq.12.8, p. 312	2nd Modified Condensed Q Formula for friction loss in 5-inch supply line	$FL = \frac{Q^2}{15}$
Eq.13.1, p. 319	Pump Discharge Pressure (PDP)	$PDP = NP + TPL$
Eq.13.2, p. 333	Net Pump Discharge Pressure	$NPDP_{PPS} = PDP — \text{intake reading}$
Eq.13.3, p. 335	Pressure Correction	$\text{Pressure correction} = \frac{\text{lift + total intake friction loss}}{2.3}$
Eq.13.4, p. 336	Net Pump Discharge Pressure at Draft	$NPDP_{Draft} = PDP + \text{intake pressure correction}$
Eq.14.1, p. 355	Total Number of Pumpers Needed	$\frac{\text{Relay distance}}{\text{Distance from Table 14.3}} + 1 = \text{Total number of Pumpers Needed}$
Eq.15.1, p. 379	The amount of water discharged	$Q = (K)(\sqrt{P})$

Appendix B

List of Constants and Other Information

Location	Reference	Equation
Ch.1, p. 6	Weight of one gallon of water	$\frac{62.4\ \text{lb/ft}^3}{7.48\ \text{gal/ft}^3} = 8.34$ pounds per gallon
Ch.3, p. 24	Alternate equation of Kinetic Energy, this case acceleration in water	$m = \frac{w}{g}$
Ch.3, p. 25	Acceleration of gravity	$g = 32.2\ \text{ft/sec}^2$
Ch.3, p. 26	Basic principles of Bernoulli's Equation	$(2.31)(P_1) + \frac{(v_1)^2}{(2)(g)} + h_1 = (2.31)(P_2) + \frac{(v_2)^2}{(2)(g)} + h_2$
Ch.5, p. 79	Constant	29.83 = A constant derived from the physical laws relating water velocity, pressure, and conversion factors
Ch.11, p. 238	Constant	29.7 = A constant
Ch.11, p. 250	Constant	1.57 = A constant
Ch.11, p. 252	Constant	0.0505 = A constant
Ch.11, p. 253	Constant	2 = A constant
Ch.12, p. 274	Steps for determining friction loss in siamesed lines	
Ch.12, p. 282	Constant	0.5 = A constant
Ch.12, p. 282	Constant	0.2 = A constant
Ch.12, p. 290	Steps for calculating the friction loss in an equal-length wyed hoseline assembly	
Ch.12, p. 297	Steps for calculating the friction loss in unequal-length wyed or manifold hoseline	

Glossary

A

Acceptance Test – A test conducted at the discretion of the purchaser to ensure that the apparatus meets the bid specifications once it is delivered.

Aerial Apparatus – An apparatus whose primary function is to provide access for firefighters to upper levels of a structure and to deploy elevated mater streams.

Apparatus Typing – Categorization of wildland apparatus by capability.

Attack Pumper – The pumper within a relay that supplies attack lines and appliances.

Automatic Drain Valve – A valve that drains piping when pressure in the pipe is relieved.

Automatic Nozzle – A nozzle that maintains a constant nozzle pressure of approximately 100 psi.

Average Daily Consumption – The average total amount of water used daily in a water distribution system over one year.

B

Bleeder Line (Bleeder Valve) – A valve that allows water to force air from a hose as the line fills; prevents air from entering the pump.

Branch Line – Piping that contains the sprinklers in an automatic sprinkler system.

British Thermal Unit (BTU) – The amount of heat required to raise the temperature of 1 pound of water 1°F.

Broken Stream – A fire stream that has been divided into coarse drops.

C

Casing – The main body of a pump; in centrifugal pumps, confines the water and converts its velocity to pressure.

Cellar Nozzle (Distributor) – A nozzle designed specifically for use on basement fires that are otherwise inaccessible for standard fire fighting operations.

Centrifugal Pump – A pump that uses centrifugal force to impart velocity to water and then transforms that velocity to pressure.

Changeover – The process of switching a two-stage centrifugal pump between pressure and volume pumping.

Check Valve – A valve that limits the flow of water to one direction.

Chemical Treatment – The use of chemicals to add or remove elements to or from water.

Chimney Nozzle – A nozzle used to attack chimney flue fires.

Circle System (Belt System) – A water supply system forming a loop that serves all customers on the system.

Clappered Siamese – A siamese that has a one-way clapper valve on each intake.

Coagulation – The introduction of chemicals to water that cause solid particles to bond forming larger pieces.

Coeffiecient of Discharge – A correction factor relating to the shape of the hydrant or nozzle discharge orifice.

Combination System – A water supply system using elements of both direct pumping and gravity systems.

Combined Agent Vehicle – A small initial attack aircraft rescue and fire fighting apparatus.

Constant Pressure Relay Method – A relay operation that establishes the maximum flow available from a particular relay setup by using a constant pressure in the system.

Cross Main – Piping that carries water from a feed main to branch lines in an automatic sprinkler system.

D

Dead-end Hydrant – A fire hydrant that receives water from only one direction.

Deluge Sprinkler System – A sprinkler system in which water discharges through all sprinklers simultaneously.

Diffuser – A device used to reduce a flowing hydrant stream's pressure.

Direct Pumping System – A water supply system using one or more pumps to move water from the primary source through the treatment processes.

Distributor – A smaller pipe serving individual fire hydrants and blocks of customers throughout a grid water supply system.

Double-Acting Piston Pump – A piston pump that can both receive and discharge water on each stroke.

Dry Automatic Standpipe System – A standpipe system in which a standpipe contains air under pressure that holds the water beneath a dry-pipe valve.

Dry-pipe Sprinkler System – A sprinkler system in which air under pressure replaces the water in the pipes above the sprinkler valve in the riser.

E

Energy – The ability to do work.

F

Feed Main – Large pipes that supply water from the riser to cross mains in an automatic sprinkler system.

Filtration – The passage of water through filters that catch and remove suspended matter.

Fire Department Connection (FDC) (Siamese Connection) – A pipe fitting to which a fire department pumper can connect hose and pump additional water into an occupancy's fixed fire protection system.

Fire Flow Test – The procedure that determines the water flow available for fire fighting.

Fire Stream – A stream of plain water used for fire fighting purposes.

Flow Hydrant – The hydrant that is opened and flowed to create a pressure drop at a test hydrant.

Flowmeter – An electronic gauge that indicates water flow in gallons per minute.

Flow Pressure (Velocity Pressure) – That forward velocity pressure created at a discharge opening while water is flowing.

Fog Stream (spray stream) – A patterned fire stream composed of many individual droplets of water.

Force – The simple measure of weight.

Friction Loss – That part of the total pressure lost while forcing water through pipe, fittings, fire hose, and adapters.

G

Gated Wye – A wye that is equipped with valves.

Gravity System – A water supply system whose primary water source is at a higher elevation than the distribution system.

Grid System – A water supply system of interlooped pipes connected at standard intervals.

H

Handline Nozzle – Any nozzle designed primarily for attachment to a single, moveable, hoseline and manual control and operation by one or more firefighters.

Head – Pressure expressed in units of feet of water instead of pounds per square inch.

Head Pressure – The amount of pressure created by the height of a column of water.

Hose Tender – An apparatus designed specifically to carry large amounts of hose for use in relay operations.

Hydraulic Calculation Method – The use of mathematical calculations to determine the required piping and other design factors for an automatic sprinkler system.

Hydrokinetics – The study of water in motion.

Hydrostatic Test – A test that determines whether the pump and pump piping can withstand normal operating pressures.

Hydrostatics – The study of water at rest.

I

Impeller – A revolving disk that throws water toward the outer edge of a centrifugal pump's casing.

Indicating Butterfly Valve – A butterfly valve with a paddle indicator or pointer arrow that shows the valve's position.

Initial Attack Fire Apparatus – An apparatus similar to a pumper but smaller and equipped with smaller pump and water tank.

Intake Pressure Relief Valve (relay relief valve) – A valve on the intake side of the pump that reduces possibility of water hammer or other significant rises in intake pressure.

J

Joule - A unit of work (1 calorie = 4.19 joules).

K

Kinetic Energy – Work being done by a substance in motion.

L

Lantern Ring – A spacer that provides cooling and lubrication between packing rings and the pump shaft.

Latent Heat of Vaporization - The amount of heat water can absorb when it changes from a liquid to a vapor.

Leader Line Wye – A smaller wye used to break one medium-diameter supply hose into two 1 3/4-inch or smaller attack hoses.

M

Major Fire Fighting Vehicle – An aircraft rescue and fire fighting apparatus with pumping capacity up to 2,000 gpm and tank capacity up to 6,000 gallons.

Manifold – A hose appliance used to divide one large-diameter supply line into smaller lines or to join several small lines into one large line.

Manual Dry Standpipe – A standpipe system that does not have an attached permanent water supply; must be charged and supplied by a fire department pumper attached to the fire department connection.

Manual Wet Pipe Standpipe – A wet standpipe system attached to a small water supply source; a fire department pumper must charge the fire department connection for the system to operate.

Master Stream Nozzle – A nozzle designed to be located and operated from a specific location and to flow larger volumes of water than firefighters could control manually.

Maximum Daily Consumption – The greatest total amount of water used during any 24-hour interval within three-years.

Maximum Distance Relay Method – A relay operation in which the volume of water to be flowed is predetermined and the pumpers are spaced at the maximum distance they can pump that amount of water through the planned hose lay.

Midipumper – A large initial attack apparatus, usually mounted on chassis over 12,000 pounds gross vehicle weight.

Minipumper – A small initial attack apparatus, usually mounted on 1-ton or 1 1/2-ton chassis.

Mobile Water Supply Apparatus (Tanker or Tender) – An apparatus with large water tank for supplying extended fire fighting operations in rural areas.

N

Necessary Fire Flow (NFF) – See Required Fire Flow.

Needed Fire Flow (NFF) – See Required Fire Flow

Net Pump Discharge Pressure – The total work done by the pump to get the water into, through, and out of the pump.

Normal Operating Pressure – Pressure in a water distribution system during normal consumption demands.

Nozzle Reaction – The counterforce that pushes back on firefighters handling the hoseline.

Nurse Tanker – A mobile water supply apparatus used as a reservoir to directly supply water to attack pumpers.

O

Outside Stem, and Yoke (OS&Y) Valve – A valve that has a yoke on the outside with a threaded stem that opens and closes the gate.

P

Packing Gland – A device that pushes packing material into the stuffing box where the pump shaft passes through the pump casing.

Packing Ring –A material used to create a seal where the pump shaft passes through the pump casing.

Peak Hourly Consumption – The maximum amount of water used in any one-hour interval over the course of a day.

Pipe Schedule Design – A method of sizing pipes in an automatic sprinkler system based upon the number of sprinklers supplied.

Piston Assembly Governor – A governor that reduces or increases engine speed under the control of a rod connected to a piston in a water chamber.

Pitot Gauge – A device used to measure a flowing stream's velocity pressure.

Positive Displacement Pump – A pump in which each positive action forces a specific amount of water and/or air from the pump body.

Post Indicator Valve (PIV) – A valve whose valve stem is inside a hollow metal post attached to the valve housing.

Potential Energy – Stored energy that can perform work once it is released.

Pounds per Square Inch Absolute – Pressure above a perfect vacuum.

Pounds per Square Inch Gauge – Pressure in addition to atmospheric pressure.

Power Take-Off (PTO) – Gear on the chassis transmission that transfers power to auxiliary equipment.

Preaction Sprinkler System – A sprinkler system that has a deluge-type valve and fire detection devices along with a dry-pipe system's closed sprinklers.

Preservice Test – A test performed before the purchasing fire department accepts the apparatus.

Pressure – Force per unit area.

Pressure Control System Test – A test conducted to ensure that dangerously high pressures do not build in the pump when discharge lines are close and the pump remains engaged.

Pressure Governor – A device that regulates centrifugal pump pressure by adjusting engine throttle.

Pressure Relief Valve – A valve intended to reduce the possibility of water hammer when valves/nozzles are closed too quickly.

Primary Feeder – The largest pipe in a grid water system.

Primer – A device that creates a partial vacuum to enable a centrifugal pump to begin drafting water from a static source.

Pump-and-Roll – The ability to pump water while moving.

Pump Certification Tests – A series of tests to determine if the pump and pump piping assembly operates properly.

Pumper – An apparatus whose main purpose is to provide water at an adequate pressure to produce an effective fire stream from a nozzle.

Pumper-Tanker – A mobile water supply apparatus equipped with large fire pump for use as attack apparatus.

Q

Quint – An apparatus equipped with an aerial device, ground ladders, fire pump, water tank, and fire hose.

R

Rapid Intervention Vehicle (RIV) – An aircraft rescue and fire fighting apparatus with pumping capacity of 1,250 gpm or less and tank capacity of no more than 1,500 gallons.

Relay – A pumper operation in which a pumper at the water supply source pumps water through one or more hoselines to the next pumper in line.

Relay Pumper (In-line Pumper) – The pumper within a relay that receives water from the source pumper or from another relay pumper.

Rescue Pumper – An apparatus that carries all the standard engine company equipment and a larger than standard amount of rescue and extrication equipment.

Required Discharge Pressure – The appropriate pressure for a hose lay or master stream device.

Required Fire Flow – The calculated amount of water needed to extinguish a fire in a given occupancy.

Required Fire Flow (Needed or Necessary Fire Flow) – The calculated amount of water needed to extinguish a fire in a given occupancy.

Residual Pressure – That part of total available pressure not used to overcome friction loss or gravity while forcing water through pipe, fittings, fire hose, and adapters.

Response Time Index (RTI) – The relative speed of a sprinkler's operation.

Riser, Automatic Sprinkler System – Vertical piping that extends up from the sprinkler valve to the first horizontal feed main.

Road Test – A manufacturer's test to ensure that a vehicle will operate safely once it is fully loaded with equipment and personnel and placed into service.

Rotary Gear Pump – A pump consisting of two tightly meshed gears that rotate inside a watertight case.

Rotary Vane Pump – A pump that contains a rotor mounted off-center inside the pump casing.

S

Secondary Feeder – An intermediate-sized pipe reinforcing grids within the loops of the primary feeder system.

Sedimentation – The use of gravity to remove particles from water.

Semiautomatic Dry Standpipe – Dry standpipe system that requires operating a remote control activation device to allow water into the piping.

Service Test – A test performed after a fire apparatus is placed into service within the department.

Siamese – A hose appliance that combines two or more hoselines into a single line.

Single-Acting Piston Pump – A piston pump in which every forward stroke causes water to be discharged and every return stroke causes the pump to fill with water again.

Single-Stage Centrifugal Pump – A centrifugal pump with only one impeller.

Solid Stream – A fire stream produced from a fixed-orifice, smoothbore nozzle.

Source Pumper (Supply Pumper) – The pumper connected to the water supply at the beginning of a relay operation.

Specific Heat – The capacity of a substance to absorb heat.

Speed – The rate of travel regardless of direction.

Sprinkler – The portion of an automatic sprinkler system that delivers the water to the fire area.

Stacked Tips – A series of threaded nozzle tips whose diameters decrease as they extend from the play.

Standpipe System – A water distribution system in a building with hose discharge connections that facilitate manual fire attack in remote parts of the structure.

Static Pressure – Stored potential energy available to force water through pipe, fittings, fire hose, and adapters.

Straight Stream – A fire stream discharged from a fog-stream nozzle to emulate a solid stream.

T

Tachometer – A gauge that records engine speed in revolutions per minute.

Tank Fill Line (Pump-to-Tank Line) – A line that connects the discharge side of the pump to the apparatus water tank allowing the driver/operator to fill the tank without making additional connections when supplying the pump from an external supply source.

Test Hydrant – The hydrant for which a test is being conducted.

Total Energy – The sum of potential energy and kinetic energy.

Trash Line – A short 1 1/2-inch preconnected line used to replace a booster reel.

Transfer Case – A split-shaft gear case that supplies power to midship-transfer-driven pumps.

Tree System – A water distribution system with a central primary supply pipe that feeds branch distribution lines.

Two-Stage Centrifugal Pump – A centrifugal pump with two impellers.

V

Vacuum – Pressure less than atmospheric pressure.

Velocity – The rate of motion of a particle in a specific direction.

Volute – A water passage in a centrifugal pump; gradually increases in size as it nears the discharge outlet.

W

Wall Post Indictor Valve (WPIV) – A valve similar to a post indicator valve and extended through the building wall.

Water – A chemical compound that results when two parts of hydrogen (H) combine with one part oxygen (O) to form H_2O.

Water Curtain Nozzle – A nozzle that produces a fan-shaped stream that acts as a water curtain between a fire and a combustible material or firefighters.

Water Shuttle – An operation in which tankers carry water from a water supply source to an incident scene and dump their loads into portable tanks.

Water Thief – A hose appliance with one intake and three gated discharges, one that is the same size as the intake and two that are smaller than the intake.

Wear Ring (Clearance Ring) – A replaceable ring that can be inserted into a centrifugal pump casing to maintain the desired spacing between the hub of the impeller and the casing.

Wet Automatic Standpipe System – A standpipe system in which the piping is filled with water at all times.

Wet-pipe Sprinkler System – A sprinkler system in which the entire piping system is filled with water at all times.

Work – The product of force multiplied by distance.

Wye – A hose appliance that divides one hoseline into two separate hoselines of equal size.

Chapter Review Question Answers

Chapter 1

1. Steam
2. 7.48 gal/ft^3
3. 8.34 lbs/gallon
4. 62.4 lbs/ft^3
5. The specific heat of any substance is the ratio between the amount of heat needed to raise the temperature of a specified quantity of a material and the amount of heat needed to raise the temperature of an identical quantity of water by the same number of degrees.
6. The amount of heat water can absorb when it changes from a liquid to a vapor state.
7. 2,335,000 BTUs/minute
8. The answers should include at least three of the following:
   - Water has a greater heat-absorbing capacity than other common extinguishing agents.
   - A relatively large amount of heat is required to change water into steam. This means that more heat will be absorbed from the fire.
   - The greater the surface area of water exposed, the more rapidly heat is absorbed. The surface area exposed can be expanded by using fog streams or deflecting solid streams off objects.
   - As a rule of thumb, water converted into steam occupies 1,700 times its original volume.
   - Water is cheap and readily available in most jurisdictions.
   - Water is incompressible and noncombustible.
9. The answers should include at least three of the following:
   - Water has a considerable amount of surface tension.
   - Violent reactions can occur when water is applied to certain water-reactive materials, whether they are on fire or not.
   - When water freezes at a temperature that is common in many jurisdictions.
   - It does not readily adhere to vertical surfaces.
   - The water that a firefighter uses conducts electricity because it contains chemicals, minerals, and organisms.

Chapter 2

1. The study of fluids, in this case water, while it is at rest.
2. Pascal's Law
3. Head is pressure expressed in units of feet of water instead of pounds per square inch (psi).
4. Potential energy is stored energy that has the ability to perform work once it is released.
5. 65 psi
6. 39 psi
7. There will be no change in the pressure at the bottom of the tank. See Principle #5.
8. 251 psi
9. 30 psi

Chapter 3

1. The science that involves study of the characteristics and physical properties of fluids (in our case, water) in motion.
2. Potential energy and kinetic energy.
3. Total energy within a system will remain constant. Any change in potential energy will be matched by a corresponding change of kinetic energy.
4. In a steady flow without friction, the sum of the velocity head, pressure head, and elevation head is constant for any incompressible fluid particle throughout its course.
5. a. 2. Head Pressure
   b. 6. Flow (Velocity) Pressure
   c. 3. Normal Operating Pressure
   d. 1. Atmospheric Pressure
   e. 5. Static Pressure
   f. 4. Residual Pressure
6. Velocity = 28.8 fps; flow = 0.98 ft^3/sec or 440 gpm
7. Velocity = 7.81 fps; flow = 2.725 ft^3/sec or 1,223 gpm
8. (0.213 psi/ft)(1,800 feet) = 384 psi
9. (0.048 psi/ft)(200 feet) = 9.5 psi

Chapter 4

1. A water supply source; Water treatment facilities; A mechanism for forcing water through the system; A system of piping to transport the water through the community.

2. These systems are typically high volume, low pressure systems that supply untreated water for major fire fighting operations. In some cases these systems are not pressurized until it is determined they are needed.
3. a. 2. Maximum Daily Consumption (MDC)
   b. 1. Average Daily Consumption (ADC)
   c. 3. Peak Hourly Consumption (PHC)
4. Surface water supply sources include rivers, streams, lakes, reservoirs, coastal waters, and ponds.
5. In general, surface water supply sources tend to:
   - Have softer water (less minerals) than ground water
   - Have more suspended solids and color than ground water
   - Have more bacterial contaminants that ground water
   - Be more rapidly depleted during periods of low rainfall.
6. In general, ground water supply sources tend to:
   - Have harder water (more minerals) than surface water
   - Be clearer (have less color) than surface water
   - Have less bacteria than surface water and require less treatment
   - Not as quickly affected by periods of decreased rainfall as surface water.
7. 1. Sedimentation
   2. Filtration
   3. Coagulation
   4. Chemical treatment
8. The main concerns regarding treatment facilities are that a maintenance error, natural disaster, loss of power supply, or fire could disable the pumping station(s) or severely hamper the purification process.
9. - *Gravity systems:* use a primary water source located at a higher elevation than the distribution system. The head pressure created by the higher elevation provides the energy to move the water through the entire system.
   - *Direct pumping systems:* are most commonly found in jurisdictions where the system's water supply source is located at the same or lower elevation than the system it supplies. Direct pumping systems use one or more pumps that take water from the primary source and discharge it through the treatment processes.
   - *Combination systems:* utilize elements of both the direct pumping and gravity systems.
10. a. 2. Grid system
    b. 1. Circle or belt system
    c. 3. Tree system

11. a. 3. Distributors
    b. 1. Primary feeders
    c. 2. Secondary feeders
12. Indicating and nonindicating.
13. Wet barrel and dry barrel.
14. - The property owner has control over the water supply source.
    - Either of the systems (fire protection or domestic/industrial) are unaffected by service interruptions to the other system.

Chapter 5

1. - To determine whether a water supply system is operating as designed.
   - To determine whether a properly functioning water supply system is capable of providing an adequate flow of water for the target hazards in the area of the test.
2. The water department.
3. Any of the following may be included:
   - A thread gauging device
   - Can of lubrication oil
   - Small, flat brush
   - Gate valve key
   - A large pail
   - Hydrant wrench
   - A static pressure gauge
   - A pitot tube and gauge
4. Any of the following may be included
   - Check for any obstructions near the hydrant, such as sign posts, utility poles, shrubbery, or fences.
   - Check the direction of the hydrant outlet(s) to ensure that they face the proper direction and that there is clearance between the outlet and the surrounding ground. The clearance between the bottom of the butt and the grade should be at least 15 inches.
   - Check for mechanical damage to the hydrant, such as dented outlets, damaged discharge threads, or rounded (stripped) stem nuts.
   - Check the condition of the paint for rust or corrosion.
   - Check for foreign objects that may have been inserted into the hydrant discharges.

- Check water flow by having the hydrant fully opened and then checking its ability to drain once closed. Drainage may be checked visually or by placing the palm of your hand over the discharge opening to feel for a slight sucking action.
- Make sure no vehicles or other property will be damaged or pedestrians injured by flowing water.
- Make sure that the caps on unused hydrant discharges are tight.

5. The test hydrant must be located between the flow hydrant and the water supply source. In other words, the flow hydrant should be downstream from the test hydrant.
6. The 2½-inch discharges.
7. 10 percent
8. 1,073 gpm
9. 940 gpm
10. 1,371 gpm is available at 20 psi
11. 4,911 gpm is available at 20 psi

Chapter 6

1. The Iowa State Formula; The National Fire Academy Formula; The Insurance Services Office (ISO) Formula
2. 1. One gallon of water will produce, with a margin of safety, 200 cubic feet of steam.
   2. One gallon of water will absorb, with a margin of safety, all the heat that can be produced with the oxygen available in 200 cubic feet of normal air.
3. 135 gallons of water
4. 648 gpm
5. 8,000 gpm
6. 2,333 gpm
7. 1,972 gpm
8. 2,268 gpm
9. Pipe Schedule Method and Hydraulic Calculation Method
10. Class I standpipe systems
11. 500 gpm minimum; 1,250 gpm maximum

Chapter 7

1. NFPA® 1901, Standard for Automotive Fire Apparatus
2. Fire Apparatus Manufacturer's Association (FAMA)
3. 750 gpm

4. A triple-combination pumper is an apparatus that carries hose, water, and a fire pump. This is primarily a West-Coast term.
5. 500 gallons
6. Minipumpers and midipumpers
7. As a nurse tanker and in a water shuttle operation
8. Any of the following are correct: terrain, bridge weight limits, budgetary constraints, compatibility with mutual aid tankers
9. A quint is a fire apparatus that is equipped with an aerial device, ground ladders, fire pump, water tank, and fire hose.
10. The primary drawback is that the pump, pump panel, and the water tank consume valuable compartment space on the vehicle.
11. - Major fire fighting vehicles
    - Rapid intervention vehicles (RIV)
    - Combined agent vehicles

Chapter 8

1. Pressure applied on a confined liquid from an external source will be transmitted equally in all directions throughout the liquid without a reduction in magnitude.
2. Piston pumps
3. Any three of the following:
    - They were subject to excessive wear on their parts.
    - They were maintenance intensive.
    - They created a pulsating fire stream.
    - Even small debris in the water increased damage to the pump.
4. Rotary vane and rotary gear
5. They are driven by either a small electric motor or through a clutch that extends off the apparatus drive shaft.
6. The centrifugal pump is considered a nonpositive displacement pump because it does not pump a definite amount of water with each revolution.
7. The centrifugal pump consists of two primary parts: the impeller and the casing.
8. Changeover
9. - Amount of water being discharged
    - Speed at which the impeller is turning
    - Pressure of water when it enters the pump from a pressurized source (hydrant, relay, etc.).

10. The single-stage pump has one impeller; the two-stage pump has two impellers mounted on a single shaft.
11. 75 psi
12. The addition of a second engine to the apparatus creates additional maintenance work for the driver and departmental mechanics. If the engine driving the pump fails to work, the apparatus may arrive at the fire scene but be unable to pump water to fight the fire.
13. The PTO connection receives its power from an idler gear in the chassis transmission.
14. Electric generators, hydraulic rescue tool power systems, and large-scale air compressors may also be powered by a PTO arrangement.
15. The primary differences between the two are the location of the pump on the apparatus (front vs. midship) and the location of the connection for the driveshaft that powers the pump. On PTO pumps the driveshaft is connected to an idler gear in the chassis transmission. On front-mount pumps the driveshaft is connected to a front PTO connection on the engine's crankshaft.
16. Power is supplied to the pump driver by a split-shaft gear case called the transfer case, which is located in the drive line between the transmission and the rear axle.
17. Midship transfer-driven fire pumps may not be used for pump-and-roll operations.
18. The two basic types of pressure control devices used on modern fire apparatus are relief valves and pressure governors.
19. The most common design is one that uses an adjustable spring-controlled pilot valve.
20. The basic operating principle of the pressure governor is that it regulates the power output of the engine to match pump discharge requirements. When the pressure in the discharge piping of the pump exceeds the pressure necessary to maintain safe fire streams, the governor reduces the excess pressure by slowing the engine speed.
21.
    - Vacuum
    - Exhaust
    - Positive displacement pumps.
22. There are two types of auxiliary coolers commonly found on fire apparatus: the marine type and the immersion type.
23. Through piping that connects the pump and the apparatus water tank and piping that is used to connect the pump to an external water supply.
24. The ball valve.
25. When pumping from a draft, the master intake gauge provides an indication of the amount of vacuum present at the intake of the pump during priming or when the pump is operating from draft.
26. Paddelwheel and spring probe

Chapter 9

1. Preservice tests include the manufacturer's test, pump certification test, and acceptance tests.
2. NFPA® 1901, Standard for Automotive Fire Apparatus and NFPA® 1906, Standard for Wildland Fire Apparatus.
3. The manufacturer and/or representatives of Underwriters Laboratories (UL) typically perform preservice tests.
4. The road test and the hydrostatic test.
5. 1. The apparatus must accelerate to 35 mph from a standing start within 25 seconds. This test must consist of two runs in opposite directions over the same surface.
   2. The apparatus must achieve a minimum top speed of 50 mph. This requirement may be dropped for specialized wildland apparatus not designed to operate on public roadways.
   3. The apparatus must come to a full stop from 20 mph within 35 feet.
   4. The apparatus parking brake must conform to the specifications listed by the braking system manufacturer.
6. NFPA® 1901 requires the pump to be hydrostatically tested at 250 psi for 3 minutes.
7. The water must be at least 4 feet deep and no more than 10 feet below the pump.
8. 3 hours
9. 100 percent at 150 psi, 70 percent at 200 psi, 50 percent at 250 psi
10. 165 psi
11. NFPA® 1901 states that the tank-to-pump piping should be sized so that pumpers with a capacity of 500 gpm or less can flow 250 gpm from their booster tank and pumpers with capacities greater than 500 gpm can flow at least 500 gpm.
12. NFPA® 1911, Standard for Service Tests of Fire Pump Systems on Fire Apparatus.
13. 6.78 or 7 psi
14. 50 minutes
15. 10 psi

Chapter 10

1. Solid streams, fog or spray streams, and broken streams.
2. The nozzle pressure, design, adjustment, and the condition of the nozzle influence the condition of a fire stream.
3. - A water supply source
   - A means to impart sufficient pressure on the water

- A means to transport the water from the source to the desired point of application
- Personnel trained in using the first three elements effectively

4. Wye
5. Siamese
6. 1. The stream has not lost continuity by breaking into showers of spray.
   2. The stream appears to shoot nine-tenths of the whole volume of water inside a circle 15 inches in diameter and three-quarters of it inside a 10-inch circle, as nearly as can be judged by the eye.
   3. The stream is stiff enough to attain the required height and/or reach, even though a moderate breeze is blowing.
7. Four
8. 32 degrees
9. 50 psi
10. A solid stream is a solid stream of water discharged from a smoothbore nozzle. A straight stream is meant to emulate a solid stream, but it is discharged from a fog stream nozzle.
11. - Gravity
    - Water velocity
    - Fire stream pattern
    - Water droplet friction with air
    - Wind
12. 100 psi
13. These two factors are fire stream pattern and water droplet friction with air.
14. Most broken stream nozzles form the stream by directing the water in the hose through a series of small holes in the end or sides of the nozzle.
15. Broken streams

Chapter 11

1. 350 gpm
2. The nozzle tip diameter should be no more than one-half the diameter of the hoseline to which it is attached.
3. 50 psi for handlines; 80 psi for master streams
4. 1,345 gpm
5. 82 gpm
6. Periphery deflection
7. 100 psi

8. This law states that for every action there is an equal and opposite reaction.
9. 60 to 120 feet per second
10. 207.3 pounds
11. 1,413 pounds
12. 267 pounds
13. 354.2 pounds
14. 63.1 pounds
15. 125 pounds

Chapter 12

1. Any additive that is introduced into the water, such as Class A foam concentrates or other wetting agents, will make water "slipperier" and reduce the friction loss in the system.
2. 1. It was based on old hose technology and newer hose does not have as much friction loss.

   2. It requires multiple steps and different formulas depending on how much water (±100 gpm) is being flowed.
3. $FL = 2Q^2 + Q$

   $FL = (2)(200/100)^2 + (200/100)$

   $FL = (2)(2)^2 + 2$

   $FL = (2)(4) + 2$

   $FL = 8 + 2$

   FL = 10 psi per 100 feet

   $FL_{total} = (10\text{ psi})(L)$

   $FL_{total} = (10\text{ psi})(300/100)$

   $FL_{total} = (10\text{ psi})(3)$

   $\mathbf{FL_{total} = 30\text{ psi}}$
4. $FL = 2Q^2 + Q$

   $FL = 2(150/100)^2 + (150/100)$

   $FL = 2(1.5)^2 + 1.5$

   $FL = (2)(2.25) + 1.5$

   $FL = 4.5 + 1.5$

   FL = 6 psi per 100 feet of 2½-inch hose (equivalent)

   In order to convert this figure to friction loss in 1 1¾-inch hose, locate the appropriate conversion factors in Table 12.1. For 1¾-inch hose we have the option of multiplying at above figure by 5.95 or dividing it by 0.16 (the reciprocal of 5.95):

   $FL_{1¾\text{-inch}}$ = (6) (5.95) = 35.7 or 36 psi per 100 feet of 1¾-inch hose

   —or—

   $FL_{1¾\text{-inch}}$ = 6÷0.16 = 37.5 or 38 psi per 100 feet of 1¾-inch hose

$FL_{total} = (FL_{1¾\text{-inch}} \text{ per 100 feet})(L)$

$FL_{total} = (36 \text{ or } 38 \text{ psi})(150/100)$

$FL_{total} = (36 \text{ or } 38 \text{ psi})(1.5)$

$FL_{total} = 54$ psi or 57 psi (depending on whether you divided or multiplied)

5. $FL = CQ^2L$

   $FL = (0.2)(600/100)^2(800/100)$

   $FL = (0.2)(6)^2(8)$

   $FL = (0.2)(36)(8)$

   $FL = 57.6$ or 58 psi

6. $FL_3 = CQ^2L$

   $FL_3 = (0.8)(225/100)^2(200/100)$

   $FL_3 = (0.8)(2.25)^2(2)$

   $FL_3 = (0.8)(5.0625)(2)$

   $FL_3 = 8.1$ or 8 psi

   $FL_{2½} = CQ^2L$

   $FL_{2½} = (2)(225/100)^2(150/100)$

   $FL_{2½} = (2)(2.25)^2(1.5)$

   $FL_{2½} = (2)(5.0625)(1.5)$

   $FL_{2½} = 15.2$ or 15 psi

   $FL_{Total} = FL_{2½} + FL_3$

   $FL_{Total} = 8 + 15$

   $FL_{Total} = 23$ psi

7. $FL = CQ^2L$

   $FL = (0.5)(500/100)^2(600/100)$

   $FL = (0.5)(5)^2(6)$

   $FL = (0.5)(25)(6)$

   $FL = 75$ psi

8. $GPM = 29.7\,(d)^2(\sqrt{NP})$

   $GPM = (29.7)(1.5)^2(\sqrt{80})$

   $GPM = (29.7)(2.25)(8.94)$

   $GPM = 598$ gpm

NOTE: Because this is greater than 350 gpm, appliance friction losses for the Siamese device and master stream nozzle will need to be factored in later.

   $FL_{3\text{-inch}} = CQ^2L$

   $FL_{3\text{-inch}} = (0.677)(598/100)^2(50/100)$

   $FL_{3\text{-inch}} = (0.677)(5.98)^2(0.5)$

   $FL_{3\text{-inch}} = (0.677)(35.76)(0.5)$

   $FL_{3\text{-inch}} = 12.1$ or 12 psi

$FL_{siamesed\ lines} = CQ^2L$

$FL_{siamesed\ lines} = (0.2)(598/100)^2(400/100)$

$FL_{siamesed\ lines} = (0.2)(5.98)^2(4)$

$FL_{siamesed\ lines} = (0.2)(35.76)(4)$

$FL_{siamesed\ lines} = 28.6$ or 27 psi

$FL_{Total} = FL_3$-inch + $FL_{siamesed\ lines}$ + Appl. Loss (Siamese) + Appl. Loss (Master Stream)

FL_{Total} = 12 psi + 27 psi + 10 psi + 25 psi

FL_{Total} = 74 psi

9. EP = 0.5H —or— EP = H/2

   EP = (0.5)(40) —or— EP = 40/2

   EP = 20 psi **EP = 20 psi**

10. EP = 5 psi x (number of stories –1)

    EP = 5 psi x (9 –1)

    EP = (5)(8)

    EP = 40 psi

11. $Q = 29.7(d)^2(\sqrt{NP})$

    $Q = (29.7)(1)^2(\sqrt{50})$

    $Q = (29.7)(1)(7.07)$

    Q = 210 gpm

    $FL_2 = CQ^2L$

    $FL_2 = (8)(210/100)^2(200/100)$

    $FL_2 = (8)(2.10)^2(2)$

    $FL_2 = (8)(4.41)(2)$

    $FL_2 = 70.6$ or 71 psi

    EP = 5 psi x (number of stories – 1)

    EP = 5 psi x (12 – 1)

    EP = (5)(11)

    EP = 55 psi

    $FL_{total} = FL^2$ + EP + Standpipe System Loss

    FL_{total} = 71 psi + 55 psi + 25 psi

    FL_{total} = 151 psi

12. *Attack Line 1 (1¾-inch)*

    $FL_{1¾} = CQ^2L$

    $FL_{1¾} = (15.5)(175/100)^2(200/100)$

    $FL_{1¾} = (15.5)(1.75)^2(2)$

    $FL_{1¾} = (15.5)(3.0625)(2)$

    $FL_{1¾}$ = 95 psi

    Attack Line 2 (2-inch)

    $FL_2 = CQ^2 L$

$FL_2 = (8)(225/100)^2(200/100)$

$FL_2 = (8)(2.25)^2(2)$

$FL_2 = (8)(5.0625)(2)$

$FL_2 = 81$ psi

EP = (5 psi)(No. of floors – 1)

EP = (5 psi)(3 – 1)

EP = (5 psi)(2)

EP = 10 psi

$FL_{2\ total} = FL_2 + EP$

$FL_{2\ total} = 81 + 10$

$\mathbf{FL_{2\ total} = 91}$ **psi**

Supply Line

$FL_{2½} = C(Q_{total})^2 L$

$$FL_{2½} = (2)\frac{(175+225)^2}{100} \text{ x } \frac{200}{100}$$

$FL_{2½} = (2)(400/100)^2(2)$

$FL_{2½} = (2)(4)^2(2)$

$FL_{2½} = (2)(16)(2)$

$\mathbf{FL_{2½} = 64}$ **psi**

Total Pressure Loss

The total pressure loss in the system is based on the highest loss of the two attack lines, which in this case would be Line 1, and the friction loss in the supply line. Because the flow rate was 350 gpm, adding in the appliance loss is also required.

$FL_{total} = FL_{1¾} + FL_{2½}$ + Appliance loss

$FL_{total} = 95 + 64 + 10$

$\mathbf{FL_{total} = 169}$ **psi**

13. Since no coefficient exits for dual 4-inch hose lines, calculate the friction loss using the flow through each line and the average of the two lengths of hose.

$FL_4 = CQ^2L_{ave}$

$$FL_4 = (0.2)\left(\frac{750}{100}\right)^2 \frac{[(350+400)\div 2]}{100}$$

$FL_4 = (0.2)(7.5)^2(375/100)$

$FL_4 = (0.2)(7.5)^2(3.75)$

$FL_4 = (0.2)(56.25)(3.75)$

$FL_4 = 42$ psi

EP = (0.5 psi/ft)(H)

EP = (0.5)(90)

EP = 45 psi

Total Pressure Loss

$FL_{total} = FL_4 + EP + \text{Appliance Loss}$

$FL_{total} = 42 + 45 + 25$

$FL_{total} = 112$ psi

14. The hose is kinked, a midline valve (such as a gated wye) may be partially closed, or a vehicle tire is parked on top of the hoseline.
15. 186 psi
16. (3)(6) = 18 psi loss/100 feet

    (18)(400/100) = **72 psi loss in the hose assembly**

Chapter 13

1. Pump discharge pressure is the pressure at which a fire pump must be set to overcome the total pressure loss in a hose layout and supply the correct nozzle pressure for optimum operation.
2. $FL = CQ^2L$

   $FL = (8)(175/100)^2(250/100)$

   $FL = (8)(1.75)^2(2.5)$

   $FL = (8)(3.0625)(2.5)$

   FL = 61.25 or 61 psi

   PDP = FL + NP

   PDP = 61 + 100

   PDP = 161 psi
3. $GPM = (29.7)(d)^2(\sqrt{NP})$

   $GPM = (29.7)(0.875)^2(\sqrt{50})$

   $GPM = (29.7)(0.7656)(\sqrt{7.07})$

   GPM = 161 gpm

   $FL = CQ^2L$

   $FL = (15.5)(161/100)^2(200/100)$

   $FL = (15.5)(1.61)^2(2)$

   FL = (15.5)(2.5921)(2)

   FL = 80.36 or 80 psi

   EP = (5 psi)(# of floors - 1) (Assume ground floor is Floor 0)

   EP = (5 psi)(0 – 1)

   EP = (5 psi)(-1)

   EP = -5 psi

   TPL = FL + EP

   TPL = 80 + (-5)

   TPL = 75 psi

   PDP = TPL + NP

   PDP = 75 + 50

   PDP = 125 psi

4. $FL_{2½\text{-inch}} = CQ^2L$

$FL_{2½\text{-inch}} = (2)(250/100)^2(150/100)$

$FL_{2½\text{-inch}} = (2)(2.5)^2(1.5)$

$FL_{2½\text{-inch}} = (2)(6.25)(1.5)$

$FL_{2½\text{-inch}} = 18.75$ or 19 psi

$FL_{4\text{-inch}} = C(Q_{total})^2L$

$FL_{4\text{-inch}} = (0.2)\ \frac{(250+250)^2}{100}\left(\frac{400}{100}\right)$

$FL_{4\text{-inch}} = (0.2)(500/100)^2(400/100)$

$FL_{4\text{-inch}} = (0.2)(5)^2(4)$

$FL_{4\text{-inch}} = (0.2)(25)(4)$

$FL_{4\text{-inch}} = 20$ psi

$TPL = FL_{2½\text{-inch}} + FL_{4\text{-inch}} + \text{Appliance Loss}$

$TPL = 19 + 20 + 25$

$TPL = 64$ psi

$PDP = TPL + NP$

$PDP = 64 + 100$

PDP = 164 psi

5. $GPM_{2½\text{-inch}} = (29.7)(d)^2(\sqrt{NP})$

$GPM_{2½\text{-inch}} = (29.7)(1.125)2(\sqrt{50})$

$GPM_{2½\text{-inch}} = (29.7)(1.2656)(7.07)$

$GPM_{2½\text{-inch}} = 266$ gpm

$FL_{2½\text{-inch}} = CQ^2L$

$FL_{2½\text{-inch}} = (2)(266/100)^2(200/100)$

$FL_{2½\text{-inch}} = (2)(2.66)^2(2)$

$FL_{2½\text{-inch}} = (2)(7.08)(2)$

$FL_{2½\text{-inch}} = 28.3$ or 28 psi

$FL_{1¾\text{-inch}} = CQ^2L$

$FL_{1¾\text{-inch}} = (15.5)(125/100)^2(150/100)$

$FL_{1¾\text{-inch}} = (15.5)(1.25)^2(1.5)$

$FL_{1¾\text{-inch}} = (15.5)(1.5625)(1.5)$

$FL_{1¾\text{-inch}} = 36.3$ or 36 psi

$EP_{1¾\text{-inch}} = (5\ \text{psi/floor})(\#\ \text{of floors} - 1)$

$EP_{1¾\text{-inch}} = (5\ \text{psi/floor})(3 - 1)$

$EP_{1¾\text{-inch}} = (5\ \text{psi/floor})(2)$

$EP_{1¾\text{-inch}} = 10$ psi

$TPL_{1¾\text{-inch}} = FL_{1¾\text{-inch}} + EP_{1¾\text{-inch}}$

$TPL_{1¾\text{-inch}} = 36 + 10$

$TPL_{1¾\text{-inch}} = 46$ psi

$FL_{3½\text{-inch}} = C(Q_{total})^2L$

$FL_{3½\text{-inch}} = (0.34)\left(\frac{266+125}{100}\right)^2\left(\frac{200}{100}\right)$

$FL_{3½\text{-inch}} = (0.34)(391/100)^2(200/100)$

$FL_{3½\text{-inch}} = (0.34)(3.91)^2(2)$

$FL_{3½\text{-inch}} = (0.34)(15.29)(2)$

$FL_{3½\text{-inch}} = 10.4$ or 10 psi

When determining which attack line should be used for calculating the PDP, one must consider both the pressure loss and the nozzle pressures for each line. The 2½-inch attack line will require 78 psi (28 + 50) and the 1¾-inch attack line will require 146 psi (46 + 100).

$PDP = FL_{3½\text{-inch}} + FL_{1¾\text{-inch}} + NP_{1¾\text{-inch}}$ + Appliance Loss

$PDP = 10 + 46 + 100 + 25$

PDP = 181 psi

6. $FL_{2½\text{-inch Siamese lines}} = CQ^2L$

   $FL_{2½\text{-inch Siamese lines}} = (0.22)(1{,}000/100)^2(250/100)$

   $FL_{2½\text{-inch Siamese lines}} = (0.22)(10)^2(2.5)$

   $FL_{2½\text{-inch Siamese lines}} = (0.22)(100)(2.5)$

   $FL_{2½\text{-inch Siamese lines}} = 55$ psi

   $TPL = FL_{2½\text{-inch Siamese lines}}$ + Appliance Loss

   $TPL = 55 + 25$

   $TPL = 80$ psi

   $PDP = TPL + NP$

   $PDP = 80 + 100$

   PDP = 180 psi

7. $FL_{1½\text{-inch}} = CQ^2L$

   $FL_{1½\text{-inch}} = (24)(95/100)^2(150/100)$

   $FL_{1½\text{-inch}} = (24)(0.95)^2(1.5)$

   $FL_{1½\text{-inch}} = (24)(0.9025)(1.5)$

   $FL_{1½\text{-inch}} = 32.49$ or 32 psi

   EP = (5 psi/floor)(# of floors –1)

   EP = (5 psi/floor)(7 – 1)

   EP = (5 psi/floor)(6)

   EP = 30 psi

$TPL = FL_{1½\text{-inch}} + EP + \text{Appliance Loss}$

$TPL = 32 + 30 + 25$

$TPL = 87$ psi

$PDP = TPL + NP$

$PDP = 87 + 100$

PDP = 187 psi

8. $GPM = (29.7)(d)^2(\sqrt{NP})$

$GPM = (29.7)(1.625)^2(\sqrt{80})$

$GPM = (29.7)(2.64)(8.94)$

$GPM = 701$ gpm

$FL_{3\text{ w/ }3} = CQ^2L$

$FL_{3\text{ w/ }3} = (0.677)(701/100)^2(100/100)$

$FL_{3\text{ w/ }3} = (0.677)(7.01)^2(1)$

$FL_{3\text{ w/ }3} = (0.677)(49.14)(1)$

$FL_{3\text{ w/ }3} = 33.3$ or 33 psi

$FL_{\text{dual 3s}} = CQ^2L$

$FL_{\text{dual 3s}} = (0.2)(701/100)^2(300/100)$

$FL_{\text{dual 3s}} = (0.2)(7.01)^2(3)$

$FL_{\text{dual 3s}} = (0.2)(49.14)(3)$

$FL_{\text{dual 3s}} = 29.48$ or 29 psi

$EP = (0.5\text{ psi/ft})(H)$

$EP = (0.5\text{ psi/ft})(60\text{ feet})$

$EP = 30$ psi

$TPL = FL_{3\text{ w/ }3} + FL_{\text{dual 3s}} + EP + \text{Siamese Loss} + \text{Ladder Pipe Loss}$

$TPL = 33 + 29 + 30 + 10 + 25$

$TPL = 127$ psi

$PDP = TPL + NP$

$PDP = 127 + 80$

PDP = 207

9. $FL_{master\ stream} = CQ^2L$

$FL_{master\ stream} = (0.5)(500/100)^2(250/100)$

$FL_{master\ stream} = (0.5)(5)^2(2.5)$

$FL_{master\ stream} = (0.5)(25)(2.5)$

$FL_{master\ stream} = 31.25$ or 31 psi

$TPL_{master\ stream} = FL_{master\ stream} + \text{Appliance Loss}$

$TPL_{master\ stream} = 31 + 25$

$TPL_{master\ stream} = 56$ psi

$PDP_{master\ stream} = TPL_{master\ stream} + NP_{master\ stream}$

$PDP_{master\ stream} = 56 + 100$

$PDP_{master\ stream} = 156$ psi

$GPM_{Line\ 1} = (29.7)(d)^2(\sqrt{NP})$

$GPM_{Line\ 1} = (29.7)(1)^2(\sqrt{50})$

$GPM_{Line\ 1} = (29.7)(1)(7.07)$

$GPM_{Line\ 1} = 210$ gpm

$FL_{Line\ 1} = CQ^2L$

$FL_{Line\ 1} = (2)(210/100)^2(150/100)$

$FL_{Line\ 1} = (2)(2.1)^2(1.5)$

$FL_{Line\ 1} = (2)(4.41)(1.5)$

$FL_{Line\ 1} = 13.23$ or 13 psi

$PDP_{Line\ 1} = FL_{Line\ 1} + NP_{\ Line\ 1}$

$PDP_{Line\ 1} = 13 + 80$

$PDP_{Line\ 1} = 93$ psi

$FL_{Line\ 2} = CQ^2L$

$FL_{Line\ 2} = (8)(150/100)^2(200/100)$

$FL_{Line\ 2} = (8)(1.5)^2(2)$

$FL_{Line\ 2} = (8)(2.25)(2)$

$FL_{Line\ 2} = 36$ psi

$EP_{Line\ 2} = (0.5\ psi/ft)(H)$

$EP_{Line\ 2} = (0.5\ psi/ft)(40\ feet)$

$EP_{Line\ 2} = 20\ psi$

$PDP_{Line\ 2} = FL_{Line\ 2} + NP_{Line\ 1} + EP_{Line\ 2}$

$PDP_{Line\ 2} = 36 + 100 + 20$

$PDP_{Line\ 2} = 156\ psi$

$GPM_{Line\ 3} = (29.7)(d)^2(\sqrt{NP})$

$GPM_{Line\ 3} = (29.7)(0.875)^2(\sqrt{50})$

$GPM_{Line\ 3} = (29.7)(0.7656)(7.07)$

$GPM_{Line\ 3} = 161\ gpm$

$FL_{Line\ 3} = CQ^2L$

$FL_{Line\ 3} = (15.5)(161/100)^2(200/100)$

$FL_{Line\ 3} = (15.5)(1.61)^2(2)$

$FL_{Line\ 3} = (15.5)(2.59)(2)$

$FL_{Line\ 3} = 80.35$ or $80\ psi$

$EP_{Line\ 3} = (5\ psi/floor)(\#\ of\ floors - 1)$

$EP_{Line\ 3} = (5\ psi/floor)(2 - 1)$

$EP_{Line\ 3} = (5\ psi/floor)(1)$

$EP_{Line\ 3} = 5\ psi$

$PDP_{Line\ 3} = FL_{Line\ 3} + NP_{Line\ 3} + EP_{Line\ 3}$

$PDP_{Line\ 3} = 80 + 80 + 5$

$PDP_{Line\ 3} = 165\ psi$

$FL_{5\text{-}inch} = C(Q_{total})^2\ L$

$$FL_{5\text{-}inch} = (0.08)\ \frac{(210+150+161)^2}{100} \left(\frac{400}{100}\right)$$

$FL_{5\text{-}inch} = (0.08)(521/100)^2(400/100)$

$FL_{5\text{-}inch} = (0.08)(5.21)^2(4)$

$FL_{5\text{-}inch} = (0.08)(27.14)(4)$

$FL_{5\text{-}inch} = 8.68$ or $9\ psi$

In order to determine the PDP for the manifold lines, the highest PDP of the three attack lines must be selected, which in this case is Line 3 at 165 psi.

$$PDP_{manifold} = FL_{5\text{-inch}} + PDP_{Line\ 3} + \text{Appliance Loss}$$

$$PDP_{manifold} = 9 + 165 + 10$$

$$PDP_{manifold} = 184 \text{ psi}$$

The final PDP for the pumper would be the higher of $PDP_{manifold}$ or $PDP_{master\ stream.}$ **This would be $PDP_{manifold}$ at 184 psi.**

10. $NPDP_{PPS} = PDP - \text{Intake reading}$

$$NPDP_{PPS} = 184 \text{ psi} - 35 \text{ psi}$$

$$\mathbf{NPDP_{PPS} = 149 \text{ psi}}$$

11. $\text{Pressure Correction} = \dfrac{\text{lift} + \text{total intake hose friction loss}}{2.3}$

$$\text{Pressure Correction} = \frac{11 + (6.5+0.5)}{2.3}$$

$$\text{Pressure Correction} = \frac{18}{2.3}$$

Intake pressure correction = 7.8 or 8 psi

$$NPDP_{Draft} = PDP + \text{Intake pressure correction}$$

$$NPDP_{Draft} = 182 \text{ psi} + 8 \text{ psi}$$

$$\mathbf{NPDP_{Draft} = 190 \text{ psi net pump discharge pressure at draft}}$$

Chapter 14

1. 1. Use the available water in apparatus water tanks to mount an attack or protect exposures and then allow the fire building to burn if the fire cannot be extinguished.
   2. Assemble and operate a water shuttle operation.
   3. Assemble and operate a relay pumping operation.
2. Any two of the following:
   - Relay pumping operations establish a constant flow of water between the water supply source and the incident scene.
   - Unless a water shuttle is run at optimum efficiency, there is the possibility of interruptions to the water supply to attack pumpers as tankers come and go from the fill site.
   - Relay pumping operations present considerably less hazards than tankers traveling back and forth on roadways;.
   - Relay pumping is more fuel-efficient than operating a water shuttle operation.
3. Tandem pumping

4. Source or supply pumper, relay pumper, and attack pumper
5. These include less hose to pick up after the incident, reduced friction loss, and increased volumes of water that may moved.
6. 1. The amount of water required at the emergency scene

   2. The distance between the emergency scene and the water source
7. 1. Increase the size of the hose in the relay

   2. Add one or more hoselines in the relay.

   3. Increase the pump discharge pressure of the pumpers operating in the relay.

   4. Increase the number of pumpers in the relay.
8. Minimum distance relay and constant pressure relay
9. $$\frac{\text{Relay distance} + 1}{\text{Distance from Table 14.3}} = \text{Total number of pumpers needed}$$

   $$\frac{4{,}000}{825} = 4.9 + 1 = 5.9\text{, or 6 pumpers needed}$$
10. At least three of the following:
    - It speeds relay activation. Each driver/operator knows exactly how much hose to lay out and how to pump it without awaiting orders.
    - It requires no calculations to be made on the emergency scene.
    - Radio traffic and confusion between pump operators are minimized.
    - The attack pumper driver/operator is able to govern fire lines with greater ease.
    - Operators in the relay only have to guide and adjust pressure to one constant figure.
11. Any two of the following:
    - Variations in relay pumper spacing (increase pressure for greater spacing, decrease for lesser spacing)
    - Severe elevation differences between source and fire (decrease discharge pressure when pumping downhill, increase when pumping uphill)
    - Increases in needed fire flow
    - Large diameter hose (requires lower discharge pressure to supply the same volume of water)
12. The pumper with the largest pump capacity; this pumper will be required to develop the highest net pump discharge pressure.
13. 10 psi
14. 50 to 75 psi
15. From the scene/attack pumper

Chapter 15

1. NFPA® 13, *Standard for the Installation of Sprinkler Systems*
2. Code requirements, insurance purposes, and general fire protection.
3. Wet-pipe systems, dry-pipe systems, deluge systems, and preaction systems.
4. $Q = (K)(\sqrt{P})$

   $Q = (9)(\sqrt{19})$

   $Q = (6)(4.36)$

   $Q = 26.1$ or 26 gpm
5. Ferrous metal, copper tubing, and plastic
6. 15 psi, 7 psi
7. 150 psi
8. - A closed valve in the water supply or some other part of the system
   - Inadequate water supply to the sprinkler system
   - Changes to the occupancy that negate the effect of the sprinkler system
9. $Q_{\text{per sprinkler}} = (K)(\sqrt{P})$

   $Q_{\text{per sprinkler}} = (5.5)(\sqrt{17})$

   $Q_{\text{per sprinkler}} = (5.5)(4.12)$

   $Q_{\text{per sprinkler}} = 22.7$ or 23 gpm

   $Q_{\text{total}} = (\text{No. of sprinklers flowing})(\text{gpm/sprinkler})$

   $Q_{\text{total}} = (10)(23)$

   $Q_{\text{total}} = 230$ gpm

   $FL_{2\frac{1}{2}\text{-inch}} = CQ^2L$

   $FL_{2\frac{1}{2}\text{-inch}} = (0.5)(230/100)^2(500/100)$

Note: These are essentially siamese lines, so C is taken from Table 12.3

   $FL_{2\frac{1}{2}\text{-inch}} = (0.5)(2.3)^2(5)$

   $FL_{2\frac{1}{2}\text{-inch}} = (0.5)(5.29)(5)$

   $FL_{2\frac{1}{2}\text{-inch}} = 13.2$ or 13 psi

   $EP = (5 \text{ psi/floor})(\text{Number of floors} - 1)$

   $EP = (5 \text{ psi/floor})(14 - 1)$

   $EP = (5 \text{ psi/floor})(13)$

   $EP = 65$ psi

   $FL_{\text{Total}} = FL_{2\frac{1}{2}\text{-inch}} + EP + \text{System Loss}$

   $FL_{\text{Total}} = 13 + 65 + 25$

   $FL_{\text{Total}} = 103$ psi

10. $PDP = TPL + NP$

    $PDP = 103 + 17$

    $PDP = 120$ psi

11. - Dry automatic standpipes
    - Wet automatic standpipes
    - Semiautomatic standpipes
    - Dry manual standpipes
    - Wet manual standpipes
12. Class II standpipes
13. $GPM = 29.7\ d^2\sqrt{NP}$

    $GPM = (29.7)(1)^2(\sqrt{50})$

    $GPM = (29.7)(1)(7.07)$

    $GPM = 210\ gpm$

    $FL_{2\text{-inch attack line}} = CQ^2L$

    $FL_{2\text{-inch attack line}} = (8)(210/100)^2(150/100)$

    $FL_{2\text{-inch attack line}} = (8)(2.1)^2(1.5)$

    $FL_{2\text{-inch attack line}} = (8)(4.41)(1.5)$

    $FL_{2\text{-inch attack line}} = 52.9$ or 53 psi

    $EP = (5\ psi/floor)(No.\ of\ floors\ -1)$

    $EP = (5)(12\ -1)$

    $EP = (5)(11)$

    $EP = 55\ psi$

    $FL_{3\text{-inch lines}} = CQ^2L$

    $FL_{3\text{-inch lines}} = (0.2)(210/100)^2(100/100)$

    $FL_{3\text{-inch lines}} = (0.2)(2.1)^2(1)$

    $FL_{3\text{-inch lines}} = (0.2)(4.41)(1)$

    $FL_{3\text{-inch lines}} = 0.88$ or 1 psi

    $TPL = FL_{2\text{-inch attack line}} + EP + FL_{3\text{-inch lines}} + System\ Loss$

    $TPL = 53 + 55 + 1 + 25$

    TPL = 134 psi
14. $PDP = TPL + NP$

    $PDP = 134 + 50$

    PDP = 184 psi

Index

B

C

D

E

F

G

H

I

J

Q

R

T

U

V

W

Index by Nancy Kopper